Dietary Patterns and Nutritional Value in Non-communicable Diseases

Dietary Patterns and Nutritional Value in Non-communicable Diseases

Editors

Andriana Kaliora
Chara Tzavara
Charalampia Amerikanou

MDPI

Basel • Beijing • Wuhan • Barcelona • Belgrade • Novi Sad • Cluj • Manchester

Editors

Andriana Kaliora
Harokopio University
Athens
Greece

Chara Tzavara
Harokopio University
Athens
Greece

Charalampia Amerikanou
Harokopio University
Athens
Greece

Editorial Office
MDPI
St. Alban-Anlage 66
4052 Basel, Switzerland

This is a reprint of articles from the Special Issue published online in the open access journal *Nutrients* (ISSN 2072-6643) (available at: https://www.mdpi.com/journal/nutrients/special_issues/diet_pattern_and_nutri_value_in_non_communicable).

For citation purposes, cite each article independently as indicated on the article page online and as indicated below:

Lastname, A.A.; Lastname, B.B. Article Title. *Journal Name* **Year**, *Volume Number*, Page Range.

ISBN 978-3-7258-0057-5 (Hbk)
ISBN 978-3-7258-0058-2 (PDF)
doi.org/10.3390/books978-3-7258-0058-2

Contents

nutrients MDPI

Editorial

Dietary Patterns and Nutritional Value in Non-Communicable Diseases

Charalampia Amerikanou, Chara Tzavara and Andriana C. Kaliora *

Department of Nutrition and Dietetics, School of Health Science and Education, Harokopio University of Athens, 70 El. Venizelou Ave., 17676 Athens, Greece; amerikanou@windowslive.com (C.A.); htzavara@med.uoa.gr (C.T.)
* Correspondence: akaliora@hua.gr; Tel.: +30-210-954-9226

Non-communicable diseases (NCDs) constitute the leading cause of mortality worldwide, with the four major contributors being cardiovascular diseases (CVDs), cancers, respiratory diseases, and diabetes [1]. Many of the risk factors that contribute to NCDs are modifiable, such as diet, physical activity, smoking, and alcohol use, with diet possessing an essential role in both prevention and management. In recent years, researchers have focused more on the study of dietary patterns instead of single nutrients and foods, suggesting that combined effects can be presented due to synergy or antagonism [2]. Various dietary patterns have been proposed as protective or detrimental for NCDs, with the Mediterranean diet standing out for its association with a lower incidence of NCDs and reduced mortality risk. However, there is still a lot to be explored in the link between dietary patterns and different disease-related factors, making this scientific field very promising.

CVDs are the most common NCDs and the leading cause of death worldwide. The role of diet in CVDs and its risk factors, such as hypertension, diabetes, and dyslipidemia, is pivotal; therefore, the investigation of dietary patterns in CVD and related comorbidities is of high importance. Recently, Dimitriou and colleagues [3] explored the association of dietary patterns, extracted from 1017 Greek individuals, with CVD risk. The Western-type pattern, which included red and processed meat, fried potatoes, and fast foods, was associated with increased CAD risk (OR = 1.20; 95% CI = 1.09–1.32, $p < 0.001$) [3]. In another Greek study, which included 146 Greek metabolically unhealthy obese adults, extracted dietary patterns were associated with several cardiometabolic and lifestyle parameters [4]. More specifically, the Western-type pattern was positively associated with anthropometric indices (fat and waist circumference) and insulin, and the Mediterranean diet-like pattern was positively associated with high-density lipoprotein (HDL) and mental health score, whereas it was negatively associated with depression score [4]. Regarding metabolic syndrome and its association with dietary patterns, a Chinese study with 5426 participants showed that two healthy dietary patterns ("dairy and fruits" and "coarse cereals and soy products") had protective effects on MetS (OR 0.81 (95% CI: 0.66, 0.98) and 0.74 (95%CI: 0.61, 0.91), respectively) [5].

Apart from dietary patterns, dietary fiber intake and dietary antioxidant capacity in cardiometabolic diseases have recently been investigated. Zhang and his colleagues [6] analyzed data from a longitudinal study (the China Nutrition and Health Database), including 24 h recalls from 2004 to 2015, examining 8307 individuals, and they computed dietary fiber intake. Contrary to other similar studies, whole-grain fiber intake was positively associated with hypertension (hazard ratio (95% confidence interval) (quartile 3 vs. quartile 1) was 1.21 (1.04, 1.40)), but not with other cardiometabolic diseases [6]. A Polish study (n = 5690) evaluated the association between dietary total antioxidant capacity (DTAC), healthy diet quality, and the risk of CVDs [7]. The dietary assessment was based on 24 h dietary recalls, and DTAC was extracted from published databases that use the ferric ion reducing antioxidant potential (FRAP) method to measure the antioxidant potential of foods. DTAC was associated with a reduced CVD odds ratio (OR = 0.593, p <0.0001) and a higher Healthy

Citation: Amerikanou, C.; Tzavara, C.; Kaliora, A.C. Dietary Patterns and Nutritional Value in Non-Communicable Diseases. *Nutrients* **2024**, *16*, 82. https://doi.org/10.3390/nu16010082

Received: 7 December 2023
Accepted: 9 December 2023
Published: 26 December 2023

Diet Indicator (HDI). HDI is an index based on the World Health Organization's recommendations, and it includes the consumption of saturated and polyunsaturated fatty acids, cholesterol, protein, fiber, free sugars, fruits, and vegetables. Also, DTAC was higher in individuals with a higher dietary intake of total polyphenols, antioxidant vitamins, and minerals ($p < 0.0001$) [7].

Another metabolic dysfunction with a high worldwide prevalence is nonalcoholic fatty liver disease (NAFLD). A prospective analysis of 128,695 UK Biobank participants recently revealed a relationship between dietary patterns and the long-term outcomes of NAFLD [8]. Patterns that were prudent and high in whole grains were associated with a lower risk of end-stage liver disease (0.74 and 0.87, respectively) and all-cause mortality (0.86 and 0.94, respectively), whereas a meat-rich dietary pattern had a U-shaped association with all-cause mortality [8]. In a cross-sectional study of 320 Lebanese NAFLD patients, high adherence to a traditional diet was associated with a lower risk of fibrosis [0.18 (0.04–0.85), $p = 0.031$]. The traditional pattern included vegetables, legumes, vegetable oils, nuts, fish, red wine, and cooked rice [9].

Although many studies have addressed the association between diet and cancer risk in recent years, there is a lack of studies on the influence of dietary patterns on cancer therapy response, as well as the combined effect of dietary and genetic factors on cancer risk. Tiberio and his colleagues [10] analyzed the eating patterns of 82 breast cancer patients undergoing chemotherapy and how these can affect positron emission tomography/computed tomography (PET/CT) outcomes related to therapy response. The colon uptake value was correlated positively with drinks (alcohol and spirits; $r = 0.33$, $p < 0.01$) and foods (red and cured meats; $r = 0.25$, $p = 0.04$) related to inflammation, and rectum uptake was correlated negatively with anti-inflammatory foods (fruits and vegetables; $r = -0.23$, $p = 0.04$) [10]. Liu and his colleagues [11] created a polygenic risk score using UK Biobank genetic and dietary data (n = 415,589). A healthy dietary score based on the consumption of fruits, vegetables, grains, fish, and meat reduced the risk of upper gastrointestinal cancer (UGI) [hazard ratio: 0.76 (0.62–0.93), $p = 0.009$]. When a healthy pattern was combined with a low genetic risk, it was associated with a lower risk of UGI cancer compared to a high genetic risk and an unhealthy dietary pattern [11].

The exploitation of dietary patterns in low- and middle-income countries has uncovered a transition of traditional diets towards westernized diets, as well as the contribution of this dietary shift to obesity, malnutrition, and other NCDs. A data analysis of 8957 adults from the 8th Philippine National Nutrition Survey (PNNS) estimated a 30% prevalence of the double burden of malnutrition (DBM), indicating the coexistence of undernutrition and overnutrition in the same population [12]. Seven dietary patterns were extracted in this study, with a rice pattern and a meat and sugar pattern in males and a protein-rich food, cereal, and sugar pattern in females being associated with lower odds of DBM. Vegetable and corn patterns showed an increased risk of DBM in women [12]. In a study aiming to explore the dietary patterns of children in South Africa, aged 1–10 years, and their relationship with socio-demographic factors, a greater adherence to unhealthy patterns in a higher socio-economic status and in the presence of an obese mother was observed [13]. The improvement of children's eating habits is pivotal not only in low- and middle-income countries but also in children from households with low incomes in developed countries [13]. In the USA, improving children's dietary intake with healthier foods and beverages has been implemented through the USDA Child and Adult Care Food Program (CACFP) [14]. A longitudinal study of childcare centers participating in CACFP evaluated their meal patterns and concluded that, although CACFP centers followed better nutrition standards than non-CACFP centers, their menu quality had not improved since 2017, when an increase in whole grains, fruits, and vegetables had been proposed [14]. The above indicates that there is a lack of effective strategies towards the improvement of children's diet quality and eating habits in both developing and developed countries.

The importance of diet is also evident in other common NCDs. Functional dyspepsia affects 10–30% of adults, with food being one of its main triggering factors. A review tried to shed light on whether specific food categories, dietary patterns, or eating habits influence the symptomatology of functional dyspepsia [15]. The authors concluded that, although several foods and patterns have been proposed as dyspepsia triggers, evidence on the association between diet and symptomatology is scarce and inconsistent, highlighting the need for more research in this field [15]. Fibromyalgia is a chronic syndrome with a prevalence of 2–3% in the general population. A systematic review that addressed the relationship between diet and pain in fibromyalgia included 12 studies (11 interventions and 1 observational) and 546 participants [16]. Interestingly, all plant-based and anti-inflammatory diets improved pain measurements, whereas not all gluten-free or elimination/restrictive diets showed statistical significance in pain improvement [16]. The significance of plant-based diets also emerged in a study of cognitive impairment in Chinese adults [17]. In this longitudinal prospective study, 1077 out of 4792 participants developed cognitive impairment, which showed a reverse J-shaped association with BMI, meaning that overweight and obese participants had a decreased risk, whereas underweight participants had an increased risk of cognitive impairment. Interestingly, overweight participants with a higher plant-based diet index had a lower hazard ratio for cognitive impairment than those with a lower index (HR = 0.74; 95% CI 0.57–0.95 vs. HR = 0.87; 95% CI 0.67–1.12). The above suggests that high adherence to plant-based diets enhances the protective effect of overweight on cognitive impairment [17].

Conclusively there is a growing interest in the relationship between diet and dietary patterns with NCD-related parameters. Our Special Issue offers some new data in this field. More research is needed to elucidate the exact mechanisms under which dietary patterns, as well as individual foods and nutrients, exert their effects on NCDs and related clinical, genetic, environmental, and behavioral factors.

Funding: This research received no external funding.

Acknowledgments: We would like to thank the authors who contributed to this Special Issue, as well as the reviewers for their valuable revisions.

Conflicts of Interest: The authors declare no conflict of interest.

References

1. Budreviciute, A.; Damiati, S.; Sabir, D.K.; Onder, K.; Schuller-Goetzburg, P.; Plakys, G.; Katileviciute, A.; Khoja, S.; Kodzius, R. Management and Prevention Strategies for Non-communicable Diseases (NCDs) and Their Risk Factors. *Front. Public Health* **2020**, *8*, 574111. [CrossRef] [PubMed]
2. Dominguez, L.J.; Di Bella, G.; Veronese, N.; Barbagallo, M. Impact of Mediterranean Diet on Chronic Non-Communicable Diseases and Longevity. *Nutrients* **2021**, *13*, 2028. [CrossRef] [PubMed]
3. Dimitriou, M.; Kalafati, I.P.; Rallidis, L.S.; Kolovou, G.; Dedoussis, G.V. A Posteriori Dietary Patterns and Coronary Artery Disease in a Greek Case–Control Study. *Nutrients* **2023**, *15*, 4733. [CrossRef] [PubMed]
4. Amerikanou, C.; Kleftaki, S.-A.; Valsamidou, E.; Tzavara, C.; Gioxari, A.; Kaliora, A.C. Dietary Patterns, Cardiometabolic and Lifestyle Variables in Greeks with Obesity and Metabolic Disorders. *Nutrients* **2022**, *14*, 5064. [CrossRef] [PubMed]
5. Wei, L.; Fan, J.; Dong, R.; Zhang, M.; Jiang, Y.; Zhao, Q.; Zhao, G.; Chen, B.; Li, J.; Liu, S. The Effect of Dietary Pattern on Metabolic Syndrome in a Suburban Population in Shanghai, China. *Nutrients* **2023**, *15*, 2185. [CrossRef] [PubMed]
6. Zhang, Z.; Chen, B.; Zeng, J.; Fan, M.; Xu, W.; Li, X.; Xing, Y.; Xu, S. Associations between Consumption of Dietary Fibers and the Risk of Type 2 Diabetes, Hypertension, Obesity, Cardiovascular Diseases, and Mortality in Chinese Adults: Longitudinal Analyses from the China Health and Nutrition Survey. *Nutrients* **2022**, *14*, 2650. [CrossRef] [PubMed]
7. Zujko, M.E.; Wa'skiewicz, A.; Witkowska, A.M.; Cicha-Mikołajczyk, A.; Zujko, K.; Drygas, W. Dietary Total Antioxidant Capacity—A New Indicator of Healthy Diet Quality in Cardiovascular Diseases: A Polish Cross-Sectional Study. *Nutrients* **2022**, *14*, 3219. [CrossRef]
8. Liu, Z.; Huang, H.; Xie, J.; Xu, C. Dietary Patterns and Long-Term Outcomes in Patients with NAFLD: A Prospective Analysis of 128,695 UK Biobank Participants. *Nutrients* **2023**, *15*, 271. [CrossRef] [PubMed]
9. Sayegh, N.F.; Heraoui, G.N.H.A.; Younes, H.; Sayegh, L.N.; Boulos, C.; Sayegh, R. Relation of Dietary Patterns and Nutritional Profile to Hepatic Fibrosis in a Sample of Lebanese Non-Alcoholic Fatty Liver Disease Patients. *Nutrients* **2022**, *14*, 2554. [CrossRef] [PubMed]
10. Tiberio, P.; Antunovic, L.; Gaudio, M.; Viganò, A.; Pastore, M.; Miggiano, C.; Jacobs, F.; Benvenuti, C.; Farina, E.; Chiti, A.; et al. The Relationship among Bowel ^{18}F-FDG PET Uptake, Pathological Complete Response, and Eating Habits in Breast Cancer Patients Undergoing Neoadjuvant Chemotherapy. *Nutrients* **2023**, *15*, 211. [CrossRef] [PubMed]

11. Liu, W.; Wang, T.; Zhu, M.; Jin, G. Healthy Diet, Polygenic Risk Score, and Upper Gastrointestinal Cancer Risk: A Prospective Study from UK Biobank. *Nutrients* **2023**, *15*, 1344. [CrossRef] [PubMed]
12. de Juras, A.R.; Hsu, W.-C.; Cheng, Y.-Y.; Ku, L.-J.E.; Yu, T.; Peng, C.-J.; Hu, S.C. Sex Differences in Dietary Patterns of Adults and Their Associations with the Double Burden of Malnutrition: A Population-Based National Survey in the Philippines. *Nutrients* **2022**, *14*, 3495. [CrossRef] [PubMed]
13. Senekal, M.; Nel, J.H.; Eksteen, G.; Steyn, N.P. Dietary Patterns, Socio-Demographic Predictors Thereof, and Associations of Dietary Patterns with Stunting and Overweight/Obesity in 1–<10-Year-Old Children in Two Economically Active Provinces in South Africa. *Nutrients* **2023**, *15*, 4136. [CrossRef] [PubMed]
14. Andreyeva, T.; Mozaffarian, R.S.; Kenney, E.L. Updated Meal Patterns in the Child and Adult Care Food Program and Changes in Quality of Food and Beverages Served: A Natural Experimental Study. *Nutrients* **2022**, *14*, 3786. [CrossRef] [PubMed]
15. Amerikanou, C.; Kleftaki, S.-A.; Valsamidou, E.; Chroni, E.; Biagki, T.; Sigala, D.; Koutoulogenis, K.; Anapliotis, P.; Gioxari, A.; Kaliora, A.C. Food, Dietary Patterns, or Is Eating Behavior to Blame? Analyzing the Nutritional Aspects of Functional Dyspepsia. *Nutrients* **2023**, *15*, 1544. [CrossRef] [PubMed]
16. Maddox, E.K.; Massoni, S.C.; Hoffart, C.M.; Takata, Y. Dietary Effects on Pain Symptoms in Patients with Fibromyalgia Syndrome: Systematic Review and Future Directions. *Nutrients* **2023**, *15*, 716. [CrossRef] [PubMed]
17. Liang, F.; Fu, J.; Turner-McGrievy, G.; Wang, Y.; Qiu, N.; Ding, K.; Zeng, J.; Moore, J.B.; Li, R. Association of Body Mass Index and Plant-Based Diet with Cognitive Impairment among Older Chinese Adults: A Prospective, Nationwide Cohort Study. *Nutrients* **2022**, *14*, 3132. [CrossRef] [PubMed]

nutrients

MDPI

Article

Dietary Patterns, Cardiometabolic and Lifestyle Variables in Greeks with Obesity and Metabolic Disorders

Charalampia Amerikanou [1], Stamatia-Angeliki Kleftaki [1], Evdokia Valsamidou [1], Chara Tzavara [1], Aristea Gioxari [1,2,*] and Andriana C. Kaliora [1]

1 Department of Nutrition and Dietetics, School of Health Science and Education, Harokopio University, 70 El. Venizelou Ave, 17676 Athens, Greece
2 Department of Nutritional Science and Dietetics, School of Health Science, University of the Peloponnese, Antikalamos, 24100 Kalamata, Greece
* Correspondence: a.gioxari@uop.gr; Tel.: +30-2721045326

Abstract: There is considerable evidence that some dietary patterns contribute to obesity and metabolic disorders but there is less data on diet's association with different health parameters. We investigated the interaction between different dietary patterns and anthropometric, biochemical, lifestyle, and psychological health parameters in a Greek population with obesity and metabolic disorders. To the best of our knowledge, this is the first study in Greece with a thorough and holistic approach in analyzing such relationships. For assessing food patterns, revealing underlying structures, and reducing the number of variables we applied exploratory factor analysis (EFA). Principal Component Analysis was chosen as the extraction method using Varimax rotation, and three regression sets were computed. The study involved 146 Greek metabolically unhealthy obese adults, both men and women. Our cohort was categorized into four dietary patterns: "Western type diet", "Mediterranean-like diet", "Healthy diet", and "Animal meat and sauces diet". Dietary patterns characterized by a high consumption of energy-dense and animal-derived foods were positively associated with anthropometric and biochemical parameters related to metabolic disorders. Plant-based, healthier dietary patterns, on the other hand, were associated with better biochemical and mental health profiles among metabolically unhealthy obese individuals.

Keywords: dietary patterns; metabolic disorders; obesity; dyslipidemia; insulin resistance; anthropometrics; mental health; plant-based diet; Mediterranean diet; Western diet

Citation: Amerikanou, C.; Kleftaki, S.-A.; Valsamidou, E.; Tzavara, C.; Gioxari, A.; Kaliora, A.C. Dietary Patterns, Cardiometabolic and Lifestyle Variables in Greeks with Obesity and Metabolic Disorders. *Nutrients* **2022**, *14*, 5064. https://doi.org/10.3390/nu14235064

Academic Editor: Jose Lara

Received: 2 November 2022
Accepted: 25 November 2022
Published: 28 November 2022

1. Introduction

Nowadays, chronic metabolic disorders are a growing public health concern, especially obesity and diabetes [1]. The prevalence of overweight adults exceeded 1.9 billion in 2016, with a great proportion (650 million) being obese [2]. Similarly, the International Diabetes Foundation (IDF) reported that prevalence of diabetes in 2019 was 463 million, and this is expected to increase further and reach 700 million by 2045 [3]. Apart from type 2 diabetes mellitus (T2DM), obesity has been associated with many other metabolic disorders, such as cardiovascular diseases (CVDs), non-alcoholic fatty liver disease (NAFLD), several types of cancers and, recently, it was linked with COVID-19 mortality [4]. Although body weight is influenced by both genetic and environmental factors, the increased prevalence of adiposity globally suggests that urbanization, followed by food marketing and access to refined and energy-dense food items, along with reduced physical activity, are the main drivers of obesity [5].

Since obesity has been associated mainly with diet quality and unhealthy dietary choices, rather than single foods or nutrients, dietary patterns analysis is one of the best approaches to examine the influence of overall diet, by including the interactive effect of individual items, macro- and micro- nutrients, and bioactive compounds. Furthermore, it offers the opportunity to associate diet with several parameters related to metabolic disorders, such as anthropometric and cardiometabolic markers [6,7]. Moreover, dietary patterns responsible for obesity could be used as a tool for its prevention [8]. The effects of several dietary patterns on obesity and related metabolic disorders have been investigated throughout the years. Most studies evidence that diets with high consumption of plant foods, such as the Mediterranean diet, are valuable in the prevention of obesity, whereas dietary patterns with a high intake of red meat and refined grains, such as the Western pattern, increase the risk of obesity and related comorbidities [9].

It is well established that several dietary patterns contribute to obesity and metabolic disorders. Nevertheless, a wide range of anthropometric, biochemical, and lifestyle parameters have seldom been linked to diets among Greeks. Therefore, in this study, we aimed to explore the interaction between different dietary patterns in a Greek population with obesity and metabolic disorders with several related cardiometabolic and lifestyle parameters. More specifically, apart from anthropometric and biochemical indices, we addressed the associations with lifestyle, physical activity, and physical and psychological health parameters as well.

2. Materials and Methods

2.1. Study Design

This observational study used the baseline data from a three-month intervention that investigated the effect of a mushroom-based snack on markers related to obesity and metabolic disorders [10]. The inclusion criteria were adult age, central obesity, and metabolic disorders, including dyslipidaemia, glucose intolerance, insulin resistance and hypertension. Non-eligible subjects were pregnant and lactating females, patients with thyroid disease, or with abuse of alcohol, or with known psychiatric or mental disorders. The Ethics Committee of Harokopio University approved the protocol of the study (ID protocol: 62/03-07-2018), which was in line with the Helsinki Declaration and the Data Protection Act 1998. Additionally, the trial acquired registration with clinicaltrials.gov (ID Number: NCT04081818). Patient recruitment took place in Harokopio University of Athens (Greece) between 2020 and 2021. All subjects were informed in detail about the study before giving their signed consent.

2.2. Anthropometric Measurements

Body weight (BW), height, body composition i.e., body fat (BF), fat free mass (FFM), and total body water (TBW), as well as waist and hip circumferences (WC, HC), and waist-to-hip ratio (WHR), were measured. Body weight was recorded to the nearest kg using a flat scale, early in the morning with light clothing and without shoes. Accordingly, height was measured to the nearest cm using a stadiometer (Seca Mode 220, Hamburg, Germany). Body mass index (BMI) was calculated as BW (kg)/Height^2 (m^2). To estimate body composition (BF, FFM, TBW) bioelectrical impedance analysis (Tanita BC-418, Tokyo, Japan) was performed. Body circumferences (WC, HC, WHR) were measured with a non-stretch, but flexible tape, on minimal clothing.

2.3. Biochemical Parameters

Blood samples, of 20 mL, were collected after an overnight fast and were centrifuged at 3000 rpm for 10 min at 20 °C for serum isolation. Serum biochemical markers were measured with an automatic biochemical analyser (Cobas 8000 analyser, Roche Diagnostics GmbH, Mannheim, Germany): glucose; insulin; total cholesterol (TC); high-density lipoprotein (HDL); low-density lipoprotein (LDL); triglycerides (TG); urea; uric acid; creatinine; alanine aminotransferase (ALT); aspartate aminotransferase (AST); γ-glutamyl transferase (γ-GT); alkaline phospatase (ALP); uric acid; lactate dehydrogenase (LDH); iron (Fe); ferritin; albumin; and C-reactive protein (CRP).

Vitamin D was measured with an automated immunoassay system, Cobas e801 (Roche Diagnostics, Mannheim, Germany).

2.4. Physical Activity and Quality of Life Assessment

We assessed physical activity, smoking status, risk for depression, self-esteem, insomnia level and related factors of quality of life. To assess physical activity, the International Physical Activity Questionnaire Short Form (IPAQ-SF) was applied. Physical activity levels were expressed as the metabolic equivalent task in minutes per week (MET-min/week), based on the IPAQ scoring system [11]. Regarding smoking status, participants were asked about their smoking habits and were classified as smokers and non-smokers. To evaluate the risk for depression, the 10-item questionnaire Center for Epidemiologic Studies Depression Scale Revised (CESD-R-10) was used. Scoring ranged between 0 and 60, while scores ≥16 were considered as depression [12]. Self-esteem was assessed by applying the 10-item Rosenberg Self-Esteem scale that includes both positive and negative feelings about oneself [13]. Sleep quality was evaluated by the Athens Insomnia Scale (AII), consisting of 10 items on nocturnal sleep problems and daytime dysfunction [14]. Finally, the impact of health on an individual's everyday life was assessed through the self-reported Short Form-12 (SF-12) questionnaire with two summary scores reporting on a mental (MCS-12) and a physical component (PCS-12).

2.5. Dietary Assesment

Dietary patterns were produced using data from a standardized semi-quantitative (69 items) food frequency questionnaire (69-FFQ) [15]. This questionnaire evaluated the long-term habitual intake of 69 food items and beverages, such as dairy, cereals, meat, fish, legumes etc. For each food item the participants were asked to choose one of the following consumption options: "never/rarely", "1–3 times/month, 1–2 times/week, 3–6 times/week, 1 times/day," to "≥2 times/day". Finally, the 69 food items were divided into 25 food groups with similar nutrient profiles, as presented in Supplementary Table S1.

2.6. Statistical Analysis

Statistical analyses were conducted using SPSS statistical software (version 24.0). We carried out Exploratory Factor Analysis (EFA) to evaluate food patterns, disclose underlying structures and reduce the number of variables. Varimax rotation was applied as the extraction method for Principal Component Analysis (PCA). For sample adequacy the Kaiser–Meyer-/–Olkin (KMO) procedure was used. Values of 0.40 and 1.00 were set as cut-off points for factor loadings and Eigen values, respectively. Calculation of Cronbach's alpha coefficient was used to determine internal consistency reliability. Scales with reliabilities equal to, or greater than, 0.70 were considered acceptable. Spearman's rho correlation coefficients were used to explore the association of dietary patterns with biochemical and the other study measurements. For associations that were found to be significant, adjustments were made via linear regression analyses. Three sets of regression were computed, one with adjustment for age, gender and BMI, one with adjustment for age, gender, BMI, physical activity and smoking and the last with adjustment for age, gender, BMI, physical, smoking, current medication and the other patterns. Adjusted regression coefficients (β) with standard errors (SE) were computed from the results of the linear regression analyses. Regression analyses were conducted after having logarithmically transformed the dependent variables. Statistical significance was set at p-value < 0.05, and all p-values were two-tailed.

3. Results

A total of 146 Greek metabolically unhealthy obese adults, men and women, were enrolled in the dietary patterns study. The descriptive characteristics of the sample are depicted in Table 1.

Table 1. Descriptive characteristics of the study population.

Characteristics of the Study Participants	
Sex, n (%)	
Men	55 (37.7)
Women	91 (62.3)
Age (years), mean (SD)	53.5 (11.4)
Educational years, mean (SD)	15.3 (3.4)
Family status, n (%)	
Married	109 (74.6)
Divorced	7 (4.8)
Single	20 (13.7)
Widowed	5 (3.4)
Other	5 (3.4)
Smoking, n (%)	
No	109 (74.7)
Yes	33 (22.6)
N/A	4 (2.7)
Antihypertensive treatment, n (%)	
No	91 (62.3)
Yes	55 (37.7)
Statins, n (%)	
No	104 (71.2)
Yes	42 (28.8)
Antidiabetic agents, n (%)	
No	111 (76.0)
Yes	35 (24.0)
Anthropometric parameters	
BMI (kg/m^2), mean (SD)	34 (6.3)
WC (cm), mean (SD)	110.4 (13.4)
HC (cm), mean (SD)	117.9 (14.9)
Lifestyle parameters	
IPAQ-SF (total MET- min/week), mean (SD)	1723 (2142)
AII, mean (SD)	5.4 (3.7)
CESD-R, mean (SD)	16 (10)
Rosenberg Self-Esteem scale, mean (SD)	31 (5)
PCS-12, mean (SD)	45.7 (9.1)
MCS-12, mean (SD)	48.6 (9.5)

Data are expressed as counts (%) or mean values (standard deviation, SD). BMI, body mass index; WC, waist circumference; HC, hip circumference; IPAQ-SF, International Physical Activity Questionnaire Short Form; AII, Athens Insomnia Scale; CESD-R, Center for Epidemiologic Studies Depression Scale Revised; PCS-12, Physical Component Score; MCS-12, Mental Composite Score.

For the PCA, the sampling adequacy of EFA was evaluated through a KMO of 0.72 and a significant Bartletts' sphericity ($p < 0.001$). The analysis of the 25 food groups revealed 4 principal components/factors, each characterizing a possible dietary pattern. All factors combined explained 44.2% of the variance. Factor 1 had 10 food groups and explained 16.7% of the variance. Factor 2 had 7 food groups and explained 13.4% of the variance, factor 3 had 3 food groups and explained 7.6% of the variance, while factor 4 also had 3 food groups and explained 6.5% of the variance. All food groups had a loading of at least 0.40, except for eggs and alcohol that did not load to any of the four factors as their highest loading was 0.29. All factors had acceptable reliability, since their alpha coefficient of Cronbach was above 0.7.

The factor loadings between the 25 food groups and the 4 dietary patterns are presented in Table 2. The first pattern was dubbed as a "Western-type pattern", which was characterized by high fat dairy, refined grains, fast food and processed meat, pies, sweets, salty snacks, soft drinks, animal and hydrogenated fats and seed oil. The second pattern, the "Mediterranean-like diet pattern" included whole grains, fish, vegetables (raw and cooked), fruit (raw, fruit juices and dried fruits), pulses and nuts. The third pattern,

"Healthy pattern", included low-fat dairy, olive oils and oil, coffee and tea. Finally, the last pattern, the "Animal meat and sauces pattern", encompassed red meat, poultry and sauces.

Table 2. Food groups and respective factor loadings for the four dietary patterns of the study.

Food Groups	Dietary Patterns			
	Western-Type Pattern	Mediterranean-Like Diet Pattern	Healthy Pattern	Animal Meat and Sauces Pattern
Dairy (High-fat)	0.57			
Dairy (Low-fat)			0.59	
Refined grains	0.66			
Whole grains		0.60		
Fast Food	0.74			
Red Meat				0.50
Processed Meat	0.57			
Fish		0.59		
Vegetables ǀ cooked mixed vegetables		0.75		
Fruits ǀ fruit juices		0.53		
Pies	0.66			
Sweets	0.56			
Salty Snacks	0.63			
Olive oil ǀ olives			0.63	
Soft drinks	0.52			
Sauces				0.43
Animal & HydrogenatedFats	0.54			
Poultry				0.70
Pulses		0.50		
Dried fruits		0.62		
Nuts		0.66		
Coffee and Tea			0.55	
Seed oil	0.53			
Alcohol			−0.29	
Eggs				0.29
Cronbach's a	0.79	0.82	0.71	0.73

The results of the correlation analysis of the dietary patterns with the parameters examined in our study are presented in Table 3. More specifically, Table 3 depicts correlations of dietary patterns with anthropometric measurements, with biochemical indices and with lifestyle parameters. Body weight, fat and waist circumference were positively correlated with "Western-type pattern" ($p = 0.030$, $p = 0.004$ and $p = 0.028$, respectively) and "Animal meat and sauces pattern" ($p = 0.001$, $p = 0.036$ and $p = 0.041$, respectively). FFM, TBW and BMI were positively correlated with "Animal meat and sauces pattern" ($p = 0.004$, $p = 0.012$ and $p = 0.044$, respectively). Regarding biochemical parameters, insulin was positively correlated with "Western-type pattern" ($p = 0.014$) and negatively with "Mediterranean-like diet pattern" ($p = 0.037$). HDL was positively correlated with "Mediterranean-like diet pattern" ($p = 0.011$). ALT and AST were positively correlated with "Western-type pattern" ($p = 0.041$ and $p = 0.029$, respectively). Fe was negatively correlated with "Western-type pattern" ($p = 0.043$) and "Animal meat and sauces pattern" ($p = 0.024$). Vitamin D was negatively correlated with "Western-type pattern" ($p = 0.013$) and CRP positively correlated with "Animal meat and sauces pattern" ($p = 0.010$). Finally, lifestyle parameters showed some significant correlations with the four patterns. More specifically, AII was positively correlated with "Western-type pattern" ($p = 0.042$). Rosenberg Self-Esteem scale, PCS-12, MCS-12 were positively correlated ($p = 0.014$, $p = 0.006$ and $p = 0.038$, respectively), whereas CESD-R negatively correlated, with "Mediterranean-like diet pattern" ($p = 0.033$). IPAQ positively correlated with "Mediterranean-like diet pattern" ($p = 0.023$) and negatively with "Animal meat and sauces pattern" ($p = 0.012$).

Table 3. Correlation between dietary patterns and anthropometrics, biochemical and lifestyle parameters.

Parameters		Western-Type Pattern	Mediterranean-Like Diet Pattern	Healthy Pattern	Animal Meat and Sauces Pattern
Anthropometrics					
BW (kg)	rho	0.19	−0.01	0.02	0.29
	p	**0.030**	0.947	0.801	**0.001**
BF (%)	rho	0.13	0.06	0.06	0.02
	p	0.136	0.470	0.504	0.844
BF (kg)	rho	0.26	0.07	0.06	0.19
	p	**0.004**	0.457	0.499	**0.036**
FFM (kg)	rho	0.05	−0.11	−0.05	0.26
	p	0.565	0.246	0.616	**0.004**
TBW (kg)	rho	0.11	−0.05	−0.08	0.23
	p	0.226	0.579	0.355	**0.012**
BMI (kg/m^2)	rho	0.13	0.04	0.08	0.18
	p	0.130	0.687	0.390	**0.044**
WC (cm)	rho	0.20	−0.08	0.03	0.18
	p	**0.028**	0.381	0.747	**0.041**
HC(cm)	rho	0.14	0.09	0.06	0.10
	p	0.102	0.312	0.470	0.236
Biochemical					
Urea (mg/dL)	Rho	−0.04	−0.11	−0.09	0.10
	p	0.648	0.214	0.284	0.265
Uricacid (mg/dL)	Rho	−0.09	0.09	0.03	0.04
	p	0.295	0.289	0.726	0.693
Creatinine (mg/dL)	Rho	−0.02	−0.12	−0.02	0.13
	p	0.863	0.190	0.805	0.159
Glucose (mg/dL)	Rho	−0.03	−0.13	0.09	−0.08
	p	0.752	0.151	0.337	0.349
Insulin (µIU/mL)	Rho	0.23	−0.20	0.00	0.11
	p	**0.014**	**0.037**	0.964	0.236
TC (mg/dL)	Rho	0.11	−0.14	−0.08	−0.03
	p	0.220	0.119	0.373	0.768
TG (mg/dL)	Rho	0.12	−0.12	−0.02	−0.05
	p	0.190	0.168	0.817	0.603
HDL (mg/dL)	Rho	−0.10	0.22	0.02	−0.09
	p	0.281	**0.011**	0.789	0.331
LDL (mg/dL)	Rho	0.10	0.11	−0.10	−0.04
	p	0.274	0.205	0.270	0.665
AST (iu/L)	Rho	0.00	0.04	0.05	−0.02
	p	0.997	0.625	0.588	0.819
ALT (iu/L)	Rho	0.19	0.02	0.13	0.06
	p	**0.041**	0.843	0.131	0.492
γ-GT (iu/L)	Rho	0.16	−0.09	0.03	0.11
	p	0.090	0.338	0.764	0.227
ALP (U/L)	Rho	0.20	0.05	−0.01	−0.02
	p	**0.029**	0.576	0.925	0.825
Fe (µg/dL)	Rho	−0.19	0.02	−0.05	−0.21
	p	**0.043**	0.831	0.579	**0.024**
Ferritin (ng/mL)	Rho	0.06	0.09	0.00	0.05
	p	0.538	0.316	0.991	0.556
Albumin (g/dL)	Rho	0.08	0.14	0.05	−0.08
	p	0.390	0.131	0.568	0.404
Vitamin D (ng/mL)	Rho	−0.22	0.01	−0.07	−0.10
	p	**0.013**	0.917	0.433	0.254
CRP (mg/L)	Rho	0.08	0.09	0.04	0.24
	p	0.409	0.326	0.655	**0.010**
LDH (U/L)	Rho	0.04	0.07	0.06	0.03
	p	0.712	0.433	0.555	0.719

Table 3. *Cont.*

Parameters		Western-Type Pattern	Mediterranean-Like Diet Pattern	Healthy Pattern	Animal Meat and Sauces Pattern
Lifestyle					
AII	Rho	0.19	−0.14	−0.12	0.17
	p	**0.042**	0.120	0.205	0.059
CESD-R	Rho	0.17	−0.21	0.00	0.09
	p	0.088	**0.033**	0.991	0.340
Rosenberg Self-Esteem scale	Rho	−0.07	0.24	0.07	0.12
	p	0.482	**0.014**	0.443	0.212
PCS-12	Rho	−0.03	0.26	−0.07	0.03
	p	0.785	**0.006**	0.444	0.759
MCS-12	Rho	−0.15	0.20	0.04	−0.12
	p	0.121	**0.038**	0.679	0.200
IPAQ-SF (total MET-min/week)	Rho	0.03	0.20	0.05	−0.22
	p	0.748	**0.023**	0.567	**0.012**

Values resulted from Spearman test. Level of significance was set as 0.05. Significant *p* are in bold. BW, body weight; BF, body fat; FFM, free fat mass; TBW, total body water; BMI, body mass index; WC, waist circumference; HC; hip circumference; TC, total cholesterol; TG, triglycerides; HDL, high-density lipoprotein; LDL, low-density lipoprotein; AST, aspartate aminotransferase; ALT, alanine aminotransferase; γ-GT, γ-glutamyl transferase; ALP, alkaline phosphatase; Fe, iron; CRP, C-reactive protein; LDH, lactate dehydrogenase; AII, Athens Insomnia Scale; CESD-R, Center for Epidemiologic Studies Depression Scale Revised; PCS-12, Physical Composite Score; MCS-12, Mental Composite Score; IPAQ-SF, International Physical Activity Questionnaire Short Form.

Then, we applied linear regression models for the associations of dietary patterns with study variables that were significant in the correlation analysis, and, thus, the Healthy pattern was not included in this analysis (Table 4). The first model was unadjusted (β^1), the second was adjusted for age, gender and BMI (β^2), the third for age, gender, BMI, physical activity and smoking status (β^3). Finally, a model adjusted for all the above parameters, plus medication and the scoring for the other three patterns, was applied (β^4). The "Western-type pattern" showed statistically significant positive association with fat ($\beta^4 = 4.76$, $p < 0.001$), WC ($\beta^4 = 4.55$, $p = 0.009$), insulin ($\beta^4 = 0.23$, $p = 0.012$), ALT ($\beta^4 = 0.26$, $p = 0.012$) and ALP ($\beta^4 = 10.91$, $p = 0.037$) and negative association with vitamin D levels ($\beta^4 = -0.49$, $p = 0.037$). The "Mediterranean-like diet pattern" was positively associated with HDL ($\beta = 4.02$, $p = 0.021$), and MCS-12 ($\beta^4 = 0.87$, $p = 0.014$) and negatively with CESD-R ($\beta^4 = -5.52$, $p = 0.008$). Finally, the "Animal meat and sauces" pattern was positively associated with TBW ($\beta^4 = 0.18$, $p = 0.008$), FAT ($\beta^4 = 6.17$, $p = 0.033$), FFM ($\beta^4 = 0.03$, $p = 0.034$).

Table 4. Association between dietary patterns scores and study parameters.

	Western-Type Pattern							
Dependent Variables	**β[1] (SE)**	*p*	**β[2] (SE)**	*p*	**β[3] (SE)**	*p*	**β[4] (SE)**	*p*
BF (kg)	9.01 (2.55)	**0.001**	5.87 (1.11)	**<0.001**	5.66 (1.14)	**<0.001**	4.38 (1.26)	**0.001**
WC(cm)	6.19 (2.99)	**0.041**	5.13 (1.56)	**0.001**	5.42 (1.60)	**0.001**	4.95 (1.77)	**0.006**
Insulin (μIU/mL)	0.21 (0.09)	**0.016**	0.23 (0.09)	**0.009**	0.23 (0.09)	**0.009**	0.24 (0.09)	**0.008**
ALT (iu/L)	0.20 (0.10)	0.035	0.24 (0.10)	**0.014**	0.24 (0.10)	**0.017**	0.25 (0.10)	**0.019**
ALP (U/L)	9.91 (4.35)	**0.025**	8.83 (4.68)	**0.062**	9.55 (4.75)	**0.047**	9.78 (4.97)	**0.041**
Fe (μg/dL)	−0.36 (0.18)	**0.040**	−0.27 (0.2)	0.176	−0.28 (0.2)	0.174	−0.19 (0.22)	0.393
Vitamin D (ng/mL)	−0.41 (0.2)	**0.046**	−0.44 (0.21)	**0.043**	−0.44 (0.21)	**0.041**	−0.55 (0.23)	**0.021**
AII	1.99 (0.85)	**0.021**	1.51 (0.87)	0.085	1.42 (0.91)	0.121	1.10 (0.98)	0.263
	Mediterranean-like diet pattern							
Insulin (μIU/mL)	−0.17 (0.09)	**0.050**	−0.17 (0.09)	**0.050**	−0.16 (0.09)	0.070	−0.16 (0.10)	0.102
HDL (mg/dL)	4.34 (1.7)	**0.012**	3.86 (1.5)	**0.011**	3.44 (1.53)	**0.026**	4.03 (1.73)	**0.022**
CESD-R	−4.13 (1.9)	**0.032**	−3.86 (1.88)	**0.043**	−4.09 (1.92)	**0.036**	−5.63 (2.07)	**0.008**
Rosenberg Self-Esteem scale	0.19 (0.08)	**0.023**	0.18 (0.08)	**0.034**	0.19 (0.08)	**0.020**	0.28 (0.35)	0.417
PCS-12	0.22 (0.08)	**0.009**	0.18 (0.08)	**0.025**	0.18 (0.08)	**0.031**	0.53 (0.34)	0.124
MCS-12	0.19 (0.08)	**0.026**	0.19 (0.08)	**0.023**	0.19 (0.08)	**0.025**	0.87 (0.35)	**0.016**
	Animal meat and sauces pattern							
BF (kg)	8.09 (2.48)	**0.001**	3.54 (1.21)	**0.004**	4.01 (1.22)	**0.001**	6.17 (2.88)	**0.033**
FFM (kg)	0.04 (0.02)	**0.048**	0.02 (0.01)	**0.033**	0.02 (0.01)	**0.040**	0.03 (0.02)	**0.034**
TBW (kg)	0.20 (0.09)	**0.033**	0.15 (0.06)	**0.016**	0.16 (0.06)	**0.010**	0.17 (0.07)	**0.011**
BMI (kg/m^2)	2.63 (1.27)	**0.041**	2.24 (1.40)	0.113	2.47 (1.44)	0.089	2.23 (1.52)	0.144[5]
WC (cm)	5.74 (2.89)	**0.049**	3.78 (1.60)	**0.020**	4.33 (1.63)	**0.009**	2.89 (1.74)	0.100
Fe (μg/dL)	−0.42 (0.2)	**0.041**	−0.41 (0.22)	0.067	−0.42 (0.23)	0.067	−0.40 (0.25)	0.114
CRP (mg/L)	2.26 (1.13)	**0.048**	2.21 (1.26)	0.082	2.23 (1.3)	0.089	2.09 (1.43)	0.146

[1] unadjusted regression coefficient of each pattern (Standard Error), [2] regression coefficients of each pattern adjusted for age, gender and BMI (Standard Error), [3] regression coefficients of each pattern adjusted for age, gender, BMI, physical activity and smoking (Standard Error), [4] regression coefficients of each pattern adjusted for age, gender, BMI, physical activity, smoking, current medication and the other three patterns (Standard Error). [5] In this analysis, BMI was not included as independent variable. Values resulted from linear regression models. Level of significance was set as 0.05. Significant *p* are in bold. BF, body fat; FFM, free fat mass; TBW, total body water; WC, waist circumference; ALT, alanine aminotransferase; ALP, alkaline phosphatase; Fe, iron; HDL, high density cholesterol; CRP, C-reactive protein; AII, Athens Insomnia Scale; CESD-R, Center for Epidemiologic Studies Depression Scale Revised; PCS-12, Physical Composite Score; MCS-12, Mental Composite Score.

4. Discussion

In this study, we investigated the associations between the four main dietary patterns identified in a Greek population of metabolically unhealthy people with obesity and several parameters related to metabolic disorders. To the best of our knowledge, this is the first study in Greece with a holistic approach in analyzing the relationships between anthropometric, biochemical, lifestyle, physical activity, and psychological health parameters with dietary patterns in a metabolically unhealthy study population.

The dietary patterns identified in our cohort were the "Western-type pattern", the "Mediterranean-like diet pattern", the "Healthy pattern" and the "Animal meat and sauces pattern". Similar patterns were previously reported on in cross-sectional studies that investigated dietary patterns associated with the presence of metabolic disorders. For example, the Western-type dietary pattern was found to have a direct association with metabolic syndrome (MS) prevalence in many studies [16–18] and higher adherence to a Western dietary pattern was associated with greater risk of developing MS in prospective cohort studies [19,20]. In a meta-analysis conducted by Rodríguez-Monforte et al., [18] which included 28 cross-sectional studies, the pooled odds ratio (OR) for MS was 0.83 for prudent/healthy patterns and 1.28 for Western/unhealthy patterns. In Greece, in the ATTICA study, the adoption of a traditional Mediterranean-like diet [21] and of a healthy food pattern (rich in low-fat products) [22] was associated with lower odds of MS. Finally, "energy-dense" dietary patterns, characterized by higher consumption of animal meat and/or sauces were usually significantly associated with MS [22–24].

Regarding the associations with cardiometabolic and other factors related to metabolic disorders, our analysis revealed some interesting findings. More specifically, the "Western-type pattern" was positively associated with fat, WC, insulin, ALT and ALP, and negatively associated with vitamin D. Western or western-like patterns have been previously associated with body weight, body fat, waist circumference and BMI in healthy, obese and patients with metabolic disorders [17,25–29]. In CARDIA, a prospective study with 5115 young adults, individuals following the prudent diet had lower risk of high WC and MS than Western diet consumers [30]. BMI and % body fat were positively associated with an "energy-dense meat" pattern, similar to western patterns in the EXPLORE study, which investigated dietary patterns and body composition profiles in premenopausal New Zealand European women [31].

Western or "westernized" dietary patterns have also been associated with several biomarkers of obesity and CVD risk. One of the first studies in this field was The Health Professionals Follow-up study, a prospective cohort study of 51529 US male health professionals, which revealed positive correlations of the Western pattern with insulin and negative with plasma folate concentrations [32]. Additionally, the Western pattern was positively associated with total cholesterol, insulin and fasting blood glucose in patients with T2DM [33]. In the NHANES analysis, which included dietary patterns determined in 13310 US adults, the Western pattern was associated positively with serum insulin, and glycated hemoglobin and negatively with red blood cell folate concentrations [34]. In our study, a positive association with insulin levels confirmed the importance of diet quality in insulin resistance; as such, it is well established that lower consumption of fruits and vegetables and higher consumption of refined grains, processed meat and sweets are associated with increased risk of T2DM [35]. We also observed a positive association with liver function enzymes (AST and ALT). Individuals with high adherence to a Western dietary pattern, were more likely to have elevated ALT and AST levels, not only in MS, but in the general population as well [36,37]. Finally, the Western-type pattern was inversely associated with vitamin D levels. Previous studies have shown that dietary patterns rich in dairy, sea food, eggs, and vegetables were positively associated with serum 25(OH)D levels, while patterns rich in sweets, alcohol, fats, and soft drinks were inversely associated with serum 25(OH)D [38–40].

In our study, the "Mediterranean-like diet pattern" was positively associated with HDL and MCS-12 and negatively with CESD-R. In the ATTICA study. A pattern rich in fish, vegetables, legumes, cereals, and fruits, comparable to our "Mediterranean-like pattern", was inversely associated with HDL-cholesterol level [22]. Nevertheless, there is a plethora of studies showing that greater adherence to the Mediterranean diet is associated with improved blood lipid profile and a significant reduction of major cardiovascular events of almost 30% [41,42].

MCS-12 and CESD-R are two very useful tools for validating mental health and depression, respectively, and, when combined, they indicate an excellent convergent validity in identifying probably clinically significant depression [43]. Excess body weight expressed as disarrayed metabolic status and depression are linked through inflammation and stress in a bidirectional way [44,45]. This vicious cycle is sustained not only through the obvious severe impact of both pathologies on one's mental health but also through intertwined biochemical pathways [44]. Herein, the "Mediterranean-like diet pattern" was positively associated with MCS-12 and negatively with CESD-R, indicating better mental health for those with a greater adherence to this dietary pattern. Similarly, in the MARK study, a longitudinal study of 500 Spanish people with intermediate cardiovascular risk, greater adherence to the Mediterranean diet was associated with higher scores on the MCS-12 [46]. In a large cohort of North Americans with osteoarthritis, a higher adherence to the Mediterranean diet was associated with a higher PCS-12 and a lower CESD [47]. The same did not occur in the Seniors-ENRICA cohort. The PREDIMED score and Trichopoulou's Mediterranean Diet score (MSD) were used to measure Mediterranean diet adherence. Only PREDIMED was associated with PCS, whereas neither PREDIMED nor MSD was associated with PCS or MCS [48]. An improved nutritional status may not be the only factor contributing to mental health when adherence to the Mediterranean diet is high. Improved mental health may also be due to a Mediterranean lifestyle that not only encourages healthy food choices but also enables friends and family to share lunchtimes, which contributes to a higher quality of life [49].

Finally, the "Animal meat and sauces pattern" was positively associated with TBW (kg), BF (kg) and FFM (kg). FFM consists of metabolically active tissues, bones, organs and TBW. The association of this pattern with both parameters was apparently due to the fact that high protein content of animal origin in the diet increases muscle mass and high muscle mass parallels high TBW due to the high content of intracellular water. On the other hand, the association of this pattern with BF could be explained by the high fat content in the diet.

This study contained several strengths, one being the use of a validated food frequency questionnaire, providing comprehensive information on eating habits, together with interviewing of the participants by experienced nutritionists and adjustment for strong potential confounders in the regression analyses. Furthermore, it was conducted in a well-characterized Greek population exploring the relationship between different dietary patterns of this Mediterranean country and health outcomes related to metabolic disorders with a statistical approach that enabled the evaluation of the overall quality of the diet [50]. Limitations might include possible under- or over- reporting of FFQs, the relatively small sample size, the fact that associations derived from an observational study did not necessarily indicate causality, and confirmation of the results in prospective studies was needed, and the disadvantages that sprout from the nature of the PCA and EFA analyses [50].

5. Conclusions

In conclusion, our results suggest that dietary patterns characterized by high consumption of energy-dense and animal derived foods are positively associated with anthropometric and biochemical parameters related to metabolic disorders. On the contrary, plant-based, healthier dietary patterns are associated with a better biochemical and mental health profile of metabolically unhealthy obese individuals.

Supplementary Materials: The following supporting information can be downloaded at: https://www.mdpi.com/article/10.3390/nu14235064/s1, Table S1: Food groups and food items

Author Contributions: Conceptualization, C.A., A.G. and A.C.K.; investigation, C.A., S.-A.K., E.V.; formal analysis, C.T.; writing—original draft preparation, C.A.; writing—review and editing, C.A., S.-A.K., E.V., C.T., A.G. and A.C.K.; supervision, A.C.K.; funding acquisition, A.C.K. All authors have read and agreed to the published version of the manuscript.

Funding: The implementation of the doctoral thesis of S.-A.K. was co-financed by Greece and the European Union (European Social Fund) through the Operational Program "Human Resource Development, Education and Lifelong Learning", 2014–2020, within the framework of the Action "Strengthening human resources through the implementation of doctoral research-Sub-Action 2: IKY grant program for doctoral candidates of Greek universities". Additionally, this work was co-financed by the European Union and Greek national funds (European Social Fund—ESF) through the Operational Program Competitiveness, Entrepreneurship, and Innovation, under the call RESEARCH-CREATE-INNOVATE (project code: T1EDK-02560).

Institutional Review Board Statement: The study was conducted in accordance with the Declaration of Helsinki, and approved by the Harokopio University Ethics Committee (ID protocol: 62/03-07-2018).

Informed Consent Statement: Informed consent was obtained from all subjects involved in the study.

Data Availability Statement: The data presented in this study are available on request from the corresponding author.

Acknowledgments: We are grateful to all participants in this study.

Conflicts of Interest: The authors declare no conflict of interest.

References

1. Cefalu, W.T.; Rodgers, G.P. COVID-19 and metabolic diseases: A heightened awareness of health inequities and a renewed focus for research priorities. *Cell Metab.* **2021**, *33*, 473–478. [CrossRef] [PubMed]
2. World Health Organization. Obesity and Overweight. Available online: https://www.who.int/news-room/fact-sheets/detail/obesity-and-overweight (accessed on 20 October 2022).
3. IDF Diabetes Atlas. Resources. Available online: https://www.diabetesatlas.org/en/resources/ (accessed on 20 October 2022).
4. Yang, M.; Liu, S.; Zhang, C. The Related Metabolic Diseases and Treatments of Obesity. *Healthcare* **2022**, *10*, 1616. [CrossRef] [PubMed]
5. Sarma, S.; Sockalingam, S.; Dash, S. Obesity as a multisystem disease: Trends in obesity rates and obesity-related complications. *Diabetes Obes. Metab.* **2021**, *23*, 3–16. [CrossRef] [PubMed]
6. Ebrahimi, S.; Leech, R.M.; McNaughton, S.A.; Abdollahi, M.; Houshiarrad, A.; Livingstone, K.M. Associations between diet quality and obesity in a nationally representative sample of Iranian households: A cross-sectional study. *Obes. Sci. Pract.* **2021**, *8*, 12–20. [CrossRef] [PubMed]
7. Latorre-Millán, M.; Rupérez, A.I.; González-Gil, E.M.; Santaliestra-Pasías, A.; Vázquez-Cobela, R.; Gil-Campos, M.; Aguilera, C.M.; Gil, Á.; Moreno, L.A.; Leis, R.; et al. Dietary Patterns and Their Association with Body Composition and Cardiometabolic Markers in Children and Adolescents: Genobox Cohort. *Nutrients* **2020**, *12*, 3424. [CrossRef]
8. Romieu, I.; Dossus, L.; Barquera, S.; Blottière, H.M.; Franks, P.W.; Gunter, M.; Hwalla, N.; Hursting, S.D.; Leitzmann, M.; Margetts, B.; et al. Energy balance and obesity: What are the main drivers? *Cancer Causes Control* **2017**, *28*, 247–258. [CrossRef]
9. Medina-Remón, A.; Kirwan, R.; Lamuela-Raventós, R.M.; Estruch, R. Dietary patterns and the risk of obesity, type 2 diabetes mellitus, cardiovascular diseases, asthma, and neurodegenerative diseases. *Crit. Rev. Food Sci. Nutr.* **2018**, *58*, 262–296. [CrossRef]
10. Kleftaki, S.A.; Amerikanou, C.; Gioxari, A.; Lantzouraki, D.Z.; Sotiroudis, G.; Tsiantas, K.; Tsiaka, T.; Tagkouli, D.; Tzavara, C.; Lachouvaris, L.; et al. A Randomized Controlled Trial on *Pleurotuseryngii* Mushrooms with Antioxidant Compounds and Vitamin D_2 in Managing Metabolic Disorders. *Antioxidants* **2022**, *11*, 2113. [CrossRef]
11. Craig, C.L.; Marshall, A.L.; Sjöström, M.; Bauman, A.E.; Booth, M.L.; Ainsworth, B.E.; Pratt, M.; Ekelund, U.; Yngve, A.; Sallis, J.F.; et al. International physical activity questionnaire: 12-country reliability and validity. *Med. Sci. Sports Exerc.* **2003**, *35*, 1381–1395. [CrossRef]
12. Björgvinsson, T.; Kertz, S.J.; Bigda-Peyton, J.S.; McCoy, K.L.; Aderka, I.M. Psychometric properties of the CES-D-10 in a psychiatric sample. *Assessment* **2013**, *20*, 429–436. [CrossRef]
13. Rosenberg, M. *Society and the Adolescent Self-Image*; Princeton University Press: Princeton, NJ, USA, 1965.
14. Soldatos, C.R.; Dikeos, D.G.; Paparrigopoulos, T.J. Athens Insomnia Scale: Validation of an instrument based on ICD-10 criteria. *J. Psychosom. Res.* **2000**, *48*, 555–560. [CrossRef] [PubMed]
15. Bountziouka, V.; Bathrellou, E.; Giotopoulou, A.; Katsagoni, C.; Bonou, M.; Vallianou, N.; Barbetseas, J.; Avgerinos, P.C.; Panagiotakos, D.B. Development, repeatability and validity regarding energy and macronutrient intake of a semi-quantitative food frequency questionnaire: Methodological considerations. *Nutr. Metab. Cardiovasc. Dis.* **2012**, *22*, 659–667. [CrossRef] [PubMed]
16. Denova-Gutiérrez, E.; Castañón, S.; Talavera, J.O.; Gallegos-Carrillo, K.; Flores, M.; Dosamantes-Carrasco, D.; Willett, W.C.; Salmerón, J. Dietary patterns are associated with metabolic syndrome in an urban Mexican population. *J. Nutr.* **2010**, *140*, 1855–1863. [CrossRef] [PubMed]
17. Hosseini, Z.; Whiting, S.J.; Vatanparast, H. Current evidence on the association of the metabolic syndrome and dietary patterns in a global perspective. *Nutr. Res. Rev.* **2016**, *29*, 152–162. [CrossRef]
18. Rodríguez-Monforte, M.; Sánchez, E.; Barrio, F.; Costa, B.; Flores-Mateo, G. Metabolic syndrome and dietary patterns: A systematic review and meta-analysis of observational studies. *Eur. J. Nutr.* **2017**, *56*, 925–947. [CrossRef]
19. Esmaillzadeh, A.; Kimiagar, M.; Mehrabi, Y.; Azadbakht, L.; Hu, F.B.; Willett, W.C. Dietary patterns, insulin resistance, and prevalence of the metabolic syndrome in women. *Am. J. Clin. Nutr.* **2007**, *85*, 910–918. [CrossRef]
20. Lutsey, P.L.; Steffen, L.M.; Stevens, J. Dietary intake and the development of the metabolic syndrome: The Atherosclerosis Risk in Communities study. *Circulation.* **2008**, *117*, 754–761. [CrossRef]
21. Panagiotakos, D.B.; Pitsavos, C.; Chrysohoou, C.; Skoumas, J.; Tousoulis, D.; Toutouza, M.; Toutouzas, P.; Stefanadis, C. Impact of lifestyle habits on the prevalence of the metabolic syndrome among Greek adults from the ATTICA study. *Am. Heart J.* **2004**, *147*, 106–112. [CrossRef]

22. Panagiotakos, D.B.; Pitsavos, C.; Skoumas, Y.; Stefanadis, C. The association between food patterns and the metabolic syndrome using principal components analysis: The ATTICA Study. *J. Am. Diet. Assoc.* **2007**, *107*, 979–997. [CrossRef]
23. He, Y.; Li, Y.; Lai, J.; Wang, D.; Zhang, J.; Fu, P.; Yang, X.; Qi, L. Dietary patterns as compared with physical activity in relation to metabolic syndrome among Chinese adults. *Nutr. Metab. Cardiovasc. Dis.* **2013**, *23*, 920–928. [CrossRef]
24. Wagner, A.; Dallongeville, J.; Haas, B.; Ruidavets, J.B.; Amouyel, P.; Ferrières, J.; Simon, C.; Arveiler, D. Sedentary behaviour, physical activity and dietary patterns are independently associated with the metabolic syndrome. *Diabetes Metab.* **2012**, *38*, 428–435. [CrossRef] [PubMed]
25. Lopez-Garcia, E.; Schulze, M.B.; Fung, T.T.; Meigs, J.B.; Rifai, N.; Manson, J.E.; Hu, F.B. Major dietary patterns are related to plasma concentrations of markers of inflammation and endothelial dysfunction. *Am. J. Clin. Nutr.* **2004**, *80*, 1029–1035. [CrossRef] [PubMed]
26. Seifu, C.N.; Fahey, P.P.; Hailemariam, T.G.; Frost, S.A.; Atlantis, E. Dietary patterns associated with obesity outcomes in adults: An umbrella review of systematic reviews. *Public Health Nutr.* **2021**, *24*, 6390–6414. [CrossRef] [PubMed]
27. Murtaugh, M.A.; Herrick, J.S.; Sweeney, C.; Baumgartner, K.B.; Guiliano, A.R.; Byers, T.; Slattery, M.L. Diet composition and risk of overweight and obesity in women living in the southwestern United States. *J. Am. Diet. Assoc.* **2007**, *107*, 1311–1321. [CrossRef] [PubMed]
28. Schulze, M.B.; Fung, T.T.; Manson, J.E.; Willett, W.C.; Hu, F.B. Dietary patterns and changes in body weight in women. *Obesity* **2006**, *14*, 1444–1453. [CrossRef]
29. Talenezhad, N.; Mirzavandi, F.; Rahimpour, S.; Amel Shahbaz, A.P.; Mohammadi, M.; Hosseinzadeh, M. Empirically derived dietary pattern and odds of non-alcoholic fatty liver diseases in overweight and obese adults: A case-control study. *BMC Gastroenterol.* **2022**, *22*, 158. [CrossRef]
30. Duffey, K.J.; Steffen, L.M.; Van Horn, L.; Jacobs, D.R., Jr.; Popkin, B.M. Dietary patterns matter: Diet beverages and cardiometabolic risks in the longitudinal Coronary Artery Risk Development in Young Adults (CARDIA) Study. *Am. J. Clin. Nutr.* **2012**, *95*, 909–915. [CrossRef]
31. Schrijvers, J.K.; McNaughton, S.A.; Beck, K.L.; Kruger, R. Exploring the Dietary Patterns of Young New Zealand Women and Associations with BMI and Body Fat. *Nutrients* **2016**, *8*, 450. [CrossRef]
32. Fung, T.T.; Rimm, E.B.; Spiegelman, D.; Rifai, N.; Tofler, G.H.; Willett, W.C.; Hu, F.B. Association between dietary patterns and plasma biomarkers of obesity and cardiovascular disease risk. *Am. J. Clin. Nutr.* **2001**, *73*, 61–67. [CrossRef]
33. Zad, N.D.; Yusof, R.M.; Esmaili, H.; Jamaluddin, R.; Mohseni, F. Association of dietary pattern with biochemical blood profiles and bodyweight among adults with Type 2 diabetes mellitus in Tehran, Iran. *J. Diabetes Metab. Disord.* **2015**, *14*, 28. [CrossRef]
34. Kerver, J.M.; Yang, E.J.; Bianchi, L.; Song, W.O. Dietary patterns associated with risk factors for cardiovascular disease in healthy US adults. *Am. J. Clin. Nutr.* **2003**, *78*, 1103–1110. [CrossRef] [PubMed]
35. Moradi, S.; Kermani, M.A.H.; Bagheri, R.; Mohammadi, H.; Jayedi, A.; Lane, M.M.; Asbaghi, O.; Mehrabani, S.; Suzuki, K. Ultra-Processed Food Consumption and Adult Diabetes Risk: A Systematic Review and Dose-Response Meta-Analysis. *Nutrients* **2021**, *13*, 4410. [CrossRef] [PubMed]
36. Lin, L.Y.; Hsu, C.Y.; Chiou, H.Y.; Lee, H.A.; Hsu, L.M.; Chang, P.Y.; Kurniawan, A.L.; Chao, J.C. Association between Dietary Patterns and Serum Hepatic Enzyme Levels in Adults with Dyslipidemia and Impaired Fasting Plasma Glucose. *Nutrients* **2021**, *13*, 987. [CrossRef]
37. Lorzadeh, E.; Akhondi-Meybodi, M.; Mozaffari-Khosravi, H.; Mirzaei, M.; Salehi-Abargouei, A. Association between empirically derived dietary patterns and liver function tests in adults: Shahedieh cohort study. *Nutrition* **2021**, *81*, 110897. [CrossRef] [PubMed]
38. Sharifan, P.; Yaghooti-Khorasani, M.; Asadi, Z.; Darroudi, S.; Rezaie, M.; Safarian, M.; Vatanparast, H.; Eslami, S.; Tayefi, M.; Pourrahim, E.; et al. Association of dietary patterns with serum vitamin D concentration among Iranian adults with abdominal obesity. *Clin. Nutr. Open Sci.* **2021**, *40*, 40–49. [CrossRef]
39. Denova-Gutiérrez, E.; Clark, P.; Muñoz-Aguirre, P.; Flores, M.; Talavera, J.O.; Chico-Barba, L.G.; Rivas, R.; Ramírez, P.; Salmerón, J. Dietary patterns are associated with calcium and vitamin D intake in an adult Mexican population. *Nutr. Hosp.* **2016**, *33*, 276. [CrossRef]
40. Grigoriou, E.; Trovas, G.; Papaioannou, N.; Dontas, I.; Makris, K.; Apostolou-Karampelis, K.; Dedoussis, G. Dietary Patterns of Greek Adults and Their Associations with Serum Vitamin D Levels and Heel Quantitative Ultrasound Parameters for Bone Health. *Nutrients* **2020**, *12*, 123. [CrossRef]
41. Guasch-Ferré, M.; Salas-Salvadó, J.; Ros, E.; Estruch, R.; Corella, D.; Fitó, M.; Martínez-González, M.A.; PREDIMED Investigators. The PREDIMED trial. Mediterranean diet and health outcomes: How strong is the evidence? *Nutr. Metab. Cardiovasc. Dis.* **2017**, *27*, 624–632. [CrossRef]
42. Bakaloudi, D.R.; Chrysoula, L.; Kotzakioulafi, E.; Theodoridis, X.; Chourdakis, M. Impact of the Level of Adherence to Mediterranean Diet on the Parameters of Metabolic Syndrome: A Systematic Review and Meta-Analysis of Observational Studies. *Nutrients* **2021**, *13*, 1514. [CrossRef]
43. Yu, D.S.; Yan, E.C.; Chow, C.K. Interpreting SF-12 mental component score: An investigation of its convergent validity with CESD-10. *Qual. Life Res.* **2015**, *24*, 2209–2217. [CrossRef]
44. Plackett, B. The vicious cycle of depression and obesity. *Nature* **2022**, *608*, 42–43. [CrossRef] [PubMed]

45. Milaneschi, Y.; Simmons, W.K.; van Rossum, E.F.C.; Penninx, B.W. Depression and obesity: Evidence of shared biological mechanisms. *Mol. Psychiatry* **2019**, *24*, 18–33. [CrossRef] [PubMed]
46. Sanchez-Aguadero, N.; Alonso-Dominguez, R.; Garcia-Ortiz, L.; Agudo-Conde, C.; Rodriguez-Martin, C.; de Cabo-Laso, A.; Sanchez-Salgado, B.; Ramos, R.; Maderuelo-Fernandez, J.A.; Gomez-Marcos, M.A.; et al. Diet and physical activity in people with intermediate cardiovascular risk and their relationship with the health-related quality of life: Results from the MARK study. *Health Qual. Life Outcomes* **2016**, *14*, 169. [CrossRef] [PubMed]
47. Veronese, N.; Stubbs, B.; Noale, M.; Solmi, M.; Luchini, C.; Maggi, S. Adherence to the Mediterranean diet is associated with better quality of life: Data from the Osteoarthritis Initiative. *Am. J. Clin. Nutr.* **2016**, *104*, 1403–1409. [CrossRef] [PubMed]
48. Pérez-Tasigchana, R.F.; León-Muñoz, L.M.; López-García, E.; Banegas, J.R.; Rodríguez-Artalejo, F.; Guallar-Castillón, P. Correction: Mediterranean Diet and Health-Related Quality of Life in Two Cohorts of Community-Dwelling Older Adults. *PLoS One* **2016**, *11*, e0155171. [CrossRef]
49. Muñoz, M.A.; Fíto, M.; Marrugat, J.; Covas, M.I.; Schröder, H.; REGICOR and HERMES investigators. Adherence to the Mediterranean diet is associated with better mental and physical health. *Br. J. Nutr.* **2009**, *101*, 1821–1827. [CrossRef]
50. Zhao, J.; Li, Z.; Gao, Q.; Zhao, H.; Chen, S.; Huang, L.; Wang, W.; Wang, T. A review of statistical methods for dietary pattern analysis. *Nutr. J.* **2021**, *20*, 37. [CrossRef]

nutrients

MDPI

Article

The Effect of Dietary Pattern on Metabolic Syndrome in a Suburban Population in Shanghai, China

Lanxin Wei [1], Jing Fan [1], Ruihua Dong [1], Mei Zhang [2], Yonggen Jiang [3], Qi Zhao [1], Genming Zhao [1], Bo Chen [1], Jing Li [2,*] and Shaojie Liu [1,*]

1 Key Laboratory of Public Health Safety of Ministry of Education, School of Public Health, Fudan University, Shanghai 200032, China; lxwei21@m.fudan.edu.cn (L.W.); 21111020063@m.fudan.edu.cn (J.F.); ruihua_dong@fudan.edu.cn (R.D.); zhaoqi@shmu.edu.cn (Q.Z.); gmzhao@shmu.edu.cn (G.Z.); chenb@fudan.edu.cn (B.C.)
2 Zhongshan Community Health Care Center, Songjiang District, Shanghai 201613, China; 18918287235@163.com
3 Songjiang District Center for Disease Control and Prevention, Shanghai 201620, China; sjjktj@hotmail.com
* Correspondence: zhongshanlijing@163.com (J.L.); liushaojie@fudan.edu.cn (S.L.)

Abstract: Metabolic syndrome (MetS) is recognized as one of the most severe non-communicable chronic diseases. Diet plays an essential role in the development and exacerbation of MetS. Thus, this study aimed to investigate the relationship between dietary patterns and MetS in a suburban population in Shanghai, China. Data were collected on the Zhongshan community from the Shanghai Suburban Adult Cohort and Biobank (SSACB) study between May and September 2017. A total of 5426 participants who completed the questionnaire investigation, physical measurements, and biological sample collection were effectively enrolled in this study. Both posteriori and priori methods were utilized to generate different dietary patterns, including the dietary approaches to stop hypertension (DASH) and Mediterranean diet (MD). The prevalence of MetS in this study was 22.47%. Compared to the reference, dietary patterns with a higher intake of "dairy and fruits" and "coarse cereals and soy products" had protective effects on MetS ($p < 0.05$). However, no significant correlation with MetS was observed for DASH and MD. Our study recommends higher consumption of fruits, coarse cereals, and soy products, which was associated with a lower prevalence of MetS in the suburban population of Shanghai. The correlation of DASH and MD with MetS in the Chinese population requires further exploration.

Keywords: metabolic syndrome; posteriori dietary pattern; priori dietary pattern; cross-section study

Citation: Wei, L.; Fan, J.; Dong, R.; Zhang, M.; Jiang, Y.; Zhao, Q.; Zhao, G.; Chen, B.; Li, J.; Liu, S. The Effect of Dietary Pattern on Metabolic Syndrome in a Suburban Population in Shanghai, China. *Nutrients* **2023**, *15*, 2185. https://doi.org/10.3390/nu15092185

Academic Editor: Hermann Toplak

Received: 7 April 2023
Revised: 25 April 2023
Accepted: 28 April 2023
Published: 4 May 2023

1. Introduction

Metabolic syndrome (MetS) is a collection of intricate syndromes typified by metabolic disorders, including being overweight, obesity, hypertension, hyperglycemia, and dyslipidemia, as defined by the diagnostic criteria of the Chinese Diabetes Society (CDS). The prevalence of MetS has increased significantly in several countries. In Korea, the prevalence of MetS increased from 24.9% in 1998 to 37% in 2012 [1]. In China, the prevalence of MetS increased from 13.7% in 2000 to 24.2% in 2012 [2,3]. In the United States, approximately one-third of adults have MetS [4], whereas in Russia, the prevalence was 23.1% and 11.0% among women and men, respectively [5]. MetS has been regarded as one of the most serious non-communicable chronic diseases and is a risk factor for type 2 diabetes and cardiovascular disease (CVD) [6,7]. Several studies have found that the prevalence of diabetes and CVD in the population with MetS was five-fold and doubled-fold higher than that without MetS, respectively [8,9]. The rapid increase in MetS prevalence may be attributed to changes in population behavior patterns in modern society [10,11]. Among these patterns, diet plays an essential and independent role in the incidence and development of MetS [12,13]. Hence, it is important and meaningful to explore the association between diet consumption and the prevalence of MetS.

Several studies have assessed the effects of single foods or food compositions on MetS [14,15]. However, given the synergistic effects of different foods, dietary patterns may be more appropriate to evaluate the association between food intake and MetS. Two methods are commonly used for evaluating dietary patterns: the a priori and a posteriori methods. A priori dietary patterns, also known as dietary indices, can intuitively provide a comprehensive explanation of complex results. Various dietary indices have been developed to investigate the relationship between MetS and dietary patterns, including the dietary approaches to stop hypertension (DASH) diet, the Mediterranean diet, and the Healthy Eating Index [16–19]. Conversely, the a posteriori method analyses dietary data using cluster analysis, factor analysis, or latent class analysis to identify and generalize different dietary patterns. This method is more flexible than the a priori method, with fixed food items and is closer to reality. Thus, both methods have individual advantages in exploring the association between diet and disease. However, most studies have only used one method to investigate the relationship between dietary patterns and MetS [20,21]. Therefore, it is necessary to adopt both methods to comprehensively assess the relationship between dietary patterns and MetS.

In this study, we employ both a priori and a posteriori methods to examine the association between dietary patterns and MetS in a suburban population in China.

2. Subjects and Methods

2.1. Study Population

The study subjects were recruited between May and September 2017 from the Zhongshan community from the Shanghai Suburban Adult Cohort and Biobank (SSACB) [22]. The overarching objective of this cohort is to identify the environmental, lifestyle, and genetic risk factors for non-communicable chronic diseases in adults residing in the suburbs of Shanghai. The cohort details have been described previously [22]. Briefly, a stratified clustered sampling design was employed to obtain data from participants aged between 20 to 74 years old, who were randomly selected through a multistage sampling method. We selected study participants from the Zhongshan community from the SSACB. Participants were excluded if: (1) they lacked a food frequency questionnaire (FFQ) or consumption data on cooking oil and condiments; (2) their self-reported energy intake was less than 800 kcal/d or more than 4000 kcal/d for males and less than 500 kcal/d or more than 3500 kcal/d for females; (3) they had abnormal intake of a single food (exceeding 1000 g/d). Ultimately, a total of 5426 participants were deemed eligible for our study. The flowchart of this study was showed in Figure 1. The Medical Research Ethics Committee of the School of Public Health, Fudan University, reviewed and approved the study (No.: IRB#2016-04-0586), and all the research subjects provided informed consent before participating in the investigation.

2.2. Questionnaire Survey

Trained investigators conducted face-to-face interviews to collect data, via questionnaire surveys, from all participants. The surveys captured general information (e.g., age, gender, education level, marital status, and retirement status), lifestyle behaviors (e.g., cigarette smoking, alcohol and tea consumption, and physical activity), as well as food consumption information via a 29 food group-based FFQ, which included rice and grain products, among other food types. The reliability and validity of the FFQ was previously established [23]. Briefly, the FFQ's reliability was assessed for 152 participants selected randomly from the Zhongshan community in the SSACB by comparing two FFQ surveys taken 12 months apart. Concurrently, the FFQ's validity was evaluated for 165 participants by comparing the FFQ data with 3-day 24-hour dietary recalls. The results demonstrated a strong correlation coefficient for both food groups (reliability: 0.36–0.54; validity: 0.20–0.41) and nutrients (reliability: 0.39–0.60; validity: 0.12–0.42).

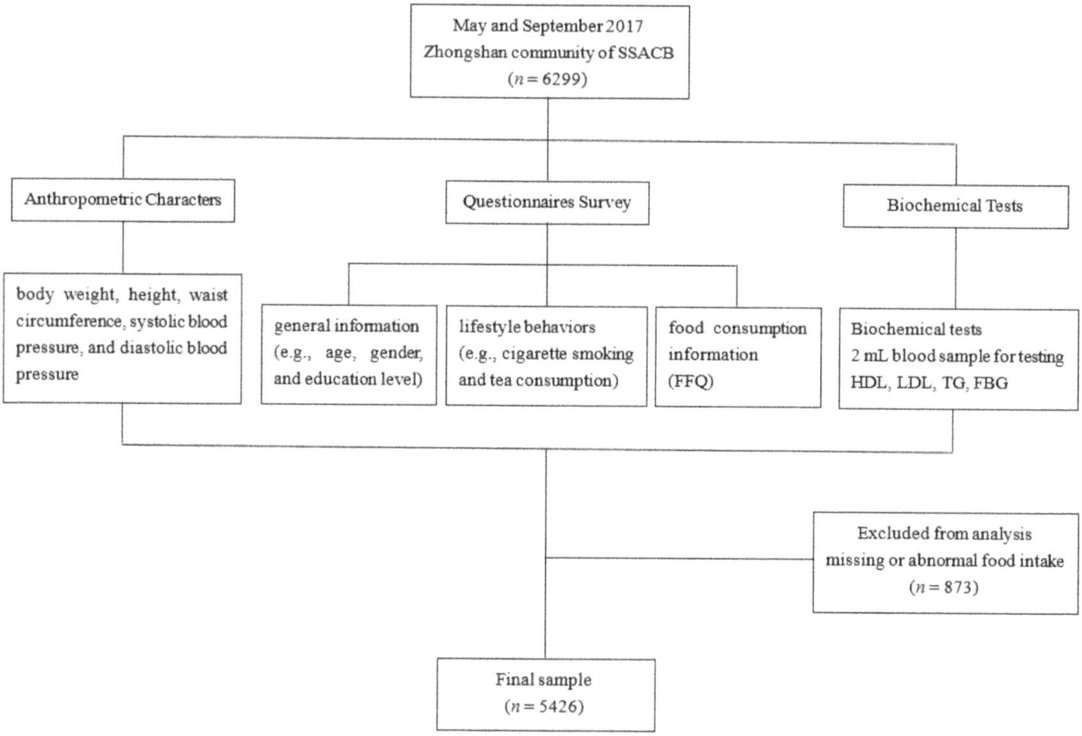

Figure 1. The flowchart of this study.

2.3. Anthropometric Characters

Trained technologists measured the anthropometric characteristics, including body weight, height, waist circumference, systolic blood pressure, and diastolic blood pressure, following the national standard WS/T424-2013. Specifically, height was measured using a column height meter with an accuracy of 0.1 cm, while weight was measured using an electronic weight meter with an accuracy of 0.1 kg. Waist circumference was measured using a flexible tape at the midpoint between the iliac crest and the last rib, with the subject at minimal respiration, and had an accuracy of 0.1 cm. Systolic and diastolic blood pressure were measured at the right arm of seated study participants after a five minute rest period, with readings recorded to the nearest 2 mmHg. To ensure accuracy, all physical measurements were performed three times, and the mean values were used for analysis.

2.4. Biochemical Tests

Trained technologists collected a 2 mL blood sample from all the study participants and drew it into tubes containing EDTA. The samples were then transported to the Shanghai Dian Diagnosis Innovation Professional clinical laboratories for testing. Biochemical tests were conducted to measure the cardiovascular-related indicators for determining the prevalence of MetS in individuals. Specifically, we measured high-density lipoprotein (HDL), low-density lipoprotein (LDL), triglyceride (TG), and fasting blood glucose (FBG).

2.5. Covariate Assessment

We calculated the total hours of metabolism equivalent (MET) per week, based on the Compendium of Physical Activity [24]. Energy intake per day was estimated using the FFQ. Body mass index (BMI, kg/m^2) was determined using the formula = weight (kg) / height^2 (m^2), and then classified into four categories based on the Working Group on Obesity in China: underweight (< 18.5 kg/m^2), normal (18.5–23.9 kg/m^2), overweight (24.0–27.9 kg/m^2), and obesity ($\geq$ 28.0 kg/m^2). Education level was categorized as primary school and below, junior high school, high school and above. Marital status was classified as married or other. To identify the presence of MetS, we used the criteria set by the Chinese Diabetes Society (CDS), which defines MetS as having three or more of the following five criteria: (1) elevated waist circumference: waist circumference $\geq$ 90 cm in males, $\geq$ 85 cm in females; (2) hyperglycemia status: fasting blood glucose (FBG) $\geq$ 6.1 mmol/L or a diabetes diagnosis; (3) hypertension status: systolic/diastolic blood pressure $\geq$ 130/85 mmHg or a hypertension diagnosis; (4) elevated triglyceridemic: triglyceridemic (TG) $\geq$ 1.7 mmol/L; (5) reduced HDL: HDL < 1.04 mmol/L [25].

2.6. *Analysis of Dietary Patterns*

2.6.1. A posteriori Dietary Pattern Analysis

Both factor analysis and cluster analysis were utilized to identify the dietary patterns of the study participants. The procedures are explained in detail below.

First, the average daily intake for each food group was subjected to a cluster analysis using the K-mean method to identify dietary patterns among the study participants. The analysis was performed with 50 iterations, centering results on zero, and the subjects were divided into four groups based on the Euclidean distance between the observations.

Second, a factor analysis was conducted to identify dietary patterns. The adequacy of the sample was first verified using the Kaiser–Meyer–Olkin (KMO) index and Bartlett's test of sphericity. Principal component analysis was then employed to extract factors, followed by orthogonal rotation. The factors with eigenvalues greater than one were retained. Only absolute values of factor loading > 0.2 were included in the analysis, as high factor loads indicated a strong relationship with the identified factors. Factor scores were further categorized into four quartiles, where high scores indicated a greater adherence to the identified dietary pattern.

2.6.2. A priori Dietary Pattern Analysis

The adherence to two diet quality scores for the dietary approaches to stop hypertension (DASH) diet and the Mediterranean diet (MD) were calculated. The DASH and MD scores ranged from 0 to 9, with higher scores indicating greater adherence to the respective dietary patterns.

The DASH diet score for each participant was determined using the formula developed by Mellen et al. [26]. The score comprised of nine nutrients (total fat, saturated fat, protein, fiber, cholesterol, calcium, magnesium, sodium, and potassium), with micronutrient goals reported per 1000 kcal. Each nutrient was assigned a score of 0, 0.5, or 1 based on whether the target was not met, partially met, or fully met, respectively. The DASH score was then calculated as the sum of all the nutrient targets met by the individual (Supplementary Table S1).

The traditional MD score was originally developed by Trichopoulou et al. [27], to assess adherence to the Mediterranean diet in a Greek population. Fung et al. [28] later revised this score and developed the alternate Mediterranean diet score, which included vegetables, fruits, nuts, whole grains, legumes, fish, the monounsaturated/saturated fat ratio, red and processed meats, and alcohol. In our study, the MD score for each participant was calculated using the formula developed by Fung et al. The total MD score was computed as the sum of the nutrient targets met, with a score of 1 assigned to the consumption of food groups considered beneficial to health at or above the sex-specific median (vegetables, fruits, nuts, whole grains, legumes, fish, and the monounsaturated/saturated fat ratio), below the median for food groups presumed to be detrimental to health (red and processed meats), and moderate ethanol consumption (5–15 g/d in females and 15–25 g/d in males) (Supplementary Table S2).

2.7. Statistical Analysis

Continuous variables were reported as mean ± SD and categorical variables were presented as frequency (%). Balance tests for continuous and categorical variables were conducted using the Student *t*-test and chi-square, respectively. Non-conditional logistic regression analysis was performed to examine the relationship between each diet pattern score and MetS. All statistical analyses were conducted using IBM SPSS Statistics software Version 20.0. Two-sided p values < 0.05 were considered statistically significant.

3. Results

3.1. Personal Characteristics and Prevalence of MetS

Table 1 displays the personal characteristics of the study participants stratified by MetS status. Out of the total sample of 5426 participants, 1219 were diagnosed with MetS, while 4207 were classified as non-MetS. The prevalence of MetS was 22.47%. Statistically significant differences ($p < 0.05$) were observed between the two groups in terms of age, BMI, physical activity, gender, retirement status, education level, cigarette smoking, alcohol drinking, and tea consumption. Furthermore, individuals with MetS exhibited higher levels for waist circumference, fasting blood glucose, systolic blood pressure, diastolic blood pressure, triglycerides, HDL-C, and LDL-C compared to those without MetS ($p < 0.05$).

Table 1. Personal characteristics of study participants between the population with and without MetS.

Variables	With MetS (n = 1219)	Without MetS (n = 4207)	p-Value
Gender [a]			< 0.001
Male	602(49.38)	1404(33.40)	
Female	617(50.62)	2803(66.63)	
Retirement status [a]			0.002
Yes	831(68.17)	2667(63.39)	
No	388(31.83)	1540(36.61)	
Marital status [a]			0.126
Married	1141(93.60)	3883(92.30)	
Other	78(6.40)	324(7.70)	
Education level [a]			< 0.001
Primary school and below	593(48.65)	1632(38.79)	
Junior high school	419(34.37)	1624(38.60)	
High school and above	207(16.98)	951(22.61)	
Cigarette smoking [a]			< 0.001
No	829(68.01)	3354(79.72)	
Yes	390(31.99)	853(20.28)	
Alcohol drinking [a]			< 0.001
No	1014(83.18)	3741(88.92)	
Yes	205(16.82)	466(11.08)	
Tea drinking [a]			< 0.001
No	710(58.24)	2858(67.93)	
Yes	509(41.76)	1349(32.07)	
Age [b]	58.31 ± 8.77	55.39 ± 10.22	< 0.001
BMI (kg/m^2) [b]	26.93 ± 3.09	23.95 ± 3.00	< 0.001
Energy intake (Kcal/d) [b]	1378.17 ± 416.41	1393.72 ± 434.84	0.267
Physical activity (MET min/week) [b]	3458.72 ± 2207.95	3670.39 ± 2157.09	0.003
Waist circumference (cm) [b]	86.73 ± 8.29	77.28 ± 7.96	< 0.001
Fasting blood glucose (mmol/L) [b]	6.64 ± 2.06	5.46 ± 1.08	< 0.001
Systolic blood pressure (mmHg) [b]	146.38 ± 20.85	132.42 ± 21.57	< 0.001
Diastolic blood pressure (mmHg) [b]	88.34 ± 11.02	81.10 ± 10.87	< 0.001
Triglycerides (mmol/L) [b]	2.94 ± 1.89	1.50 ± 0.83	< 0.001
HDL cholesterol (mmol/L) [b]	1.06 ± 0.27	1.43 ± 0.29	< 0.001
LDL cholesterol (mmol/L) [b]	2.71 ± 0.98	2.80 ± 0.78	0.001

Abbreviations: MetS: metabolic syndrome; SD: standard deviation; MET: metabolic equivalent task. [a] Categorical variables were presented as frequency (%); [b] continuous variables were reported as mean ± SD.

3.2. A posteriori Dietary Pattern Analysis

We classified 29 food items from the FFQ into 19 predefined food groups for the purpose of cluster analysis (Supplementary Table S3). As indicated in Table 2, the resulting four clusters were labeled as follows: "snacks and beverages" dietary pattern (cluster I with $n = 291$), "grains and vegetables" dietary pattern (cluster II with $n = 1051$), "balanced" dietary pattern (cluster III with $n = 2722$), and "dairy and fruits" dietary pattern (cluster IV with $n = 1362$).

Table 2. Identifying a posteriori dietary pattern based on food intake of study participants by cluster analysis.

Food Groups	Cluster I ($n = 291$)	Cluster II ($n = 1051$)	Cluster III ($n = 2722$)	Cluster IV ($n = 1362$)
Grain	341.84	402.63	311.99	277.44
Coarse cereals	19.01	14.83	10.92	32.72
Starchy vegetables	20.23	19.14	14.41	32.56
Vegetables	220.41	378.58	165.81	235.99
Dark vegetables	129.09	244.55	65.91	115.80
Fruits	137.86	109.17	95.11	168.65
Dairy	125.05	56.24	56.04	149.97
Meats	80.79	81.39	48.75	78.19
Fish	41.88	46.52	27.12	57.66
Other seafood	20.04	16.41	10.14	25.17
Soy products	6.58	5.25	3.18	8.15
Eggs	32.52	37.25	25.45	41.28
Sugared beverages	212.81	8.90	6.20	7.57
Snacks	42.18	27.84	18.94	42.19
Salted food	17.55	18.01	8.56	14.24

Cluster I: snacks and beverages dietary pattern; cluster II: grains and vegetables dietary pattern; cluster III: balanced dietary pattern; cluster IV: dairy and fruits dietary pattern.

Both the Kaiser–Meyer–Olkin index (0.711) and Bartlett's test ($p < 0.001$) confirmed the suitability of the data for factor analysis. We identified five distinct dietary patterns among the study participants and labeled them as follows: "high protein" dietary pattern (factor I), "grains and vegetables" dietary pattern (factor II), "coarse cereals and soy products" dietary pattern (factor III), "snacks and beverages" dietary pattern (factor IV), and "dairy and fruits" dietary pattern (factor V), based on the highest factor loading for food items and interpretability (Figure 2 and Supplementary Table S4). The variance explained by factor I, factor II, factor III, factor IV, and factor V was 10.933%, 9.869%, 9.556%, 9.062%, and 8.533%, respectively.

3.3. A priori Dietary Pattern Analysis

As presented in Table 3, the present results revealed that study participants had a priori dietary pattern score of 1.48 (SD: 1.14) for DASH and 3.93 (SD: 1.59) for MD. Notably, no statistically significant differences in the DASH and MD scores were observed between the study populations with and without MetS ($p > 0.05$).

3.4. Association between Dietary Patterns and MetS

Table 4 presents the results of the non-conditional logistic regression analysis examining the association between the a posteriori dietary patterns and MetS. The results of the cluster analysis revealed that the study participants with the "dairy and fruits" dietary pattern (cluster IV) had a protective effect against MetS compared to those with the "balanced" dietary pattern (cluster I) in both unadjusted and adjusted models, with OR values of 0.66 (95% CI: 0.56, 0.78) and 0.81 (95% CI: 0.66, 0.98), respectively. For factor analysis, in the unadjusted model, a higher score for the "coarse cereals and soy products" and "dairy and fruits" dietary pattern was associated with a significantly lower risk of MetS, with OR values of 0.67 (95% CI: 0.56, 0.80; $p_{trend} < 0.001$) and 0.60 (95% CI: 0.50, 0.72; $p_{trend} < 0.001$), respectively. After adjusting for covariates, a higher score for the "coarse cereals and soy products" dietary pattern was still associated with a significantly lower risk of MetS, with an OR value of 0.74 (95%CI: 0.61, 0.91; $p_{trend} = 0.007$).

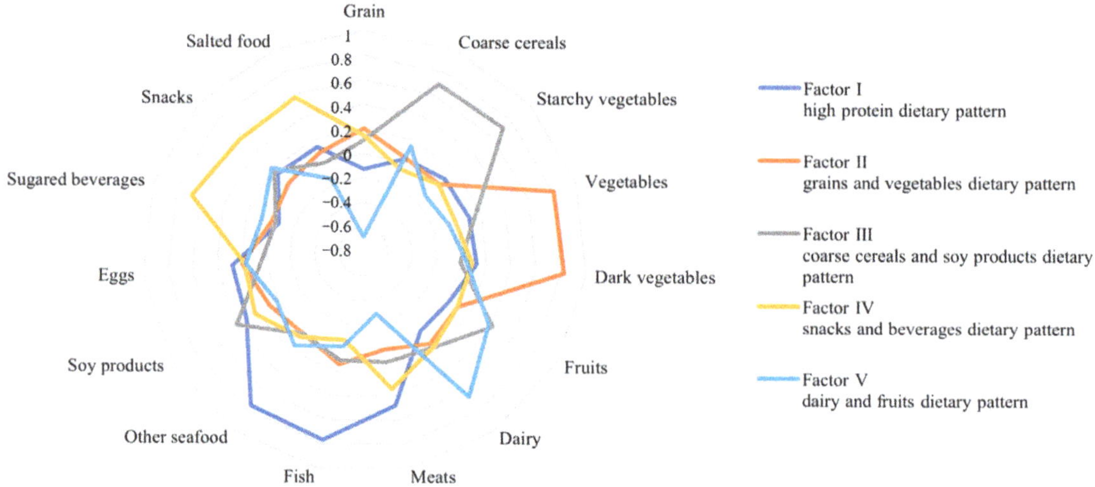

Figure 2. Radar chart for the different dietary patterns from the factor analysis.

Table 3. A priori dietary pattern scores of study participants between the population with and without MetS (n = 5426).

Dietary Pattern Score, Mean ± SD	Total (n = 5426)	With MetS (n = 1219)	Without MetS (n = 4207)	p-Value [a]
DASH	1.48 ± 1.14	1.46 ± 1.13	1.48 ± 1.14	0.488
MD	3.93 ± 1.59	3.89 ± 1.57	3.95 ± 1.59	0.248

Abbreviations: DASH: dietary approaches to stop hypertension; MD: Mediterranean diet. [a] p-value for t-test.

Table 4. A posteriori dietary patterns in association with the risk of MetS using non-conditional logistic regression models.

Method	Dietary Pattern	Model 1		Model 2	
		OR (95 % CI)	p-Value	OR (95 % CI)	p-Value
Cluster analysis	Balanced pattern	Reference		Reference	
	Snacks and beverages pattern	0.92 (0.69, 1.23)	0.580	1.06 (0.77, 1.46)	0.718
	Grains and vegetables pattern	1.03 (0.88, 1.22)	0.699	0.97 (0.81, 1.17)	0.764
	Dairy and fruits pattern	0.66 (0.56, 0.78)	<0.001	0.81 (0.66, 0.98)	0.032
Factor analysis [a]	High protein pattern				
	Q1	Reference		Reference	
	Q2	0.99 (0.82, 1.18)	0.864	1.02 (0.85, 1.22)	0.838
	Q3	0.85 (0.70, 1.01)	0.070	0.91 (0.75, 1.10)	0.318
	Q4	0.99 (0.83, 1.19)	0.936	1.12 (0.93, 1.36)	0.239

Table 4. *Cont.*

Method	Dietary Pattern	Model 1		Model 2	
		OR (95 % CI)	*p*-Value	OR (95 % CI)	*p*-Value
	p-value for trend [b]	0.660		0.382	
	Grains and vegetables pattern				
	Q1	Reference		Reference	
	Q2	0.94 (0.78, 1.12)	0.467	0.95 (0.79, 1.15)	0.616
	Q3	1.05 (0.88, 1.26)	0.579	1.07 (0.89, 1.29)	0.453
	Q4	0.93 (0.78, 1.12)	0.433	0.92 (0.76, 1.12)	0.406
	p-value for trend [b]	0.622		0.553	
	Coarse cereals and soy products pattern				
	Q1	Reference		Reference	
	Q2	0.81 (0.68, 0.97)	0.022	0.86 (0.72, 1.03)	0.098
	Q3	0.79 (0.66, 0.94)	0.009	0.86 (0.71, 1.03)	0.105
	Q4	0.67 (0.56, 0.80)	<0.001	0.74 (0.61, 0.91)	0.005
	p-value for trend [b]	<0.001		0.007	
	Snacks and beverages pattern				
	Q1	Reference		Reference	
	Q2	1.09 (0.91, 1.31)	0.333	1.08 (0.90, 1.30)	0.417
	Q3	0.99 (0.82, 1.18)	0.890	1.02 (0.85, 1.24)	0.822
	Q4	0.99 (0.82, 1.19)	0.898	1.09 (0.88, 1.34)	0.422
	p-value for trend [b]	0.593		0.537	
	Dairy and fruits pattern				
	Q1	Reference		Reference	
	Q2	0.82 (0.69, 0.98)	0.025	0.97 (0.80, 1.16)	0.707
	Q3	0.69 (0.57, 0.82)	<0.001	0.95 (0.78, 1.15)	0.598
	Q4	0.60 (0.50, 0.72)	<0.001	0.97 (0.79, 1.19)	0.753
	p-value for trend [b]	<0.001		0.722	

Note: [a] factor scores were divided into Q1, Q2, Q3, and Q4 according to the quartiles; [b] test for trend based on variable containing median value for each quartile. Model 1: unadjusted model; model 2: adjusted for age, energy intake, physical activity, gender, education level, retirement status, smoking behavior, alcohol drinking, and tea drinking.

The non-conditional logistic regression analysis examining the relationship between the a priori dietary patterns and MetS is presented in Table 5. Our findings revealed no significant associations of the DASH and MD scores with MetS ($p > 0.05$).

Table 5. A priori dietary patterns in association with the risk of MetS using non-conditional logistic regression models.

Dietary Pattern Score	Model 1		Model 2	
	OR (95 % CI)	*p*-Value	OR (95 % CI)	*p*-Value
DASH	0.98 (0.93, 1.04)	0.488	1.02 (0.96, 1.08)	0.541
MD	0.98 (0.94, 1.02)	0.247	1.01 (0.97, 1.06)	0.675

Abbreviations: DASH: dietary approaches to stop hypertension; MD: Mediterranean diet. Model 1: unadjusted model; model 2: adjusted for age, energy intake, physical activity, gender, education level, retirement status, smoking behavior, alcohol drinking, and tea drinking.

4. Discussion

In this cross-sectional study, we firstly utilized both a posteriori and a priori dietary patterns to investigate the associations between dietary patterns and MetS in a suburban population in Shanghai, China. Our findings demonstrated the inverse associations of the "dairy and fruits" and "coarse cereals and soy products" dietary patterns with MetS. In addition, two a priori dietary pattern scores for DASH and MD were calculated. However, we did not observe any significant correlations between DASH or MD and MetS in the Chinese population.

The prevalence of MetS in our study population was 22.47%, which is similar to the 24.5% reported in a systematic review and meta-analysis of 226,653 Chinese individuals [29], and consistent with the global prevalence of 20–25% [30]. The high prevalence of MetS may be attributed to lifestyle changes in modern society, among which dietary pattern may pose an independent and essential factor in influencing the levels of MetS-related indicators [10,31]. Especially for the Western-style diet high in fat and animal-based foods, a meta-analysis revealed that "Western" dietary patterns were significantly associated with increased MetS risk [32]. Over the past few decades, there has been a significant shift in Chinese dietary patterns and behaviors, with a move from a predominantly plant-based diet to a Western-style diet [33]. Therefore, we generated current dietary patterns in a suburban population in Shanghai and investigated the relationship between these patterns and MetS.

In this study, we observed a protective effect from the "dairy and fruits" dietary pattern on MetS following cluster analysis. High consumption of fruits, egg, and dairy was a crucial component of this pattern. Numerous studies have shown the beneficial effects of fruits and vegetables on chronic diseases, such as MetS and CVD [34,35]. These foods are rich in vitamins, dietary fiber, minerals, and phytochemicals, which contribute to the body's antioxidant, anti-inflammatory, and electrolyte properties [35]. Similarly, some studies have indicated a negative correlation between egg consumption and the risk of MetS [36–38]. Eggs are a source of high-quality protein, unsaturated fatty acids, vitamins, minerals, and bioactive components that can effectively regulate lipid absorption, hepatic lipid metabolism, increase HDL-C levels [39], and improve insulin sensitivity [40]. However, the relationship between dairy consumption and MetS is not consistent. One meta-analysis revealed that a higher intake of dairy products significantly reduced the risk of MetS by 17% in cross-sectional studies and by 14% in cohort studies [41]. Another meta-analysis also reported similar findings for milk and yogurt consumption, but not for cheese consumption [42]. Conversely, Babio et al. suggested that higher cheese consumption was associated with a greater risk of MetS [43]. These inconsistent results may be related to the type of dairy product. Milk and yogurt contain various minerals, such as calcium and potassium, which can reduce fat absorption, resulting in weight and fat loss [44,45], and can also decrease sodium retention, thereby lowering blood pressure [46,47]. Furthermore, whey protein in milk and yoghurt may contribute to reducing endogenous fat, leading to a decline in plasma triglycerides, total cholesterol, and LDL [44]. However, cheese is a high-fat dairy product that loses its whey protein during production, and its phosphorus content, energy density, and sodium are higher than other dairy products. These variations in nutrient content may result in different effects compared to milk or yogurt [43]. Nonetheless, in our study, we were unable to differentiate between the types of dairy products consumed, and this issue requires further investigation.

In our study, we observed a significant negative correlation between the "coarse cereals and soy products" dietary pattern and MetS after factor analysis. Notably, high consumption of coarse cereals is a vital component of this dietary pattern. This finding is consistent with previous research indicating that a low intake of coarse cereals increases the risk of MetS in males, as reported by Cheng et al., based on the China Health and Nutrition Survey [48]. The consumption of coarse cereals has also been shown to promote a healthy gut microbiome, supporting the growth and activity of probiotics [49]. Furthermore, the consumption of coarse cereals has been linked to cardiovascular disease prevention through various pathways, including CaMKII/p-BFAF-3, NF-κB, MAPK, and PI3K/Akt [50]. Another key characteristic of the "coarse cereals and soy products" dietary pattern is the high consumption of soy products and starchy vegetables. Studies have demonstrated that soy products can prevent or improve some symptoms associated with MetS [51–53].

Soy products contain essential nutrients, such as protein, polyunsaturated fatty acid, fiber, sterol, and soybean isoflavone, which can reduce glycemic markers (fasting blood glucose and serum insulin levels) [54], improve serum lipids (TG, TC, and LDL-C) [55], blood pressure [56], and flow-mediated dilation [57]. However, the relationship between starchy vegetables, such as potatoes, and MetS is not yet clear. Some studies have suggested that high consumption of potatoes increases the risk of MetS-related diseases, such as type 2 diabetes [58] and hypertension [59], while others have found no significant association between potato or starchy vegetable intake and the risk of obesity, type 2 diabetes, CVD, or MetS [60,61]. This inconsistency may be attributed to differences in cooking methods for starchy vegetables, particularly in Western diets where they are often fried or baked, leading to negative health effects [62,63]. Starchy vegetables are rich in micronutrients and other healthy biological compounds, such as phenolic acid, carotenoid, and resistant starch, which are essential for macronutrient metabolism, antioxidant protection, and chronic disease status [64,65]. Further research is needed to clarify the relationship between starchy vegetable consumption and MetS.

In our study, we evaluated the two a priori dietary patterns, DASH and MD, to explore the association between dietary pattern and MetS. The DASH diet was developed to regulate blood pressure and can also improve various chronic diseases, such as cardiovascular diseases, diabetes, and kidney ailments [66–68]. Previous studies have revealed the positive effects of the DASH diet on MetS and its components [69–71]. Nonetheless, our findings do not align with these conclusions. Specifically, we found no connection between DASH and MetS, which may be due to differences in ethnic backgrounds across populations. The DASH diet was originally developed in a U.S. population, and the Chinese population is demographically and culturally different from the U.S. population. As reported by Joyce et al. [72], the correlation between DASH and MetS varies across different heritage groups. DASH may be more effective in capturing diet–MetS associations in certain Hispanic/Latino subpopulations, such as central/south Americans. Further investigations are required to identify the underlying reasons. The Mediterranean-style diet is a conventional dietary pattern in countries such as Greece, Italy, and other Mediterranean coastal regions. It emphasizes the use of olive oil as the primary source of fat and moderate wine consumption [73]. Numerous studies have indicated that greater adherence to the Mediterranean diet is associated with a lower risk of MetS in Western populations [74–77]. However, this association is not as apparent in Asian populations [78]. In our study, we also did not observe a significant correlation between MD and MetS in the Chinese population, which may be due to differences in dietary practices between Western and Eastern countries. Particularly, in our studied population, the consumption of olive oil and grape wine, which are typical components of the Mediterranean diet, was relatively infrequent.

Our study has several limitations that need to be considered. Firstly, in a posterior methods, the final presentation of data does not rely on any prior knowledge, which may introduce subjectivity in the interpretation and naming, and limit the reproducibility and validity of the data. Secondly, seasonal variations may influence the intake of various foods, potentially compromising the reliability of the FFQ data. Furthermore, distinguishing between milk, yogurt, and cheese consumption can be challenging when analyzing the effects of dairy products on MetS using the FFQ. Similar challenges exist for other foods, including fruits. This limitation precludes an in-depth investigation into the impacts of specific types of these foods on MetS. Thirdly, the sample size of this study was relatively small, and study participants were only from a single region, which may not be representative of the wider population. Therefore, future investigations with an expanded sample size and enrolled population from diverse regions are necessary to examine the relationship between dietary patterns and MetS. Finally, the cross-sectional design of the study limits our ability to establish causality between dietary patterns and MetS. Future studies using longitudinal data and intervention trials are expected to better elucidate the causal relationship.

5. Conclusion

Our findings suggest that study participants following the "dairy and fruits" and "coarse cereals and soy products" dietary pattern had a protective effect on MetS. Our study recommends increasing the consumption of fruits, coarse cereals, and soy products to reduce the risk of MetS. However, we did not observe any significant correlations between DASH or MD and MetS in the Chinese population. Further exploration is needed to verify the correlation of DASH and MD with MetS in the Chinese population.

Supplementary Materials: The following supporting information can be downloaded at: https://www.mdpi.com/article/10.3390/nu15092185/s1, Table S1: DASH diet scoring system; Table S2: MD scoring system; Table S3: food groupings used in dietary pattern analysis; Table S4: factor loadings and dietary patterns for the 15 food groups derived from factor analysis.

Author Contributions: L.W., S.L., M.Z., J.F. and Y.J. designed the research; L.W., S.L. and J.F. performed the statistical analyses; L.W., S.L. and B.C. wrote the paper; S.L., R.D., Q.Z., B.C., G.Z. and J.L. reviewed and edited the manuscript; L.W., S.L., B.C. and J.L. had primary responsibility for the final content. All authors have read and agreed to the published version of the manuscript.

Funding: The work was supported by the Key Science and Technology Program of the Songjiang District (20SJKJGG215).

Institutional Review Board Statement: The study was conducted in accordance with the Declaration of Helsinki, and the protocol was approved by the ethical review board of the School of Public Health, Fudan University (IRB#2016-04-0586) on 5 April 2016.

Informed Consent Statement: Informed consent was obtained from all subjects involved in the study. Written informed consent has been obtained from the study population to publish this paper.

Data Availability Statement: The data presented in this study are available on request from the corresponding author. The data are not publicly available due to privacy of the study participants.

Acknowledgments: We thank all the participants who participated in the study, the investigators who enrolled the study subjects. We also thank all the members in the Shanghai Suburban Adult Cohort and Biobank study group.

Conflicts of Interest: The authors declare no conflict of interest.

References

1. Won, K.B.; Chang, H.J.; Niinuma, H.; Niwa, K.; Jeon, K.; Cho, I.J.; Shim, C.Y.; Hong, G.R.; Chung, N. Inverse association between central obesity and arterial stiffness in Korean subjects with metabolic syndrome: A cross-sectional cohort study. *Diabetol. Metab. Syndr.* **2015**, *7*, 3. [CrossRef] [PubMed]
2. Song, Q.B.; Zhao, Y.; Liu, Y.Q.; Zhang, J.; Xin, S.J.; Dong, G.H. Sex difference in the prevalence of metabolic syndrome and cardiovascular-related risk factors in urban adults from 33 communities of China: The CHPSNE study. *Diabetes Vasc. Dis. Res.* **2015**, *12*, 189–198. [CrossRef] [PubMed]
3. Li, Y.; Zhao, L.; Yu, D.; Wang, Z.; Ding, G. Metabolic syndrome prevalence and its risk factors among adults in China: A nationally representative cross-sectional study. *PLoS ONE* **2018**, *13*, e0199293. [CrossRef] [PubMed]
4. Saklayen, M.G. The Global Epidemic of the Metabolic Syndrome. *Curr. Hypertens. Rep.* **2018**, *20*, 12. [CrossRef] [PubMed]
5. Sidorenkov, O.; Nilssen, O.; Brenn, T.; Martiushov, S.; Arkhipovsky, V.L.; Grjibovski, A.M. Prevalence of the metabolic syndrome and its components in Northwest Russia: The Arkhangelsk study. *BMC Public Health* **2010**, *10*, 23. [CrossRef]
6. Hadaegh, F.; Ghasemi, A.; Padyab, M.; Tohidi, M.; Azizi, F. The metabolic syndrome and incident diabetes: Assessment of alternative definitions of the metabolic syndrome in an Iranian urban population. *Diabetes Res. Clin. Pract.* **2008**, *80*, 328–334. [CrossRef]
7. Sacco, S.; Comelli, M.; Molina, V.; Montrasio, P.L.; Giani, E.; Cavanna, F. A simplified indication of metabolic syndrome to recognize subjects with a moderate risk to develop type 2 diabetes mellitus in a large Italian sample. *Acta Diabetol.* **2014**, *51*, 35–41. [CrossRef]
8. Lorenzo, C.; Williams, K.; Hunt, K.J.; Haffner, S.M. The National Cholesterol Education Program—Adult Treatment Panel III, International Diabetes Federation, and World Health Organization definitions of the metabolic syndrome as predictors of incident cardiovascular disease and diabetes. *Diabetes Care* **2007**, *30*, 8–13. [CrossRef]
9. Alberti, K.G.; Eckel, R.H.; Grundy, S.M.; Zimmet, P.Z.; Cleeman, J.I.; Donato, K.A.; Fruchart, J.C.; James, W.P.; Loria, C.M.; Smith, S.C., Jr. Harmonizing the metabolic syndrome: A joint interim statement of the International Diabetes Federation Task Force on Epidemiology and Prevention; National Heart, Lung, and Blood Institute; American Heart Association; World Heart Federation;

International Atherosclerosis Society; and International Association for the Study of Obesity. *Circulation* **2009**, *120*, 1640–1645. [CrossRef]
10. Xu, H.; Li, X.; Adams, H.; Kubena, K.; Guo, S. Etiology of Metabolic Syndrome and Dietary Intervention. *Int. J. Mol. Sci.* **2018**, *20*, 128. [CrossRef]
11. Song, P.; Yu, J.; Chang, X.; Wang, M.; An, L. Prevalence and Correlates of Metabolic Syndrome in Chinese Children: The China Health and Nutrition Survey. *Nutrients* **2017**, *9*, 79. [CrossRef] [PubMed]
12. Hosseini, Z.; Whiting, S.J.; Vatanparast, H. Current evidence on the association of the metabolic syndrome and dietary patterns in a global perspective. *Nutr. Res. Rev.* **2016**, *29*, 152–162. [CrossRef]
13. Calton, E.K.; James, A.P.; Pannu, P.K.; Soares, M.J. Certain dietary patterns are beneficial for the metabolic syndrome: Reviewing the evidence. *Nutr. Res.* **2014**, *34*, 559–568. [CrossRef] [PubMed]
14. Carlson, J.J.; Eisenmann, J.C.; Norman, G.J.; Ortiz, K.A.; Young, P.C. Dietary fiber and nutrient density are inversely associated with the metabolic syndrome in US adolescents. *J. Am. Diet Assoc.* **2011**, *111*, 1688–1695. [CrossRef] [PubMed]
15. Chiva-Blanch, G.; Urpi-Sarda, M.; Ros, E.; Valderas-Martinez, P.; Casas, R.; Arranz, S.; Guillén, M.; Lamuela-Raventós, R.M.; Llorach, R.; Andres-Lacueva, C.; et al. Effects of red wine polyphenols and alcohol on glucose metabolism and the lipid profile: A randomized clinical trial. *Clin. Nutr.* **2013**, *32*, 200–206. [CrossRef]
16. Godos, J.; Zappalà, G.; Bernardini, S.; Giambini, I.; Bes-Rastrollo, M.; Martinez-Gonzalez, M. Adherence to the Mediterranean diet is inversely associated with metabolic syndrome occurrence: A meta-analysis of observational studies. *Int. J. Food Sci. Nutr.* **2017**, *68*, 138–148. [CrossRef]
17. Siervo, M.; Lara, J.; Chowdhury, S.; Ashor, A.; Oggioni, C.; Mathers, J.C. Effects of the Dietary Approach to Stop Hypertension (DASH) diet on cardiovascular risk factors: A systematic review and meta-analysis. *Br. J. Nutr.* **2015**, *113*, 1–15. [CrossRef]
18. He, D.H.; Yang, M.; Zhang, R.H.; Ma, X.G.; Huang, L.C.; Huang, E.S.; Gu, W.; Zhu, Y.B.; Zhao, D.; Zhu, X.H.; et al. Dietary Patterns Associated Metabolic Syndrome in Chinese Adults. *Biomed. Environ. Sci.* **2015**, *28*, 370–373. [CrossRef]
19. Gadgil, M.D.; Anderson, C.A.; Kandula, N.R.; Kanaya, A.M. Dietary patterns are associated with metabolic risk factors in South Asians living in the United States. *J. Nutr.* **2015**, *145*, 1211–1217. [CrossRef]
20. Shab-Bidar, S.; Golzarand, M.; Hajimohammadi, M.; Mansouri, S. A posteriori dietary patterns and metabolic syndrome in adults: A systematic review and meta-analysis of observational studies. *Public Health Nutr.* **2018**, *21*, 1681–1692. [CrossRef]
21. Rodríguez-Monforte, M.; Sánchez, E.; Barrio, F.; Costa, B.; Flores-Mateo, G. Metabolic syndrome and dietary patterns: A systematic review and meta-analysis of observational studies. *Eur. J. Nutr.* **2017**, *56*, 925–947. [CrossRef] [PubMed]
22. Zhao, Q.; Chen, B.; Wang, R.; Zhu, M.; Shao, Y.; Wang, N.; Liu, X.; Zhang, T.; Jiang, F.; Wang, W.; et al. Cohort profile: Protocol and baseline survey for the Shanghai Suburban Adult Cohort and Biobank (SSACB) study. *BMJ Open* **2020**, *10*, e035430. [CrossRef] [PubMed]
23. Wang, Y.; Huang, Y.; Wu, H.; He, G.; Li, S.; Chen, B. Association between Dietary Patterns and Frailty Prevalence in Shanghai Suburban Elders: A Cross-Sectional Study. *Int. J. Environ. Res. Public Health* **2021**, *18*, 852. [CrossRef] [PubMed]
24. Ainsworth, B.E.; Haskell, W.L.; Whitt, M.C.; Irwin, M.L.; Swartz, A.M.; Strath, S.J.; O'Brien, W.L.; Bassett, D.R., Jr.; Schmitz, K.H.; Emplaincourt, P.O.; et al. Compendium of physical activities: An update of activity codes and MET intensities. *Med. Sci. Sport. Exerc.* **2000**, *32*, S498–S504. [CrossRef]
25. Chiness Diabetes Society. Guideline for the prevention and treatment of type 2 diabetes mellitus in China (2020 edition). *Chin. J. Diabetes Mellit.* **2021**, *13*, 315–409.
26. Mellen, P.B.; Gao, S.K.; Vitolins, M.Z.; Goff, D.C., Jr. Deteriorating dietary habits among adults with hypertension: DASH dietary accordance, NHANES 1988-1994 and 1999-2004. *Arch. Intern. Med.* **2008**, *168*, 308–314. [CrossRef]
27. Trichopoulou, A.; Kouris-Blazos, A.; Wahlqvist, M.L.; Gnardellis, C.; Lagiou, P.; Polychronopoulos, E.; Vassilakou, T.; Lipworth, L.; Trichopoulos, D. Diet and overall survival in elderly people. *BMJ* **1995**, *311*, 1457–1460. [CrossRef]
28. Fung, T.T.; Rexrode, K.M.; Mantzoros, C.S.; Manson, J.E.; Willett, W.C.; Hu, F.B. Mediterranean diet and incidence of and mortality from coronary heart disease and stroke in women. *Circulation* **2009**, *119*, 1093–1100. [CrossRef]
29. Li, R.; Li, W.; Lun, Z.; Zhang, H.; Sun, Z.; Kanu, J.S.; Qiu, S.; Cheng, Y.; Liu, Y. Prevalence of metabolic syndrome in Mainland China: A meta-analysis of published studies. *BMC Public Health* **2016**, *16*, 296. [CrossRef]
30. Vidigal Fde, C.; Ribeiro, A.Q.; Babio, N.; Salas-Salvadó, J.; Bressan, J. Prevalence of metabolic syndrome and pre-metabolic syndrome in health professionals: LATINMETS Brazil study. *Diabetol. Metab. Syndr.* **2015**, *7*, 6. [CrossRef]
31. Bovolini, A.; Garcia, J.; Andrade, M.A.; Duarte, J.A. Metabolic Syndrome Pathophysiology and Predisposing Factors. *Int. J. Sport. Med.* **2021**, *42*, 199–214. [CrossRef]
32. Fabiani, R.; Naldini, G.; Chiavarini, M. Dietary Patterns and Metabolic Syndrome in Adult Subjects: A Systematic Review and Meta-Analysis. *Nutrients* **2019**, *11*, 2056. [CrossRef]
33. Bu, T.; Tang, D.; Liu, Y.; Chen, D. Trends in Dietary Patterns and Diet-related Behaviors in China. *Am. J. Health Behav.* **2021**, *45*, 371–383. [CrossRef]
34. Tian, Y.; Su, L.; Wang, J.; Duan, X.; Jiang, X. Fruit and vegetable consumption and risk of the metabolic syndrome: A meta-analysis. *Public Health Nutr.* **2018**, *21*, 756–765. [CrossRef] [PubMed]
35. Alissa, E.M.; Ferns, G.A. Dietary fruits and vegetables and cardiovascular diseases risk. *Crit. Rev. Food Sci. Nutr.* **2017**, *57*, 1950–1962. [CrossRef]

36. Wang, H.; Wang, W.; Shen, M.; Yang, Z.; Wang, N.; Zhu, Z.; Wu, Z.; Xie, D. Association between egg consumption and metabolic syndrome in Chinese population: A cross-sectional study. *BMJ Open* **2021**, *11*, e050317. [CrossRef]
37. Ding, J.; Zhang, Y. Relationship between Egg Consumption and Metabolic Syndrome. A Meta-Analysis of Observational Studies. *J. Nutr. Health Aging* **2022**, *26*, 373–382. [CrossRef] [PubMed]
38. Shin, S.; Lee, H.W.; Kim, C.E.; Lim, J.; Lee, J.K.; Lee, S.A.; Kang, D. Egg Consumption and Risk of Metabolic Syndrome in Korean Adults: Results from the Health Examinees Study. *Nutrients* **2017**, *9*, 687. [CrossRef] [PubMed]
39. Blesso, C.N.; Andersen, C.J.; Bolling, B.W.; Fernandez, M.L. Egg intake improves carotenoid status by increasing plasma HDL cholesterol in adults with metabolic syndrome. *Food Funct.* **2013**, *4*, 213–221. [CrossRef]
40. Blesso, C.N.; Andersen, C.J.; Barona, J.; Volek, J.S.; Fernandez, M.L. Whole egg consumption improves lipoprotein profiles and insulin sensitivity to a greater extent than yolk-free egg substitute in individuals with metabolic syndrome. *Metabolism* **2013**, *62*, 400–410. [CrossRef] [PubMed]
41. Chen, G.C.; Szeto, I.M.; Chen, L.H.; Han, S.F.; Li, Y.J.; van Hekezen, R.; Qin, L.Q. Dairy products consumption and metabolic syndrome in adults: Systematic review and meta-analysis of observational studies. *Sci. Rep.* **2015**, *5*, 14606. [CrossRef]
42. Jin, S.; Je, Y. Dairy Consumption and Risk of Metabolic Syndrome: Results from Korean Population and Meta-Analysis. *Nutrients* **2021**, *13*, 1574. [CrossRef] [PubMed]
43. Babio, N.; Becerra-Tomás, N.; Martínez-González, M.; Corella, D.; Estruch, R.; Ros, E.; Sayón-Orea, C.; Fitó, M.; Serra-Majem, L.; Arós, F.; et al. Consumption of Yogurt, Low-Fat Milk, and Other Low-Fat Dairy Products Is Associated with Lower Risk of Metabolic Syndrome Incidence in an Elderly Mediterranean Population. *J. Nutr.* **2015**, *145*, 2308–2316. [CrossRef] [PubMed]
44. Rice, B.H.; Cifelli, C.J.; Pikosky, M.A.; Miller, G.D. Dairy components and risk factors for cardiometabolic syndrome: Recent evidence and opportunities for future research. *Adv. Nutr.* **2011**, *2*, 396–407. [CrossRef] [PubMed]
45. Christensen, R.; Lorenzen, J.K.; Svith, C.R.; Bartels, E.M.; Melanson, E.L.; Saris, W.H.; Tremblay, A.; Astrup, A. Effect of calcium from dairy and dietary supplements on faecal fat excretion: A meta-analysis of randomized controlled trials. *Obes. Rev.* **2009**, *10*, 475–486. [CrossRef]
46. Zemel, M.B. Calcium modulation of hypertension and obesity: Mechanisms and implications. *J. Am. Coll. Nutr.* **2001**, *20*, 428S–435S, discussion 440S–442S. [CrossRef] [PubMed]
47. Young, D.B.; Lin, H.; McCabe, R.D. Potassium's cardiovascular protective mechanisms. *Am. J. Physiol.* **1995**, *268*, R825–R837. [CrossRef]
48. Cheng, M.; Wang, H.; Wang, Z.; Du, W.; Ouyang, Y.; Zhang, B. Relationship between dietary factors and the number of altered metabolic syndrome components in Chinese adults: A cross-sectional study using data from the China Health and Nutrition Survey. *BMJ Open* **2017**, *7*, e014911. [CrossRef]
49. Ren, G.; Fan, X.; Teng, C.; Li, Y.; Everaert, N.; Blecker, C. The Beneficial Effect of Coarse Cereals on Chronic Diseases through Regulating Gut Microbiota. *Foods* **2021**, *10*, 2891. [CrossRef]
50. Fu, J.; Zhang, Y.; Hu, Y.; Zhao, G.; Tang, Y.; Zou, L. Concise review: Coarse cereals exert multiple beneficial effects on human health. *Food Chem.* **2020**, *325*, 126761. [CrossRef]
51. Ruscica, M.; Pavanello, C.; Gandini, S.; Gomaraschi, M.; Vitali, C.; Macchi, C.; Morlotti, B.; Aiello, G.; Bosisio, R.; Calabresi, L.; et al. Effect of soy on metabolic syndrome and cardiovascular risk factors: A randomized controlled trial. *Eur. J. Nutr.* **2018**, *57*, 499–511. [CrossRef] [PubMed]
52. Mohammadifard, N.; Sajjadi, F.; Haghighatdoost, F. Effects of soy consumption on metabolic parameters in patients with metabolic syndrome: A systematic review and meta-analysis. *EXCLI J.* **2021**, *20*, 665–685. [CrossRef] [PubMed]
53. Yamagata, K.; Yamori, Y. Potential Effects of Soy Isoflavones on the Prevention of Metabolic Syndrome. *Molecules* **2021**, *26*, 5863. [CrossRef]
54. Glisic, M.; Kastrati, N.; Musa, J.; Milic, J.; Asllanaj, E.; Portilla Fernandez, E.; Nano, J.; Ochoa Rosales, C.; Amiri, M.; Kraja, B.; et al. Phytoestrogen supplementation and body composition in postmenopausal women: A systematic review and meta-analysis of randomized controlled trials. *Maturitas* **2018**, *115*, 74–83. [CrossRef]
55. Jenkins, D.J.; Mirrahimi, A.; Srichaikul, K.; Berryman, C.E.; Wang, L.; Carleton, A.; Abdulnour, S.; Sievenpiper, J.L.; Kendall, C.W.; Kris-Etherton, P.M. Soy protein reduces serum cholesterol by both intrinsic and food displacement mechanisms. *J. Nutr.* **2010**, *140*, 2302s–2311s. [CrossRef] [PubMed]
56. Nasca, M.M.; Zhou, J.R.; Welty, F.K. Effect of soy nuts on adhesion molecules and markers of inflammation in hypertensive and normotensive postmenopausal women. *Am. J. Cardiol.* **2008**, *102*, 84–86. [CrossRef]
57. Li, S.H.; Liu, X.X.; Bai, Y.Y.; Wang, X.J.; Sun, K.; Chen, J.Z.; Hui, R.T. Effect of oral isoflavone supplementation on vascular endothelial function in postmenopausal women: A meta-analysis of randomized placebo-controlled trials. *Am. J. Clin. Nutr.* **2010**, *91*, 480–486. [CrossRef]
58. Guo, F.; Zhang, Q.; Jiang, H.; He, Y.; Li, M.; Ran, J.; Lin, J.; Tian, L.; Ma, L. Dietary potato intake and risks of type 2 diabetes and gestational diabetes mellitus. *Clin. Nutr.* **2021**, *40*, 3754–3764. [CrossRef]
59. Borgi, L.; Rimm, E.B.; Willett, W.C.; Forman, J.P. Potato intake and incidence of hypertension: Results from three prospective US cohort studies. *BMJ* **2016**, *353*, i2351. [CrossRef]
60. Borch, D.; Juul-Hindsgaul, N.; Veller, M.; Astrup, A.; Jaskolowski, J.; Raben, A. Potatoes and risk of obesity, type 2 diabetes, and cardiovascular disease in apparently healthy adults: A systematic review of clinical intervention and observational studies. *Am. J. Clin. Nutr.* **2016**, *104*, 489–498. [CrossRef]

61. Li, Z.; Wang, D.; Ruiz-Narváez, E.A.; Peterson, K.E.; Campos, H.; Baylin, A. Starchy Vegetables and Metabolic Syndrome in Costa Rica. *Nutrients* **2021**, *13*, 1639. [CrossRef] [PubMed]
62. Nahab, F.; Pearson, K.; Frankel, M.R.; Ard, J.; Safford, M.M.; Kleindorfer, D.; Howard, V.J.; Judd, S. Dietary fried fish intake increases risk of CVD: The REasons for Geographic And Racial Differences in Stroke (REGARDS) study. *Public Health Nutr.* **2016**, *19*, 3327–3336. [CrossRef]
63. Provido, S.M.P.; Abris, G.P.; Hong, S.; Yu, S.H.; Lee, C.B.; Lee, J.E. Association of fried food intake with prehypertension and hypertension: The Filipino women's diet and health study. *Nutr. Res. Pract.* **2020**, *14*, 76–84. [CrossRef]
64. King, J.C.; Slavin, J.L. White potatoes, human health, and dietary guidance. *Adv. Nutr.* **2013**, *4*, 393s–401s. [CrossRef] [PubMed]
65. Robert, L.; Narcy, A.; Rock, E.; Demigne, C.; Mazur, A.; Rémésy, C. Entire potato consumption improves lipid metabolism and antioxidant status in cholesterol-fed rat. *Eur. J. Nutr.* **2006**, *45*, 267–274. [CrossRef]
66. Phillips, C.M.; Harrington, J.M.; Perry, I.J. Relationship between dietary quality, determined by DASH score, and cardiometabolic health biomarkers: A cross-sectional analysis in adults. *Clin. Nutr.* **2019**, *38*, 1620–1628. [CrossRef]
67. Mozaffari, H.; Ajabshir, S.; Alizadeh, S. Dietary Approaches to Stop Hypertension and risk of chronic kidney disease: A systematic review and meta-analysis of observational studies. *Clin. Nutr.* **2020**, *39*, 2035–2044. [CrossRef]
68. Chiavaroli, L.; Viguiliouk, E.; Nishi, S.K.; Blanco Mejia, S.; Rahelić, D.; Kahleová, H.; Salas-Salvadó, J.; Kendall, C.W.; Sievenpiper, J.L. DASH Dietary Pattern and Cardiometabolic Outcomes: An Umbrella Review of Systematic Reviews and Meta-Analyses. *Nutrients* **2019**, *11*, 338. [CrossRef] [PubMed]
69. Saneei, P.; Fallahi, E.; Barak, F.; Ghasemifard, N.; Keshteli, A.H.; Yazdannik, A.R.; Esmaillzadeh, A. Adherence to the DASH diet and prevalence of the metabolic syndrome among Iranian women. *Eur. J. Nutr.* **2015**, *54*, 421–428. [CrossRef]
70. Ghorabi, S.; Salari-Moghaddam, A.; Daneshzad, E.; Sadeghi, O.; Azadbakht, L.; Djafarian, K. Association between the DASH diet and metabolic syndrome components in Iranian adults. *Diabetes Metab. Syndr.* **2019**, *13*, 1699–1704. [CrossRef]
71. Lari, A.; Sohouli, M.H.; Fatahi, S.; Cerqueira, H.S.; Santos, H.O.; Pourrajab, B.; Rezaei, M.; Saneie, S.; Rahideh, S.T. The effects of the Dietary Approaches to Stop Hypertension (DASH) diet on metabolic risk factors in patients with chronic disease: A systematic review and meta-analysis of randomized controlled trials. *Nutr. Metab. Cardiovasc. Dis.* **2021**, *31*, 2766–2778. [CrossRef]
72. Joyce, B.T.; Wu, D.; Hou, L.; Dai, Q.; Castaneda, S.F.; Gallo, L.C.; Talavera, G.A.; Sotres-Alvarez, D.; Van Horn, L.; Beasley, J.M.; et al. DASH diet and prevalent metabolic syndrome in the Hispanic Community Health Study/Study of Latinos. *Prev. Med. Rep.* **2019**, *15*, 100950. [CrossRef]
73. Martínez-González, M.A.; Sánchez-Villegas, A. The emerging role of Mediterranean diets in cardiovascular epidemiology: Monounsaturated fats, olive oil, red wine or the whole pattern? *Eur. J. Epidemiol.* **2004**, *19*, 9–13. [CrossRef]
74. Velázquez-López, L.; Santiago-Díaz, G.; Nava-Hernández, J.; Muñoz-Torres, A.V.; Medina-Bravo, P.; Torres-Tamayo, M. Mediterranean-style diet reduces metabolic syndrome components in obese children and adolescents with obesity. *BMC Pediatr.* **2014**, *14*, 175. [CrossRef] [PubMed]
75. Papadaki, A.; Nolen-Doerr, E.; Mantzoros, C.S. The Effect of the Mediterranean Diet on Metabolic Health: A Systematic Review and Meta-Analysis of Controlled Trials in Adults. *Nutrients* **2020**, *12*, 3342. [CrossRef]
76. Montemayor, S.; Mascaró, C.M.; Ugarriza, L.; Casares, M.; Llompart, I.; Abete, I.; Zulet, M.; Martínez, J.A.; Tur, J.A.; Bouzas, C. Adherence to Mediterranean Diet and NAFLD in Patients with Metabolic Syndrome: The FLIPAN Study. *Nutrients* **2022**, *14*, 3186. [CrossRef] [PubMed]
77. Esposito, K.; Marfella, R.; Ciotola, M.; Di Palo, C.; Giugliano, F.; Giugliano, G.; D'Armiento, M.; D'Andrea, F.; Giugliano, D. Effect of a mediterranean-style diet on endothelial dysfunction and markers of vascular inflammation in the metabolic syndrome: A randomized trial. *Jama* **2004**, *292*, 1440–1446. [CrossRef] [PubMed]
78. Veissi, M.; Anari, R.; Amani, R.; Shahbazian, H.; Latifi, S.M. Mediterranean diet and metabolic syndrome prevalence in type 2 diabetes patients in Ahvaz, southwest of Iran. *Diabetes Metab. Syndr.* **2016**, *10*, S26–S29. [CrossRef]

nutrients MDPI

Article

A Posteriori Dietary Patterns and Coronary Artery Disease in a Greek Case–Control Study

Maria Dimitriou [1,*], Ioanna Panagiota Kalafati [2], Loukianos S. Rallidis [3], Genovefa Kolovou [4] and George V. Dedoussis [2]

1 Department of Nutritional Science and Dietetics, School of Health Sciences, University of Peloponnese, 24100 Kalamata, Greece
2 Department of Nutrition and Dietetics, School of Health and Education, Harokopio University of Athens, 17676 Athens, Greece; nkalafati@gmail.com (I.P.K.)
3 Second Department of Cardiology, Attikon Hospital, School of Medicine, National and Kapodistrian University of Athens, 11527 Athens, Greece; lrallidis@gmail.com
4 First Cardiology Department, Onassis Cardiac Surgery, 17674 Athens, Greece; genovefa@kolovou.com
* Correspondence: dimitrioumelina@gmail.com

Abstract: Introduction: Diet is one of the most important modifiable risk factors associated with cardiovascular health (CH). Research identifying dietary patterns (DPs) through data-driven analysis and reporting associations between DPs and coronary artery disease (CAD) outcomes is rather limited. Objective: The aim of the present report was to generate DPs through factor analysis (FA) and to examine their association with CAD risk. Methods: Participants (n = 1017) consisted of cases diagnosed with CAD (n = 356) and controls (n = 661) drawn from the THISEAS study. Demographic, anthropometric and lifestyle data were collected. Dietary components were generated through FA. Logistic regression analysis was performed to estimate CAD relative risks. Results: FA generated seven dietary components, explaining 53.5% of the total variation in intake. The Western-type DP showed a modest significant association with CAD risk, after controlling for confounders (OR = 1.20; 95% CI = 1.09–1.32, $p < 0.001$). The vegetarian-type DP was not significantly associated with the likelihood of CAD (OR = 0.95; 95% CI = 0.84–1.04, $p = 0.259$). Discussion: The Western-type DP was positively associated with CAD risk and the odds were further increased after controlling for confounders. This finding is in concordance with previously reported positive associations between Western patterns and CAD risk. Limited data exist regarding a posteriori DPs and their effect on CAD risk.

Keywords: coronary artery disease; cardiovascular disease; dietary patterns; Western-type dietary pattern; factor analysis

Citation: Dimitriou, M.; Kalafati, I.P.; Rallidis, L.S.; Kolovou, G.; Dedoussis, G.V. A Posteriori Dietary Patterns and Coronary Artery Disease in a Greek Case–Control Study. *Nutrients* **2023**, *15*, 4733. https://doi.org/10.3390/nu15224733

Academic Editor: Benjamin D. Horne

Received: 22 September 2023
Revised: 3 November 2023
Accepted: 7 November 2023
Published: 9 November 2023

1. Introduction

Global statistics place cardiovascular disease (CVD) mortality first, since CVD deaths represent 32% of all deaths worldwide and 38% of premature deaths due to noncommunicable diseases (NCDs). Coronary artery disease (CAD) is the most common type of CVDs and it is estimated that the number of people that will die due to CAD will rise up to 23.6 million by the year 2030 [1].

Low and moderate levels of cardiovascular health (CH) are attributable to the majority of CVD events in the United States, according to the American Heart Association (AHA)'s updated statistics report [2]. Therefore, premature mortality could be reduced or even avoided when focusing on CH level improvements.

Diet is one of the most important CAD risk factors, depicts health behavior and can modify CAD risk and overall CVD risk [3,4]. According to AHA, diet is among the seven approaches for better CH [2]. Research investigating the impact of specific dietary nutrients of foods on CAD may highlight their protective effects against CAD and is still ongoing [5–9]. However, the nutrients and bioactive chemicals of food items are intercorrelated and research for associations between a single nutrient and a chronic disease may underestimate the impact of the overall diet on health outcomes [10]. In recognition of the synergy of nutrients, research was directed towards the study of dietary patterns (DPs) in the prevention or treatment of disease [11,12]. Therefore, apart from focusing on the protective potential of individual nutrients, it is also important to study the impact of DPs on the disease. A DP conceptualizes the nutritional intake, the quality and the variety of the overall diet.

Although there is supportive evidence regarding the protective effect of the Mediterranean-type diet on CAD risk, data regarding other known DPs such as, vegetarian-type, and Western-type diets remain limited. Furthermore, research identifying DPs through data-driven analysis and reporting associations between DPs and CAD outcomes is scarce, being even more scarce in the Greek population, and the results remain inconclusive. In addition, the posteriori approach is independent of current nutrition-model knowledge and is an important tool with which to identify novel DPs that may substantially increase or decrease CAD risk.

In order to bridge the gap and provide more evidence to the current literature regarding DPs on disease risk, the aim of the present report was to identify DPs through data-driven analysis and to evaluate their association with CAD risk.

2. Materials and Methods

2.1. Study Design and Population

Details regarding the materials and methods used for the study population, along with demographic, anthropometric, clinical and lifestyle assessments, have been previously published [13]. The study population comprised up to 1017 subjects of Greek origin drawn from the THISEAS database, constituting a case–control study.

Cases were coronary patients presenting with acute coronary syndrome or stable CAD defined as >50% stenosis in at least one of the three main coronary vessels. All patients had undergone coronary angiography. Controls were individuals free of CAD. Exclusion criteria from both study groups were the presence of renal or hepatic disease. Subjects from the THISEAS database with incomplete/missing dietary data and missing data regarding other parameters tested were excluded from current analyses.

Therefore, the analysis of the present report was restricted to 356 cases diagnosed with first-time CAD at the time of recruitment and 661 controls, depending on the dietary data availability of the cohort. The study protocol was approved by the Ethics Committee of Harokopio University of Athens. The flow chart of the present study is depicted in Figure 1.

2.2. Demographic and Lifestyle Characteristics

All participants were interviewed regarding their origins to ensure their Greek ancestry. Data regarding educational status were collected and in the present work educational status was measured via years of schooling.

Physical activity level (PAL) was assessed through the Harokopio Physical Activity Questionnaire (HAPAQ), which evaluates the frequency, duration and intensity of occupational, household and leisure time activities [14]. Physical activity (PA) adoption was assessed as a categorical variable categorizing the participants into two groups, based on whether they reported leisure time activities in a regular basis or not. Volunteers that reported no leisure time activities were categorized as physically inactive.

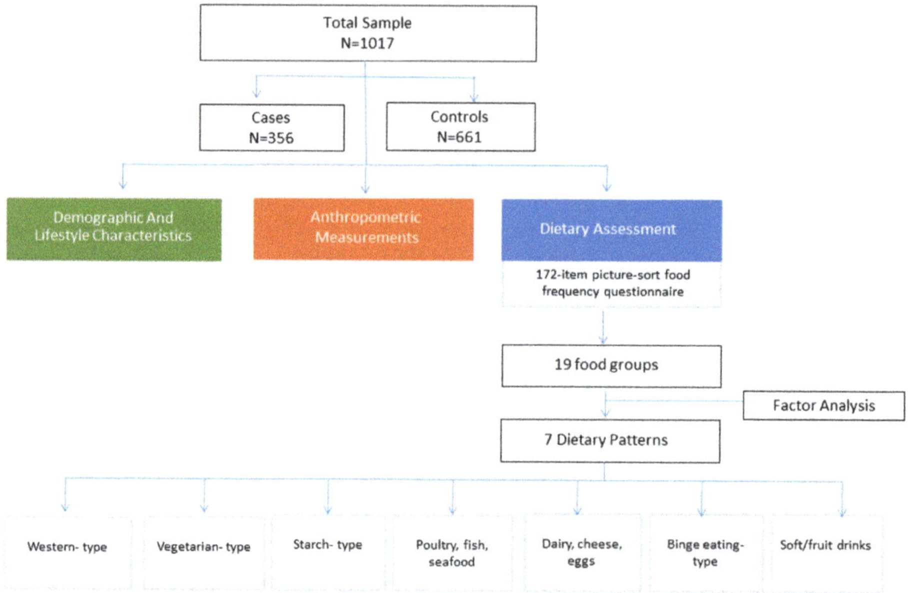

Figure 1. Study flow chart.

Participants were classified into current, never or former smokers. Information regarding the average daily number of cigarettes, the duration of smoking and time of cessation was also obtained. Current smokers were defined as those who smoked at least one cigarette per day, non-smokers those who had never smoked in their life and former smokers those who had stopped smoking for at least six months. In the present analysis, former smokers were combined together into one group with never smokers.

2.3. Anthropometric Measurements

Body weight (BW) and height (Ht) were measured in all participants, who were wearing light clothing, without shoes. Weight was measured to the nearest 0.5 kg using a leveled platform scale. Height was measured to the nearest 0.5 cm using a wall-mounted stadiometer. Body mass index (BMI) was computed as weight (kg)/height^2 (m) (Quetelet's equation):

$$\text{BMI} = \text{BW (kg)}/[\text{Ht (m)}]^2$$

2.4. Dietary Assessment

Nutritional information was recorded through a 172-item picture-sort food frequency questionnaire (FFQ). Dietary data were manually entered into an Excel spreadsheet database that translated the queried foods and beverages into food group equivalents. Regarding combinations of individual foods into one food item, the researcher referred to the ingredients, nutrient information and recipe from reference lists. In order to calculate portion sizes, the dietary guidelines for adults in Greece were used [15]. In total, 26 food groups were estimated (Supplementary Material). For analysis, the number of food groups was further narrowed down to 21 by including two similar food groups in one (e.g., dairy, full fat, and cheese, full fat).

2.5. Statistical Analysis

Continuous variables are presented as mean values and SD, while categorical variables are presented as relative frequencies. Differences between categorical variables and groups of the study were assessed using the χ^2 test. P–P plots were applied to assess the normality of the distribution of the continuous variables. Student's t test or the Mann–Whitney test was applied to evaluate differences in continuous variables between the two study groups.

The factor analysis (FA) technique was used to identify and generate DPs. Data preparation was completed prior to performing FA. This step included the quality control of the dataset by checking for outliers regarding consumption, removing missing data and ensuring continuous variables. Given that most studies have demonstrated a U-shaped relationship between CAD risk and coffee or alcohol consumption, alcohol and coffee were dropped from further analysis [16,17]. In total, 19 food groups were coded as servings per day and underwent FA. Exploratory FA was carried out to evaluate validity, disclose underlying structures and reduce the number of variables. FA was chosen as the extraction method using orthogonal rotation (Varimax rotation) in order to generate non-correlated components (namely, non-correlated DPs). The food variables that were highly correlated showed factor loadings (correlation coefficients) greater that $|0.4|$. The cut-off point for Eigen values was greater than 1.0 [18].

Logistic regression models (unadjusted and adjusted for major confounders) were used in order to estimate the relative risks of developing CAD via the calculation of ORs and their corresponding 95% confidence interval. Model 2 logistic regression was adjusted for main covariates, namely age, sex and BMI. In order to control for more covariates (beyond Model 2 covariates) related to the dependent variable and to alleviate concerns regarding a loss of study power due to multiple testing, we tested the selected set of covariates one at a time. Covariates retained as significant were the presence of hypertension, diabetes mellitus, dyslipidemia and current smoking. Physical inactivity and years of education that changed the effect size by less than 10% and were non-significant were removed. Therefore, we concluded on using Model 3 adjusted for age, sex, BMI, presence of hypertension, diabetes mellitus, dyslipidemia and current smoking.

Analyses were based on 2-sided tests, while statistical significance was set at $p \leq 0.05$. The statistical software package IBM SPSS Statistics 21.0 (SPSS Inc., Frisco, TX, USA) was used for all statistical calculations, where appropriate.

3. Results

Table 1 presents the descriptive characteristics of the study (namely, the demographic, lifestyle and clinical characteristics). The two study groups significantly differed regarding age; specifically, cases with CAD were older compared to controls ($p < 0.001$). A higher proportion of males comprised the case groups, compared with that in the control group ($p < 0.001$). As expected, the prevalence of arterial hypertension, hypercholesterolemia, type 2 diabetes mellitus (T2DM), physical inactivity and cigarette smoking was higher in cases compared to that in controls.

Table 2 depicts the score coefficients (factor loadings) derived from FA. Absolute values greater than 0.4 indicate that the food variables are highly correlated and contribute more to the development of a dietary component. In this report, seven components were generated from the initial food groups, and explained 53.5% of the total variation in intake. Specifically, the derived components from the analysis were as follows:

(1) Component 1, a Western-type pattern, which included red meat, processed meat, fried potatoes and fast foods;
(2) Component 2, a vegetarian-type pattern, which is mainly characterized by vegetables, legumes and potatoes (boiled, baked or smashed);
(3) Component 3, a starch pattern, which was loaded with refined and unrefined starch (with the unrefined starch prevailing)l
(4) Component 4, a pattern that was characterized by the consumption of poultry, fish and seafood;
(5) Component 5, a pattern that included dairy and eggs,
(6) Component 6, a binge eating-type pattern that included the intake of sweets and nuts;
(7) Component 7, a pattern that included soft drinks and fruit drinks. Fruits and fresh fruit juice were not loaded.

Table 1. Descriptive characteristics of the participants.

	Controls (N = 661)		Cases (N = 356)		
	Mean or Frequency	**±SD ***	**Mean or Frequency**	**±SD**	***p*-Value**
Demographic & Lifestyle characteristics					
Age (years)	54.1	±14.1	62.5	±10.1	<0.001
Male sex (%)	49.5		82.1		0.000
Years of education	12.3	±4.6	11.5	±4.9	0.007
	Relative Frequency (%)		**Relative Frequency (%)**		***p*-Value**
Physical inactivity	79.5%		90.9%		0.000
Current smokers	26.4%		46.7%		0.000
Clinical characteristics					
	Mean	**±SD**	**Mean**	**±SD**	***p*-Value**
Body mass index	28.4	±4.9	27.8	±3.8	0.040
Systolic blood pressure (mmHg)	134	±18	134	±20	0.944
Diastolic blood pressure (mmHg)	80	±11	80	±13	0.658
Total cholesterol (mg/dL)	210	±39	192	±48	0.000
Low-density lipoprotein cholesterol (mg/dL)	133	±35	123	±42	0.000
Triglyceride (mg/dL)	114	±64	148	±103	0.000
Blood glucose (mg/dL)	98	±23	113	±35	0.000
	Relative Frequency (%)		**Relative Frequency (%)**		***p*-Value**
Prevalence of hypertension	47.5%		90.3%		0.000
Use of antihypertensive medication	29.0%		85.1%		0.000
Prevalence of hypercholesterolemia	69.5%		88.4%		0.000
Use of lipid lowering medication	21.5%		79.4%		0.000
Prevalence of diabetes mellitus	10.6%		35.0%		0.000
Use of anti-diabetic medication	5.7%		21.8%		0.000

* SD = standard deviation.

Table 2. Loadings from principal component analysis regarding food groups consumed by the participants from the THISEAS study.

	Component [a]						
	1	2	3	4	5	6	7
Red meat	**0.584**	0.300	−0.117	0.043	0.087	0.033	−0.122
Processed meat	**0.671**	−0.076	0.097	−0.006	0.031	0.066	0.082
Potatoes, fried	**0.549**	0.135	−0.175	0.007	0.189	−0.189	0.039
Fast foods	**0.630**	−0.083	−0.165	0.037	0.152	0.149	0.173
Vegetables	−0.105	**0.625**	0.164	0.268	0.089	−0.097	0.198
Legumes	−0.036	**0.666**	−0.147	−0.010	−0.046	0.297	−0.145
Potatoes, boiled/baked/smashed	0.208	**0.615**	−0.054	−0.034	0.082	−0.031	−0.010
Refined starch	0.139	0.256	**−0.645**	−0.057	0.056	0.056	0.124
Unrefined starch	−0.030	0.096	**0.813**	0.020	0.054	−0.041	−0.092
Fish	−0.007	0.208	0.031	**0.606**	−0.048	−0.004	−0.168
Seafood	0.204	−0.108	−0.059	**0.688**	0.065	−0.028	0.135
Poultry	*0.537*	0.082	0.098	**0.431**	−0.331	−0.026	−0.099
Dairy, full fat	0.096	0.298	0.003	0.089	**0.565**	−0.193	0.041
Dairy, semi/non fat	−0.033	−0.020	*0.443*	−0.201	**−0.440**	−0.214	0.292
Eggs	0.121	−0.026	0.019	−0.084	**0.624**	0.193	−0.011
Sweets	0.214	−0.038	0.032	0.132	0.336	**0.498**	0.292
Nuts	0.024	0.095	0.024	−0.029	−0.029	**0.739**	−0.070
Soft drinks	0.180	0.112	−0.023	−0.255	−0.058	−0.225	**0.701**
Fruit drinks	−0.02	0.081	−0.158	0.224	−0.037	0.184	**0.590**
Fruits	−0.219	0.243	0.0201	0.366	−0.020	0.279	0.215

Numbers in bold indicate loadings with an absolute value of >0.4 (a higher correlation of the food group with the component); total %variance explained equals 53.5. [a] Component description: Component 1 = a Western-type pattern; Component 2 = a vegetarian-type pattern; Component 3 = a starch-type pattern; Component 4 = a pattern that is mainly characterized by the consumption of poultry, seafood and fish; Component 5 = a pattern that is mainly characterized by the consumption of dairy and eggs; Component 6 = a binge eating-type pattern that is mainly characterized by the consumption of sweets and nuts; Component 7 = a pattern that is mainly characterized by the consumption of soft drinks and fruit drinks.

In order to evaluate the associations between each extracted dietary component and CAD risk, logistic regression models were performed without adjustments (Model 1) or after controlling for main covariates (Model 2, adj. for age, sex and BMI). In addition, Model 3 was adjusted for more covariates (Model 2 + presence of arterial hypertension, dyslipidemia, diabetes mellitus and current smoking). Table 3 depicts the results from logistic regression, which evaluated the association between each dietary component and CAD likelihood. The unadjusted regression showed that Component 1 (OR = 1.10; 95% CI = 1.01–1.10, p = 0.034), Component 4 (OR = 1.03; 95% CI = 1.02–1.04, p = 0.000) and Component 5 (OR = 1.09; 95% CI = 1.03–1.16, p = 0.003) were positively associated with CAD risk. On the other hand, Component 3, Component 4 and Component 7 were inversely associated with CAD risk. After adjusting for Models 2 and 3, only the association of Component 1 (Western-type diet pattern) remained significant (Model 2: OR = 1.20; 95% CI = 1.09–1.32, $p < 0.001$) (Model 3: OR = 1.13; 95% CI = 1.02–1.24, $p < 0.017$). Model 2 revealed a modest effect of Component 1 on CAD risk. The effect size was attenuated after Model 3 analysis, although it remained significant. The positively associated Component 5, along with the inversely associated Components 3, 4 and 7 lost their significance in adjusted analyses. In addition, Component 6 was inversely associated with CAD risk (Model 1 and 2) but lost significance in Model 3 analysis. Component 2 did not demonstrate significant associations with the likelihood of CAD, in both unadjusted and adjusted analyses.

Table 3. Results from logistic regression, which evaluated the association between dietary components and the likelihood of having coronary artery disease.

	Odds Ratio	95% CI	*p*-Value
Component 1: [a] Western-type dietary pattern (DP)			
* Model 1	1.10	1.01–1.10	0.034
** Model 2	1.20	1.09–1.32	<0.001
*** Model 3	1.13	1.02–1.24	0.017
Component 2: [b] Vegetarian-type DP			
Model 1	0.97	0.90–1.05	0.48
Model 2	0.95	0.84–1.04	0.26
Model 3	0.94	0.85–1.04	0.22
Component 3: [c] Starch-type DP			
Model 1	0.95	0.91–0.99	0.007
Model 2	0.98	0.94–1.03	0.45
Model 3	1.00	0.95–1.05	0.97
Component 4: Poultry, fish and seafood DP			
Model 1	0.75	0.61–0.92	0.005
Model 2	0.85	0.67–1.07	0.15
Model 3	0.80	0.61–1.04	0.10
Component 5: Dairy, cheese and eggs DP			
Model 1	1.09	1.03–1.16	0.003
Model 2	1.06	1.00–1.14	0.06
Model 3	1.06	0.98–1.15	0.12
Component 6: [d] Binge eating-type DP			
Model 1	0.79	0.71–0.89	<0.001
Model 2	0.84	0.75–0.95	0.005
Model 3	0.88	0.77–1.01	0.07

Table 3. *Cont.*

	Odds Ratio	95% CI	*p*-Value
Component 7: [e] Soft/fruit drinks DP			
Model 1	0.66	0.52–0.83	0.001
Model 2	0.77	0.77–1.00	0.052
Model 3	0.83	0.62–1.12	0.23

* No adjustments; ** adjustments = age, sex and body mass index; *** adjustments = age, sex, body mass index, presence of arterial hypertension, dyslipidemia, diabetes mellitus and current smoking. [a] A pattern mainly characterized by the consumption of meat, processed meat, fast foods, fried potatoes and fast-food; [b] a pattern mainly characterized by the consumption of legumes, vegetables and potatoes; [c] a pattern mainly characterized by the consumption of unrefined starch; [d] a pattern mainly characterized by the consumption of sweets and nuts; [e] a pattern mainly characterized by the consumption soft drinks and fruit drinks.

Furthermore, in an attempt to apply multiple regression analysis to model the associations between DPs, dependent variables and CAD (total cholesterol, low-density-lipoprotein cholesterol, triglycerides {TG} and systolic blood pressure) in the pooled sample or in controls did not reveal significant results.

4. Discussion

The present report attempted to create posteriori DPs and assess their associations with the likelihood of having CAD. Dietary components were identified from food groups that underwent the FA, a data-driven statistical method. In total, seven components (DPs) were generated.

The first component could be described as a Western-type DP and depicted an unhealthy pattern; it was mainly characterized by the consumption of red meat, processed meat, fried potatoes and fast foods. This pattern was positively associated with CAD risk. Although the odds were further increased after controlling for main confounders (Model 2) compared to those under the unadjusted analysis (Model 1), the effect was attenuated when controlling for more confounding variables (Model 3). However, in all models the association of the pattern remained significant. This finding is in concordance with the previously reported positive association between Western-type patterns and CAD in studies with high methodological quality [19]. Western-type DPs are mainly characterized by saturated fatty acid intake, and this type of fat has been implicated as a CAD risk factor. In another Greek study, a pattern characterized by meat intake and meat products has been associated with higher waist circumference and lower levels of high-density-lipoprotein cholesterol (HDL-C) [20]. Nowadays, there is evidence in the literature indicating that the Western-type diet is associated with CVD, as well as other NCDs (such as obesity, T2DM and cancer) due to their regulation of the gut microbiota–immune system interaction [21]. In the present report, the Western-type pattern that was generated from the analysis contained ultra-processed foods (meat products and fast food) which are associated with an increased prevalence of CVD [22–24].

On the other hand, Component 2, which is a vegetarian-type DP, is mainly characterized by the consumption of vegetables, legumes and potatoes (fried potatoes were excluded) and depicts a healthy DP. Vegetables and legumes are usually components of prudent patterns, which are inversely associated with CAD risk [19]. In our findings, this component showed an expected directional effect on CAD risk, although it was not significant. In general, there is some evidence in the literature suggesting a protective effect of vegetarian type patterns on primary prevention [25]. Vegetarian-type diets seem to have a protective effect against CVD risk and improve overall CH. However, more consistent associations regarding cardiovascular benefits need to be observed [26].

The inverse association of the seventh pattern (sweets and nuts) (Model 1 and 2) is in line with the results of a case–control Norwegian study. The latter demonstrated an inverse association of sweets with myocardial infarction [27]. The authors mention that although this was an unexpected outcome, this food group contained food items, such as, nuts, almond paste and chocolate, which have been associated with reduced CAD risk. Similarly, in our report, this component also included nuts; in addition, among other sweets, almond chocolate and dark chocolate were also included. We cannot obviate the possibility that the latter food items affected the direction of the association with CAD risk. This association lost significance after Model 3 analysis.

It has been demonstrated that diets characterized by foods with high glycemic index scores have been associated with high TG and low HDL-C levels [28]. In addition, beverages with added sugar are associated with CAD risk [29]. However, in this report, soft drinks and fruit drinks (Component 7) did not reveal associations with CAD risk, after confounding. The third, fourth and fifth components also lacked significant associations, after confounding.

Potential limitations of this report are the recall bias of food intake, which may have resulted in underreported or overreported dietary intakes. Many cases had received dietary advice by the time of the interview, so we cannot rule out the possibility that this may have influenced their diet report towards the intake of favorable foods indicated in the dietary advice. This is a case–control investigation that cannot support causality. In addition, the two study groups significantly differed in gender proportions. Although logistic regression Model 2 was adjusted for the sex variable, this does not preclude the possibility of a sex misbalance affecting the outcome of some results. FA as a technique also has some limitations, since the extracted components are based on subjective decisions.

We assume that some of the non-significant findings may well have been significant if the sample size was larger and had a larger discriminatory power.

Despite the limitations of this report, we assume that these results are noteworthy, given the limited evidence in the literature identifying DPs through FA and examining DPs for CAD risk. Most studies investigate possible associations between a priori DPs and cardiovascular disease risk [30]. Research in the Greek population consistently reveals a protective effect of the Mediterranean diet on CAD risk, examined as an a priori DP [31,32]. However, research on posteriori DPs and CAD outcomes is scarce, if there is any, meaning we are unable to compare our results with others. Therefore, more research is needed in this direction that could eventually identify the optimal diet for disease prevention.

5. Conclusions

Regarding an important environmental parameter, diet, research using DPs highlights the importance of food diversity in health outcomes. Although dietary assessment is approached using different methods, many studies have consistent results regarding DPs and CAD. These results are important and may contribute to the implementation of programs and services supporting overall healthy patterns against disease prevention and development instead of the avoidance of certain foods or nutrients. Policy health promotion can encourage the improvement of eating behaviors at an individual and population level.

Supplementary Materials: The following supporting information can be downloaded at https://www.mdpi.com/article/10.3390/nu15224733/s1, Table S1. Assessment of food groups.

Author Contributions: M.D. participated in data collection, carried out data manipulation, performed statistical analyses and interpretation and drafted the manuscript; I.P.K. participated in sample recruitment and performed data entry; L.S.R. participated in sample recruitment; G.K. participated in sample recruitment; G.V.D. participated in the coordination of the study. All authors critically revised the manuscript. All authors have read and agreed to the published version of the manuscript.

Funding: Volunteer recruitment for the study was partially funded by the General Secretary of Research and Technology (PENED, 03EΔ474).

Institutional Review Board Statement: The study was approved by the Ethics Committee of Harokopio University of Athens. Approval protocol number and approval date: 10/9-6-2004, 14 June 2004.

Informed Consent Statement: Informed consent was obtained from all subjects involved in the study.

Data Availability Statement: Data are available on request from chief investigators due to ethical and privacy restrictions.

Acknowledgments: The authors thank all dietitians, clinicians and volunteers for their contribution to the study.

Conflicts of Interest: The authors declare no conflict of interest.

References

1. World Health Organization. *Global Atlas on Cardiovascular Diseases Prevention and Control*; World Health Organization: Geneva, Switzerland, 2011. Available online: http://www.who.int/cardiovascular_diseases/publications/atlas_cvd/en/ (accessed on 12 November 2017).
2. Tsao, C.W.; Aday, A.W.; Almarzooq, Z.I.; Alonso, A.; Beaton, A.Z.; Bittencourt, M.S.; Boehme, A.K.; Buxton, A.E.; Carson, A.P.; Commodore-Mensah, Y.; et al. Heart Disease and Stroke Statistics-2022 Update: A Report From the American Heart Association. *Circulation* **2022**, *145*, e153–e639. [CrossRef] [PubMed]
3. Petersen, K.S.; Kris-Etherton, P.M. Diet Quality Assessment and the Relationship between Diet Quality and Cardiovascular Disease Risk. *Nutrients* **2021**, *13*, 4305. [CrossRef] [PubMed]
4. Grech, E.D. Pathophysiology and investigation of coronary artery disease. *BMJ* **2003**, *326*, 1027–1030. [CrossRef]
5. Sesso, H.D.; Manson, J.E.; Aragaki, A.K.; Rist, P.M.; Johnson, L.G.; Friedenberg, G.; Copeland, T.; Clar, A.; Mora, S.; Moorthy, M.V.; et al. Effect of cocoa flavanol supplementation for the prevention of cardiovascular disease events: The COcoa Supplement and Multivitamin Outcomes Study (COSMOS) randomized clinical trial. *Am. J. Clin. Nutr.* **2022**, *115*, 1490–1500. [CrossRef] [PubMed]
6. Virtanen, J.K.; Nurmi, T.; Aro, A.; Bertone-Johnson, E.R.; Hyppönen, E.; Kröger, H.; Lamberg-Allardt, C.; Manson, J.E.; Mursu, J.; Mäntyselkä, P.; et al. Vitamin D supplementation and prevention of cardiovascular disease and cancer in the Finnish Vitamin D Trial: A randomized controlled trial. *Am. J. Clin. Nutr.* **2022**, *115*, 1300–1310. [CrossRef]
7. Dehghan, M.; Mente, A.; Rangarajan, S.; Mohan, V.; Lear, S.; Swaminathan, S.; Wielgosz, A.; Seron, P.; Avezum, A.; Lopez-Jaramillo, P.; et al. Association of egg intake with blood lipids, cardiovascular disease, and mortality in 177,000 people in 50 countries. *Am. J. Clin. Nutr.* **2020**, *111*, 795–803. [CrossRef]
8. Tindall, A.M.; McLimans, C.J.; Petersen, K.S.; Kris-Etherton, P.M.; Lamendella, R. Walnuts and Vegetable Oils Containing Oleic Acid Differentially Affect the Gut Microbiota and Associations with Cardiovascular Risk Factors: Follow-up of a Randomized, Controlled, Feeding Trial in Adults at Risk for Cardiovascular Disease. *J. Nutr.* **2020**, *150*, 806–817. [CrossRef]
9. Trautwein, E.A.; McKay, S. The Role of Specific Components of a Plant-Based Diet in Management of Dyslipidemia and the Impact on Cardiovascular Risk. *Nutrients* **2020**, *12*, 2671. [CrossRef]
10. Tucker, K. Dietary patterns, approaches, and multicultural perspective. *Appl. Physiol. Nutr. Metab.* **2010**, *35*, 211–218. [CrossRef]
11. Nestel, P.J.; Mori, T.A. Dietary patterns, dietary nutrients and cardiovascular disease. *Rev. Cardiovasc. Med.* **2022**, *23*, 17. [CrossRef]
12. Shan, Z.; Li, Y.; Baden, M.Y.; Bhupathiraju, S.N.; Wang, D.D.; Sun, Q.; Rexrode, K.M.; Rimm, E.B.; Qi, L.; Willett, W.C.; et al. Association between Healthy Eating Patterns and Risk of Cardiovascular Disease. *JAMA Intern. Med.* **2020**, *180*, 1090–1100. [CrossRef] [PubMed]
13. Dimitriou, M.; Rallidis, L.S.; Theodoraki, E.V.; Kalafati, I.P.; Kolovou, G.; Dedoussis, G.V. Exclusive olive oil consumption has a protective effect on coronary artery disease; overview of the THISEAS study. *Public Health Nutr.* **2016**, *19*, 1081–1087. [CrossRef] [PubMed]
14. Kavouras, S.A.; Maraki, M.I.; Kollia, M.; Gioxari, A.; Jansen, L.; Sidossis, L. Development, reliability and validity of a physical activity questionnaire for estimating energy expenditure in Greek adults. *Sci. Sport.* **2016**, *31*, e47–e53. [CrossRef]
15. Supreme Scientific Health Council & Ministry of Health and Welfare of Greece. Dietary guidelines for adults in Greece. *Arch. Hell. Med.* **1999**, *16*, 516–524.
16. Zhao, Y.; Wu, K.; Zheng, J.; Zuo, R.; Li, D. Association of coffee drinking with all-cause mortality: A systematic review and meta-analysis. *Public Health Nutr.* **2015**, *18*, 1282–1291. [CrossRef]
17. Hill, J.A. In vino veritas: Alcohol and heart disease. *Am. J. Med. Sci.* **2005**, *329*, 124–135. [CrossRef] [PubMed]
18. Wirfalt, E.; Drake, I.; Wallstrom, P. What do review papers conclude about food and dietary patterns? *Food Nutr. Res.* **2013**, *57*, 20523. [CrossRef]
19. Mente, A.; de Koning, L.; Shannon, H.S.; Anand, S.S. A systematic review of the evidence supporting a causal link between dietary factors and coronary heart disease. *Arch. Intern. Med.* **2009**, *169*, 659–669. [CrossRef]
20. Panagiotakos, D.B.; Pitsavos, C.; Skoumas, Y.; Stefanadis, C. The association between food patterns and the metabolic syndrome using principal components analysis: The ATTICA Study. *J. Am. Diet. Assoc.* **2007**, *107*, 979–987. [CrossRef]
21. García-Montero, C.; Fraile-Martínez, O.; Gómez-Lahoz, A.M.; Pekarek, L.; Castellanos, A.J.; Noguerales-Fraguas, F.; Coca, S.; Guijarro, L.G.; García-Honduvilla, N.; Asúnsolo, A.; et al. Nutritional Components in Western Diet Versus Mediterranean Diet at the Gut Microbiota-Immune System Interplay. Implications for Health and Disease. *Nutrients* **2021**, *13*, 699. [CrossRef]
22. Chen, X.; Chu, J.; Hu, W.; Sun, N.; He, Q.; Liu, S.; Feng, Z.; Li, T.; Han, Q.; Shen, Y. Associations of ultra-processed food consumption with cardiovascular disease and all-cause mortality: UK Biobank. *Eur. J. Public Health* **2022**, *32*, 779–785. [CrossRef] [PubMed]

23. Juul, F.; Vaidean, G.; Lin, Y.; Deierlein, A.L.; Parekh, N. Ultra-Processed Foods and Incident Cardiovascular Disease in the Framingham Offspring Study. *J. Am. Coll. Cardiol.* **2021**, *77*, 1520–1531. [CrossRef] [PubMed]
24. Srour, B.; Fezeu, L.K.; Kesse-Guyot, E.; Allès, B.; Méjean, C.; Andrianasolo, R.M.; Chazelas, E.; Deschasaux, M.; Hercberg, S.; Galan, P.; et al. Ultra-processed food intake and risk of cardiovascular disease: Prospective cohort study (NutriNet-Santé). *BMJ* **2019**, *365*, l1451. [CrossRef] [PubMed]
25. Rees, K.; Al-Khudairy, L.; Takeda, A.; Stranges, S. Vegan dietary pattern for the primary and secondary prevention of cardiovascular diseases. *Cochrane Database Syst. Rev.* **2021**, *2*, CD013501. [CrossRef]
26. Satija, A.; Hu, F.B. Plant-based diets and cardiovascular health. *Trends Cardiovasc. Med.* **2018**, *28*, 437–441. [CrossRef]
27. Lockheart, M.S.K.; Steffen, L.M.; Møklebust Rebnord, H.; Fimreite, R.L.; Ringstad, J.; Thelle, D.S.; Pedersen, J.I.; Jacobs, D.R. Dietary patterns, food groups and myocardial infarction: A case-control study. *Br. J. Nutr.* **2007**, *98*, 380–387. [CrossRef]
28. Jeppesen, J.; Schaaf, P.; Jones, C.; Zhou, M.Y.; Chen, Y.D.; Reaven, G.M. Effects of low-fat, high-carbohydrate diets on risk factors for ischemic heart disease in postmenopausal women. *Am. J. Clin. Nutr.* **1997**, *65*, 1027–1033. [CrossRef]
29. Fung, T.T.; Malik, V.; Rexrode, K.M.; Manson, J.E.; Willett, W.C.; Hu, F.B. Sweetened beverage consumption and risk of coronary heart disease in women. *Am. J. Clin. Nutr.* **2009**, *89*, 1037–1042. [CrossRef]
30. Wong, M.M.H.; Louie, J.C.Y. A priori dietary patterns and cardiovascular disease incidence in adult population-based studies: A review of recent evidence. *Crit. Rev. Food Sci. Nutr.* **2022**, *62*, 6153–6168. [CrossRef]
31. Critselis, E.; Kontogianni, M.D.; Georgousopoulou, E. Comparison of the Mediterranean diet and the Dietary Approach Stop Hypertension in reducing the risk of 10-year fatal and non-fatal CVD events in healthy adults: The ATTICA Study (2002–2012). *Public Health Nutr.* **2021**, *24*, 2746–2757. [CrossRef]
32. Fillipatos, T.D.; Panagiotakos, D.P.; Georgousopoulo, E.N. Mediterranean Diet and 10-year (2002–2012) Incidence of Diabetes and Cardiovascular Disease in Participants with Prediabetes: The ATTICA study. *Rev. Diabet. Stud.* **2016**, *13*, 226–235. [CrossRef] [PubMed]

nutrients

MDPI

Article

Dietary Total Antioxidant Capacity—A New Indicator of Healthy Diet Quality in Cardiovascular Diseases: A Polish Cross-Sectional Study

Małgorzata Elżbieta Zujko [1,*,†], Anna Waśkiewicz [2,†], Anna Maria Witkowska [1], Alicja Cicha-Mikołajczyk [2], Kinga Zujko [3] and Wojciech Drygas [2,4]

1 Department of Food Biotechnology, Faculty of Health Sciences, Medical University of Białystok, Szpitalna 37, 15-295 Białystok, Poland; anna.witkowska@umb.edu.pl
2 Department of Epidemiology, Cardiovascular Disease Prevention and Health Promotion, National Institute of Cardiology, Alpejska 42, 04-628 Warsaw, Poland; awaskiewicz@ikard.pl (A.W.); acicha@ikard.pl (A.C.-M.); wdrygas@ikard.pl (W.D.)
3 Department of Cardiology, Medical University of Białystok, M. Skłodowskiej-Curie 24a, 15-276 Białystok, Poland; z_kinga@wp.pl
4 Department of Social and Preventive Medicine, Faculty of Health Sciences, Medical University of Łódź, Hallera 1, 90-001 Łódź, Poland
* Correspondence: malgorzata.zujko@umb.edu.pl
† These authors contributed equally to this work.

Abstract: This study aimed to assess the relationship between the dietary total antioxidant capacity (DTAC) and the occurrence of cardiovascular diseases (CVDs), as well as healthy diet quality, in a representative sample (n = 5690) of the whole Polish adult population (WOBASZ II study). Daily food consumption was estimated by the single 24 h dietary recall method. Antioxidant vitamins (C, E, and β-carotene) and minerals (Zn, Fe, Mn, and Cu) from the diet and supplements were calculated using 5D Diet software, and dietary total polyphenol intake (DTPI) was determined using the Phenol-Explorer database and our database. Total diet quality was measured by the Healthy Diet Indicator (HDI) based on World Health Organization (WHO) recommendations for the prevention of CVD. DTAC was calculated using the data on food consumption and the antioxidant potential of foods measured by the FRAP (ferric ion reducing antioxidant potential) method. It was shown that higher DTAC was associated with a higher intake of polyphenols, antioxidant vitamins, and minerals. Moreover, a higher quartile of DTAC was associated with a reduced odds ratio for cardiovascular diseases in a Polish population, as well as with a higher HDI. Therefore, dietary recommendations for the prevention and therapy of CVDs should take into account a high DTAC. DTAC, measured by the FRAP method, can be considered an indicator of healthy diet quality.

Keywords: cardiovascular disease; dietary total antioxidant capacity; FRAP; population

Citation: Zujko, M.E.; Waśkiewicz, A.; Witkowska, A.M.; Cicha-Mikołajczyk, A.; Zujko, K.; Drygas, W. Dietary Total Antioxidant Capacity—A New Indicator of Healthy Diet Quality in Cardiovascular Diseases: A Polish Cross-Sectional Study. *Nutrients* **2022**, *14*, 3219. https://doi.org/10.3390/nu14153219

Academic Editors: Chara Tzavara, Charalampia Amerikanou and Andriana Kaliora

Received: 14 July 2022
Accepted: 5 August 2022
Published: 6 August 2022

1. Introduction

According to the World Health Organization (WHO) report, cardiovascular diseases (CVDs) are the leading cause of death worldwide. It is estimated that in 2019, deaths from CVDs accounted for 32% of all deaths. Most cardiovascular diseases can be prevented by changing modifiable risk factors, such as unhealthy diet, obesity, tobacco use, physical inactivity, and extensive alcohol consumption [1].

All of the recommended dietary patterns for CVD prevention, including the Mediterranean diet and the Dietary Approaches to Stop Hypertension model, emphasize the importance of high diet quality. A common feature of such dietary patterns is the high consumption of healthy foods, i.e., vegetables, fruits, whole grains, legumes, nuts and seeds, low-fat dairy, fish, and unprocessed lean meats and poultry, and low consumption of energy-dense, nutrient-poor foods rich in saturated fat, trans fat, added sugar and salt [2,3].

Epidemiological research has globally demonstrated that plant-based foods, which contain polyphenols, vitamins, minerals, and fiber, with proven antioxidant activity, can prevent CVDs [4,5]. Our previous study showed that individual diet modifications in terms of a higher intake of polyphenols (flavonoids and anthocyanins), fiber, and polyunsaturated fatty acids (PUFA) and a lower intake of saturated fatty acids (SFA) had a significant impact on the improvement of some metabolic syndrome risk factors, such as waist circumference, fasting glucose, and HDL cholesterol [6]. However, no association was found between lignan intake and the prevalence of CVDs [7].

A new approach to a healthy diet is the assessment of dietary total antioxidant capacity (DTAC). The whole diet contains various antioxidants (vitamins: C, E, and carotenoids; minerals: Zn, Fe, Mn, Cu, and Se; polyphenols) with additional or synergistic effects. Several assays are available to measure antioxidants in foods, but the largest database is based on the FRAP (ferric ion reducing antioxidant potential) method [8]. The current FRAP database includes more than 3100 different foods, beverages, spices, herbs, and supplements purchased at local stores and markets around the world [9]. Some authors suggest that dietary FRAP may be considered an appropriate measure of dietary quality because it positively correlates with well-known indicators of a healthy diet [10].

The epidemiological evidence of a relationship between DTAC measured by the FRAP method and CVDs is limited. Findings from a meta-analysis of prospective cohort studies revealed that higher dietary FRAP was associated with a lower risk of CVD mortality [11]. In some studies, DTAC was inversely associated with cardiovascular events and cardiometabolic risk factors [12], blood pressure and diabetes [13], prediabetes and insulin resistance [14,15], and lipid biomarkers [15]. In the other study, no association was found between DTAC and waist circumference, glucose level, insulin resistance, or lipid biomarkers. However, the study was conducted on a small group of patients with nonalcoholic fatty liver disease [16].

This study aimed to assess the relationship between the dietary total antioxidant capacity and the occurrence of cardiovascular diseases, as well as healthy diet quality, in the Polish adult population. This cross-sectional study supplements the existing knowledge in this field.

2. Materials and Methods

2.1. Ethical Approval

The study was conducted in accordance with the Helsinki Declaration and Good Clinical Practice. Approval for the WOBASZ II study was obtained from the Bioethics Committee at the National Institute of Cardiology (No. 1344), which also approved the current study (No. 1837). Written informed consent was obtained from all participants.

2.2. Study Group

The subjects (5690) were participants of the National Multicenter Health Survey II (WOBASZ II) conducted by the National Institute of Cardiology in Warsaw in 2013–2014. It was a cross-sectional study aimed at investigating the determinants of chronic non-communicable diseases in a representative sample of Polish adults aged 20 years and older. The design and methods of the WOBASZ II study have been described in detail elsewhere [17].

Baseline information about participants (smoking, educational level, physical activity, and diseases) was collected using a standardized questionnaire developed for the WOBASZ II study. Physical activity was classified as follows: low level—physical activity for at least 30 min a day once a week or less; middle level—physical activity for at least 30 min a day 2–3 times a week; and high level—physical activity for at least 30 min a day $\geq$ 4 times a week.

Subjects were classified as having CVD if they had: coronary heart disease, myocardial infarction, stroke, atrial fibrillation and/or other cardiac arrhythmias, peripheral vascular disease of the lower limbs, heart failure, coronary angioplasty or coronary artery bypass grafting, and implanted pacemaker or cardioverter-defibrillator, as was reported previously [18].

2.3. Clinical Measurements

The measurements of body mass, height, and waist circumference were performed by trained nurses using standardized procedures. Body mass index (BMI) was calculated as body mass in kilograms divided by squared height in meters, and a BMI of 18.5–24.9 kg/m^2 was defined as normal body mass [19]. Blood pressure (BP) was measured three times on the right arm after 5 min of rest in a sitting position at 1 min intervals using automatic devices (AND UA-631), and the final BP was reported as the mean of the second and third measurements. Hypertension was diagnosed when systolic blood pressure (BPs) $\geq$ 140 mm Hg and/or diastolic blood pressure (BPd) $\geq$ 90 mm Hg and/or when antihypertensive drugs were used [20]. Biochemical analyses were performed at the Diagnostyka Central Laboratory of the National Institute of Cardiology in Warsaw. Diabetes was diagnosed when the fasting glucose (FG) level was $\geq$126 mg/dL or participants were taking antidiabetic drugs [21]. Hypercholesterolemia was defined when fasting total cholesterol (TC) was $\geq$190 mg/dL or low-density lipoprotein cholesterol (LDL-C) was $\geq$115 mg/dL or subjects were taking lipid-lowering medication, and hypertriglyceridemia was defined when the fasting triglyceride level (TG) was $\geq$150 mg/dL or participants used lipid-lowering drugs [20].

A diagnosis of metabolic syndrome (MetS) was made when at least three of five risk factors were identified: waist circumference ($\geq$94 cm for men and $\geq$80 cm for women), TG ($\geq$150 mg/dL), HDL-C (<40 mg/dL for men and <50 for women), BPs $\geq$ 130 mm Hg and/or BPd $\geq$ 85 mm Hg), and FG ($\geq$100 mg/dL) [22].

2.4. Dietary Assessment

Dietary and supplement intake assessment was performed by trained interviewers using a single 24 h dietary recall method. Food portion sizes were estimated using an Album of Photographs of Food Products and Dishes [23]. Subjects were asked if they had taken any form of a dietary supplement on the recall day, and the supplement type, name brand, and dose were recorded. Nutrient intake (including vitamins and minerals) from the diet was calculated based on the amount of food consumed with the use of Polish Food Composition Tables [24].

2.5. Assessment of Dietary Antioxidants

Antioxidant vitamin (C, E, and β-carotene) and mineral (Zn, Fe, Mn, and Cu) intake from both food and dietary supplements were taken into account. In the case of vitamins, losses during technological processes and food preparation were deducted. The amounts of vitamins and minerals from diet and supplements were estimated using the NFNI (National Food and Nutrition Institute) 5D Diet software (IŻŻ Diet 5D).

Dietary total polyphenol intake (DTPI) was estimated using the online Phenol-Explorer database [25] and our database [26,27].

Dietary total antioxidant capacity (DTAC) was estimated using the FRAP method published by Carlsen [28] and our database [26,27].

2.6. Assessment of Healthy Diet Indicator (HDI)

Total diet quality was measured using the Healthy Diet Indicator (HDI) based on World Health Organization (WHO) recommendations for the prevention of CVDs [29] and described in [30]. The HDI includes the consumption of six nutrients (saturated fatty acids, polyunsaturated fatty acids, cholesterol, protein, fiber, and free sugars) and the sum of fruits and vegetables within the recommended range, with 1 point awarded for the recommended intake of a given nutrient, and 0 points if intake was not consistent with guidelines. The maximum score for an optimal diet was 7 points.

In the present study, an HDI score equal to at least 5 points was arbitrarily assumed as an indicator of a proper diet.

2.7. Statistical Analysis

The study participants were divided into four subgroups according to the quartile distribution of DTAC for men and women separately and for the total group.

Quantitative variables are presented as the mean (standard deviation) and/or median (interquartile range), while qualitative variables are reported as the number and percentage. The Wilcoxon test and the chi-square test were used for comparisons between CVD and non-CVD subgroups, if appropriate.

Mean values of antioxidants and/or polyphenol intake from food and supplements, as well as the HDI score, with a 95% confidence interval (95% CI) were calculated by a general linear model with the Tukey–Kramer adjustment for multiple comparisons. The Cochran–Armitage test for trend was applied to assess for an association between the prevalence of a proper diet (HDI $\geq$ 5) and DTAC over quartiles.

Logistic regression models were used to assess the relationship between DTAC and the prevalence of CVDs. Three models were applied: model 1: crude in men and women or adjusted for sex in the total group; model 2: adjusted for age, BMI, HDI, smoking status, and alcohol intake in men and women and additionally for sex in the total group; model 3: adjusted for age, BMI, HDI, smoking status, alcohol intake, educational level, and physical activity in men and women and additionally for sex in the total group. The first quartile (Q1) in each model was adopted as a reference. The results of logistic regression analysis are presented as odds ratios (ORs) and 95% confidence intervals.

All statistical analyses were performed using SAS software version 9.4 (SAS Institute Inc., Cary, NC, USA). A p-value less than 0.05 was considered statistically significant.

3. Results

Baseline characteristics of the study population according to CVDs for both sexes are presented in Table 1. Among 5690 analyzed participants (2554 men and 3136 women), 1138 (20%) had CVD (494 men and 644 women). It was shown that CVD participants were older (respectively: 62.3 $\pm$ 14.1 years vs. 46.4 $\pm$ 15.4 years), had higher BMI (respectively: 28.8 $\pm$ 5.1 kg/m^2 vs. 26.8 $\pm$ 5.1 kg/m^2), consumed less fiber (respectively: 18.3 $\pm$ 8.2 g/day vs. 19.3 $\pm$ 8.7 g/day), and less frequently had a higher educational level (respectively: 11.52% vs. 21.93%). On the other hand, people with CVDs were less likely to smoke cigarettes (respectively: 16.9% vs. 24.87%) and consumed less energy (respectively: 1755 $\pm$ 729 kcal vs. 2017 $\pm$ 867 kcal) and alcohol (respectively: 1.5 $\pm$ 10.2 g/day vs. 2.7 $\pm$ 15.0 g/day). Over 50% of participants had a low level of physical activity. The recommended moderate physical activity was more common in non-CVD in comparison to CVD patients (16.14% vs. 11.89%). Moreover, CVD compared to non-CVD participants more often had hypertension (68.7% vs. 39.3%), diabetes mellitus (22.84% vs. 7.81%), hypercholesterolemia (79.33% vs. 66.25%), hypertriglyceridemia (32.17% vs. 27.39%), and metabolic syndrome (52.2% vs. 31.5%).

Overall, CVD participants were older and less educated than those without CVD, which could have resulted in poorer lifestyles (food choices and exercise) and the development of obesity, diabetes, hypertension, and dyslipidemia. Health-promoting behaviors of people with CVDs were limited to smoking and drinking less alcohol and consuming less energy compared to people without CVDs. Therefore, further analyses were adjusted for confounding factors.

A comparison of antioxidant intake from food and supplements between CVD and non-CVD participants is shown in Table 2. The results were adjusted for age in men and women and for sex and age in the total group. It was detected that the diet of non-CVD subjects contained significantly more DTAC, DTPI, Fe, and Cu in comparison to CVD subjects for men and the total group and more Zn compared to the total group. However, after recalculation of the results per 1000 kcal of the diet, DTAC was significantly higher only in non-CVD relative to CVD men. Moreover, consumption of β-carotene and Mn per 1000 kcal of the diet was significantly lower in non-CVD than in CVD men and the total group. In women, no significant differences were found between CVD and non-CVD participants.

Table 1. Baseline characteristics of the study population according to CVDs for both sexes and overall.

Variables	Men (n = 2554)		p	Women (n = 3136)		p	Total (n = 5690)		p
	CVD	Non-CVD		CVD	Non-CVD		CVD	Non-CVD	
	n = 494	n = 2060		n = 644	n = 2492		n = 1138	n = 4552	
Age [years], Mean ± SD	62.4 ± 13.2	45.5 ± 15.2	<0.0001	62.3 ± 14.8	47.1 ± 15.5	<0.0001	62.3 ± 14.1	46.4 ± 15.4	<0.0001
BMI [kg/m^2], Mean ± SD	28.5 ± 4.7	27.2 ± 4.5	<0.0001	29.0 ± 5.3	26.4 ± 5.6	<0.0001	28.8 ± 5.1	26.8 ± 5.1	<0.0001
Current smoking, n (%)	93 (18.90)	645 (31.36)	<0.0001	99 (15.37)	486 (19.51)	0.0163	192 (16.90)	1131 (24.87)	<0.0001
Educational level, n (%)									
Under middle	118 (23.94)	258 (12.54)	<0.0001	221 (34.32)	376 (15.11)	<0.0001	339 (29.82)	634 (13.95)	<0.0001
Vocational	156 (31.64)	642 (31.20)		110 (17.08)	464 (18.65)		266 (23.39)	1106 (24.33)	
Middle	166 (33.67)	775 (37.66)		235 (36.49)	1034 (41.56)		401 (35.27)	1809 (39.79)	
Higher	53 (10.75)	383 (18.61)		78 (12.11)	614 (24.68)		131 (11.52)	997 (21.93)	
Physical activity, n (%)									
Low level	257 (52.34)	1142 (55.60)	0.0039	363 (56.37)	1314 (52.96)	0.0403	620 (54.63)	2456 (54.16)	0.0017
Middle level	57 (11.61)	320 (15.58)		78 (12.11)	412 (16.61)		135 (11.89)	732 (16.14)	
High level	165 (33.60)	535 (26.05)		187 (29.04)	705 (28.42)		352 (31.01)	1240 (27.34)	
Seasonally	12 (2.44)	57 (2.78)		16 (2.48)	50 (2.02)		28 (2.47)	107 (2.36)	
Energy [kcal/day], Mean ± SD	2034 ± 829	2385 ± 947	<0.0001	1542 ± 555	1713 ± 654	<0.0001	1755 ± 729	2017 ± 867	<0.0001
Dietary fiber [g/day], Mean ± SD	20.3 ± 8.7	21.1 ± 9.2	0.0733	16.8 ± 7.4	17.7 ± 7.8	0.0019	18.3 ± 8.2	19.3 ± 8.7	0.0005
Alcohol intake [g/day], Mean ± SD	3.1 ± 15.1	5.0 ± 21.0	0.0111	0.3 ± 2.7	0.7 ± 6.4	0.0792	1.5 ± 10.2	2.7 ± 15.0	0.0014
Hypertension, n (%)	359 (73.27)	890 (43.84)	<0.0001	418 (65.21)	874 (35.56)	<0.0001	777 (68.70)	1764 (39.30)	<0.0001
Diabetes mellitus, n (%)	129 (27.22)	161 (8.17)	<0.0001	119 (19.44)	178 (7.51)	<0.0001	248 (22.84)	339 (7.81)	<0.0001
Hypercholesterolemia, n (%)	401 (82.68)	1376 (68.97)	<0.0001	493 (76.79)	1563 (64.03)	<0.0001	894 (79.33)	2939 (66.25)	<0.0001
Hypertriglyceridemia, n (%)	181 (38.19)	699 (35.50)	0.2742	168 (27.50)	494 (20.70)	0.0003	349 (32.17)	1193 (27.39)	0.0018
Metabolic syndrome, n (%)	272 (55.06)	728 (35.34)	<0.0001	322 (50.00)	706 (28.33)	<0.0001	594 (52.20)	1434 (31.50)	<0.0001

CVD—cardiovascular disease; SD—standard deviation; n—number. Differences between quantitative and qualitative variables were tested by Wilcoxon rank sum test or Chi2 test, respectively.

Antioxidant intake from food and supplements by quartiles of DTAC in men, in women, and in the total group is listed in Table 3. The results were adjusted for age in men and women and for sex and age in the total group. It was shown that higher DTAC was associated with higher DTPI, intake of antioxidant vitamins (C, E, and β-carotene), and minerals (Zn, Fe, Mn, and Cu) in men, women, and the total group. After recalculation of the results per 1000 kcal of the diet, significant positive associations were found only for DTPI, vitamin C, and Cu in men and women and for DTPI, vitamin C, Fe, Mn, and Cu in the total group. However, inverse associations were found for Zn per 1000 kcal of diet in men and women and β-carotene and Zn per 1000 kcal of the diet in the total group.

The prevalence and OR (95% CI) of CVDs by quartiles of DTAC are shown in Table 4. Among men, the mean percentage of CVD prevalence was 23.1% in the first quartile (Q1), 21.8% in the second quartile (Q2), 17.2% in the third quartile (Q3), and 15.2% in the fourth quartile (Q4). Among women, it was: Q1—23.7%; Q2—21.1%; Q3—18.3%; and Q4—19.0%. In the total group, the prevalence of CVD was: Q1—23.3%; Q2—21.4%; Q3—18.1%; and Q4—17.2%. The prevalence of CVDs decreased significantly with increasing DTAC.

Table 2. Comparison of antioxidant intake from food and supplements between CVD and non-CVD participants according to sex and overall.

Variables	Men [1]			Women [1]			Total [2]		
	CVD	Non-CVD	*p*	CVD	Non-CVD	*p*	CVD	Non-CVD	*p*
	Mean (95% CI)			Mean (95% CI)			Mean (95% CI)		
DTAC [mmol/d]	11.06 (10.35–11.77)	12.69 (12.36–13.02)	<0.0001	12.15 (11.55–12.74)	12.29 (12.00–12.58)	0.6776	11.70 (11.24–12.15)	12.47 (12.25–12.69)	0.0033
DTAC/1000 kcal	5.42 (5.09–5.75)	5.81 (5.66–5.97)	0.0388	8.18 (7.77–8.59)	7.78 (7.58–7.98)	0.0931	6.86 (6.59–7.13)	6.79 (6.66–6.92)	0.6511
DTPI [mg/d]	1948 (1858–2037)	2101 (2060–2143)	0.0029	1935 (1866–2004)	2000 (1967–2034)	0.1046	1947 (1892–2002)	2050 (2023–2076)	0.0013
DTPI/1000 kcal	945 (905–986)	954 (936–973)	0.7086	1298 (1249–1347)	1255 (1232–1279)	0.1302	1126 (1094–1159)	1104 (1088–1120)	0.2346
Vitamin C [mg/d]	85.9 (78.2–93.7)	87.9 (84.3–91.5)	0.6611	92.2 (84.1–100.3)	96.1 (92.2–100.1)	0.4020	88.9 (83.2–94.6)	92.0 (89.3–94.7)	0.3437
Vitamin C/1000 kcal	43.5 (39.5–47.4)	40.5 (38.6–42.3)	0.1851	63.5 (57.5–69.4)	61.2 (58.3–64.1)	0.5097	53.5 (49.7–57.3)	50.8 (49.0–52.6)	0.2218
Vitamin E [mg/d]	12.38 (11.60–13.16)	12.67 (12.30–13.03)	0.5262	10.47 (9.28–11.65)	11.11 (10.54–11.69)	0.3495	11.38 (10.63–12.13)	11.90 (11.54–12.25)	0.2366
Vitamin E/1000 kcal	5.71 (5.38–6.04)	5.52 (5.36–5.67)	0.3043	6.82 (5.90–7.74)	6.86 (6.41–7.31)	0.9438	6.25 (5.71–6.79)	6.19 (5.93–6.45)	0.8444
β-Carotene [μg/d]	2980 (2685–3275)	2972 (2836–3109)	0.9652	3210 (2936–3484)	2923 (2789–3056)	0.0716	3109 (2908–3310)	2944 (2848–3041)	0.1577
β-Carotene/1000 kcal	1601 (1443–1759)	1396 (1322–1469)	0.0246	2135 (1933–2337)	1927 (1829–2025)	0.0764	1870 (1737–2003)	1661 (1597–1725)	0.0067
Zinc [mg/d]	11.34 (10.86–11.82)	11.62 (11.40–11.85)	0.3089	8.26 (7.91–8.61)	8.64 (8.47–8.81)	0.0570	9.78 (9.49–10.07)	10.13 (10.00–10.27)	0.0326
Zinc/1000 kcal	5.28 (5.12–5.44)	5.10 (5.03–5.18)	0.0506	5.39 (5.19–5.59)	5.29 (5.19–5.39)	0.4099	5.33 (5.20–5.46)	5.20 (5.13–5.26)	0.0809
Iron [mg/d]	11.99 (11.40–12.57)	12.70 (12.43–12.97)	0.0348	10.05 (9.38–10.71)	10.41 (10.09–10.73)	0.3369	11.03 (10.58–11.48)	11.55 (11.33–11.77)	0.0480
Iron/1000 kcal	5.61 (5.40–5.82)	5.59 (5.49–5.69)	0.8788	6.55 (6.16–6.93)	6.35 (6.16–6.53)	0.3675	6.10 (5.86–6.33)	5.96 (5.85–6.08)	0.3302
Manganese [mg/d]	4.75 (4.53–4.96)	4.66 (4.56–4.76)	0.4996	3.90 (3.74–4.06)	4.06 (3.98–4.14)	0.0842	4.31 (4.18–4.44)	4.36 (4.30–4.43)	0.4498
Manganese/1000 kcal	2.31 (2.21–2.41)	2.14 (2.09–2.18)	0.0032	2.65 (2.54–2.76)	2.56 (2.51–2.62)	0.1539	2.48 (2.40–2.55)	2.35 (2.32–2.39)	0.0038
Copper [mg/d]	1.21 (1.15–1.26)	1.27 (1.24–1.29)	0.0429	1.03 (0.99–1.07)	1.07 (1.05–1.09)	0.0873	1.12 (1.08–1.15)	1.17 (1.15–1.18)	0.0076
Copper/1000 kcal	0.568 (0.549–0.587)	0.562 (0.553–0.571)	0.5883	0.675 (0.650–0.700)	0.656 (0.644–0.668)	0.1971	0.623 (0.606–0.639)	0.609 (0.601–0.617)	0.1450

[1]—Adjusted for age; [2]—adjusted for sex and age; DTAC—dietary total antioxidant capacity; DTPI—dietary total polyphenol intake; CVD—cardiovascular disease; CI—confidence interval. The general linear model was applied to identify differences in all variables between adjusted means of CVD and non-CVD subgroups.

Table 3. Antioxidant intake from food and supplements by quartiles of DTAC according to sex and overall.

Variables	Men n = 2554				p	Women n = 3136				p	Total n = 5690				p
	Q1	Q2	Q3	Q4		Q1	Q2	Q3	Q4		Q1	Q2	Q3	Q4	
	n = 638	n = 639	n = 638	n = 639		n = 784	n = 784	n = 784	n = 784		n = 1423	n = 1422	n = 1423	n = 1422	
								DTAC (mmol/d)							
Mean ± SD	5.22 ± 1.75	9.44 ± 1.02	13.17 ± 1.20	21.67 ± 8.52	<0.0001	5.52 + 1.77	9.72 ± 1.00	13.14 ± 1.11	20.68 ± 9.07	<0.0001	5.38 ± 1.77	9.59 ± 1.01	13.15 ± 1.15	21.13 ± 8.84	<0.0001
Me, IQR	5.50	9.43	13.14	19.10		5.78	9.77	13.08	18.50		5.67	9.62	13.11	18.73	
	(3.99–6.67)	(8.52–10.34)	(12.12–14.18)	(16.90–23.03)		(4.37–7.08)	(8.85–10.61)	(12.15–14.11)	(16.52–21.56)		(4.18–6.87)	(8.71–10.47)	(12.13–14.16)	(16.76–22.20)	
Range	0.47–7.70	7.71–11.16	11.16–15.38	15.39–95.69		0.32–7.91	7.94–11.35	11.35–15.19	15.19–191.82		0.32–7.85	7.85–11.31	11.31–15.27	15.27–191.82	
								DTAC/1000 kcal							
Mean ± SD	3.15 ± 1.78	4.99 ± 2.07	6.15 ± 2.46	8.66 ± 4.34	<0.0001	4.55 ± 2.31	6.80 ± 2.69	8.50 ± 3.10	11.62 ± 7.16	<0.0001	3.88 ± 2.16	6.02 ± 2.59	7.48 ± 3.06	10.27 ± 6.24	<0.0001
Me, IQR	2.83	4.65	5.74	7.80		4.09	6.19	7.92	10.07		3.50	5.51	6.99	8.97	
	(2.02–3.90)	(3.65–5.84)	(4.55–7.21)	(5.89–9.98)		(3.03–5.55)	(4.98–7.99)	(6.43–10.02)	(8.16–13.05)		(2.46–4.83)	(4.26–7.14)	(5.42–8.92)	(6.89–11.65)	
Range	0.39–21.18	1.51–21.98	2.25–26.35	2.30–33.51		0.38–19.06	2.05–22.01	2.90–29.93	3.31–105.64		0.38–21.18	1.51–22.01	2.25–29.93	2.30–105.64	
								Mean [1] (95% CI)							
DTPI [mg/d]	1135	1738	2226	3187	<0.0001	1107	1714	2165	2962	<0.0001	1125	1729	2199	3067	<0.0001
	(1091–1179)	(1694–1782)	(2182–2270)	(3143–3231)		(1071–1143)	(1678–1750)	(2129–2200)	(2927–2998)		(1097–1153)	(1701–1757)	(2171–2227)	(3039–3095)	
DTPI/ 1000 kcal	654	886	1019	1251	<0.0001	876	1168	1372	1640	<0.0001	761	1028	1197	1450	<0.0001
	(626–682)	(858–914)	(991–1047)	(1223–1279)		(839–913)	(1131–1205)	(1335–1409)	(1603–1677)		(737–785)	(1004–1052)	(1173–1221)	(1426–1474)	
Vitamin C [mg/d]	60.1	79.3	92.8	117.7	<0.0001	67.2	85.7	94.8	133.7	<0.0001	64.2	82.1	93.5	125.8	<0.0001
	(54.0–66.2)	(73.2–85.4)	(86.7–98.9)	(111.6–123.8)		(60.5–74.0)	(79.0–92.4)	(88.0–101.5)	(126.9–140.4)		(59.6–68.8)	(77.5–86.7)	(88.9–98.1)	(121.2–130.4)	
Vitamin C/ 1000 kcal	34.9	40.6	42.1	46.5	<0.0001	54.4	57.9	60.2	74.0	<0.0001	44.9	49.1	50.9	60.5	<0.0001
	(31.7–38.1)	(37.4–43.8)	(38.9–45.3)	(43.3–49.7)		(49.3–59.5)	(52.9–63.0)	(55.1–65.3)	(69.0–79.1)		(41.8–48.1)	(45.9–52.2)	(47.7–54.0)	(57.4–63.7)	
Vitamin E [mg/d]	10.11	11.63	12.86	15.84	<0.0001	8.62	10.43	11.42	13.45	<0.0001	9.43	11.03	12.10	14.62	<0.0001
	(9.49–10.73)	(11.02–12.25)	(12.24–13.48)	(15.22–16.45)		(7.62–9.63)	(9.42–11.44)	(10.41–12.42)	(12.44–14.46)		(8.81–10.05)	(10.41–11.66)	(11.48–12.72)	(13.99–15.24)	

Table 3. *Cont.*

Variables	Men				*p*	Women				*p*	Total				*p*
	n = 2554					*n* = 3136					*n* = 5690				
	Q1	Q2	Q3	Q4		Q1	Q2	Q3	Q4		Q1	Q2	Q3	Q4	
	n = 638	*n* = 639	*n* = 638	*n* = 639		*n* = 784	*n* = 784	*n* = 784	*n* = 784		*n* = 1423	*n* = 1422	*n* = 1423	*n* = 1422	
Vitamin E/ 1000 kcal	5.44	5.54	5.44	5.80	0.1965	6.44	6.81	7.10	7.05	0.6479	5.95	6.17	6.26	6.43	0.5209
	(5.17–5.70)	(5.27–5.80)	(5.18–5.71)	(5.53–6.06)		(5.66–7.23)	(6.03–7.60)	(6.31–7.88)	(6.26–7.84)		(5.50–6.40)	(5.72–6.62)	(5.81–6.71)	(5.98–6.88)	
β-Carotene [μg/d]	2313	2971	3194	3416	<0.0001	2598	2871	3086	3373	<0.0001	2488	2894	3142	3387	<0.0001
	(2075–2552)	(2733–3210)	(2955–3432)	(3178–3655)		(2364–2832)	(2638–3105)	(2853–3320)	(3139–3607)		(2320–2655)	(2727–3062)	(2974–3310)	(3219–3555)	
β-Carotene/ 1000 kcal	1428	1546	1432	1334	0.1568	2164	1967	1907	1841	0.0574	1812	1745	1666	1586	0.0309
	(1299–1558)	(1417–1676)	(1303–1562)	(1205–1463)		(1992–2337)	(1794–2139)	(1735–2080)	(1668–2013)		(1701–1924)	(1634–1857)	(1555–1778)	(1475–1698)	
Zinc [mg/d]	9.96	10.99	11.94	13.38	<0.0001	7.36	8.11	8.75	10.04	<0.0001	8.71	9.52	10.34	11.68	<0.0001
	(9.58–10.34)	(10.61–11.37)	(11.56–12.32)	(13.00–13.76)		(7.07–7.65)	(7.82–8.40)	(8.46–9.04)	(9.75–10.33)		(8.48–8.94)	(9.29–9.76)	(10.11–10.58)	(11.45–11.92)	
Zinc/ 1000 kcal	5.35	5.20	5.07	4.93	<0.0001	5.51	5.18	5.26	5.28	0.0459	5.43	5.19	5.16	5.12	0.0004
	(5.22–5.48)	(5.07–5.33)	(4.94–5.20)	(4.80–5.06)		(5.34–5.68)	(5.01–5.35)	(5.09–5.43)	(5.11–5.45)		(5.32–5.54)	(5.08–5.30)	(5.05–5.28)	(5.00–5.23)	
Iron [mg/d]	10.26	11.58	13.29	15.12	<0.0001	8.50	9.42	10.65	12.79	<0.0001	9.39	10.50	11.91	13.99	<0.0001
	(9.81–10.72)	(11.13–12.04)	(12.83–13.74)	(14.67–15.58)		(7.94–9.05)	(8.86–9.97)	(10.10–11.20)	(12.24–13.34)		(9.03–9.76)	(10.14–10.87)	(11.54–12.28)	(13.62–14.35)	
Iron/ 1000 kcal	5.52	5.56	5.68	5.62	0.5867	6.30	6.07	6.48	6.70	0.0567	5.89	5.82	6.06	6.19	0.0411
	(5.35–5.69)	(5.39–5.73)	(5.51–5.85)	(5.45–5.79)		(5.97–6.63)	(5.75–6.40)	(6.15–6.80)	(6.37–7.03)		(5.69–6.09)	(5.63–6.02)	(5.86–6.26)	(5.99–6.39)	
Manganese [mg/d]	3.67	4.40	4.85	5.80	<0.0001	3.18	3.79	4.21	4.92	<0.0001	3.43	4.09	4.54	5.36	<0.0001
	(3.50–3.83)	(4.24–4.57)	(4.69–5.02)	(5.63–5.96)		(3.05–3.31)	(3.66–3.92)	(4.08–4.34)	(4.79–5.05)		(3.33–3.53)	(3.99–4.20)	(4.43–4.64)	(5.25–5.46)	
Manganese/ 1000 kcal	2.10	2.19	2.17	2.23	0.1706	2.50	2.55	2.62	2.65	0.1216	2.29	2.37	2.39	2.45	0.0067
	(2.02–2.18)	(2.10–2.27)	(2.09–2.25)	(2.15–2.31)		(2.41–2.60)	(2.46–2.65)	(2.52–2.71)	(2.56–2.75)		(2.23–2.36)	(2.31–2.44)	(2.33–2.46)	(2.39–2.51)	
Copper [mg/d]	0.96	1.13	1.32	1.61	<0.0001	0.79	0.96	1.10	1.39	<0.0001	0.87	1.05	1.21	1.50	<0.0001
	(0.92–0.99)	(1.09–1.17)	(1.28–1.36)	(1.58–1.65)		(0.75–0.82)	(0.93–1.00)	(1.07–1.13)	(1.36–1.43)		(0.85–0.90)	(1.02–1.07)	(1.18–1.23)	(1.48–1.53)	

Table 3. *Cont.*

Variables	Men				p	Women				p	Total				p
	n = 2554					n = 3136					n = 5690				
	Q1	Q2	Q3	Q4		Q1	Q2	Q3	Q4		Q1	Q2	Q3	Q4	
	n = 638	n = 639	n = 638	n = 639		n = 784	n = 784	n = 784	n = 784		n = 1423	n = 1422	n = 1423	n = 1422	
Copper/ 1000 kcal	0.523	0.548	0.575	0.607	<0.0001	0.594	0.626	0.674	0.746	<0.0001	0.557	0.587	0.624	0.679	<0.0001
	(0.507–0.538)	(0.532–0.563)	(0.560–0.591)	(0.592–0.623)		(0.573–0.615)	(0.605–0.647)	(0.653–0.695)	(0.725–0.767)		(0.543–0.570)	(0.573–0.600)	(0.611–0.638)	(0.666–0.693)	

[1]—Adjusted for age in men and women and for sex and age in the total group; n—number; DTAC—dietary total antioxidant capacity; DTPI—dietary total polyphenol intake; Me—median; IQR—interquartile range; Q1–Q4—quartiles of DTAC; SD—standard deviation. Differences in DTAC and DTAC adjusted for energy were tested by nonparametric Kruskal–Wallis test, while other variables were compared using a general linear model.

Table 4. Prevalence and OR (95% CI) of CVDs by quartiles of DTAC according to sex and overall.

Variables	Men				Women				Total			
	n = 2554				n = 3136				n = 5690			
	Q1	Q2	Q3	Q4	Q1	Q2	Q3	Q4	Q1	Q2	Q3	Q4
	n = 638	n = 639	n = 638	n = 639	n = 784	n = 784	n = 784	n = 784	n = 1423	n = 1422	n = 1423	n = 1422
CVD Prevalence %	23.1	21.8	17.2	15.2	23.7	21.1	18.3	19.0	23.3	21.4	18.1	17.2
(95% CI) [A]	(20.4–25.9)	(19.0–24.6)	(14.4–20.0)	(12.4–18.0)	(21.1–26.3)	(18.5–23.8)	(15.7–21.0)	(16.4–21.6)	(21.4–25.2)	(19.5–23.4)	(16.2–20.0)	(15.3–19.1)
p *	-	-	-	0.0001	-	-	-	0.0205	-	-	-	<0.0001
OR	1	0.921	0.643	0.563	1	0.796	0.650	0.617	1	0.852	0.666	0.593
(95% CI) [1]	-	(0.710–1.196)	(0.487–0.848)	(0.424–0.749)	-	(0.629–1.006)	(0.510–0.829)	(0.483–0.788)	-	(0.716–1.015)	(0.555–0.799)	(0.492–0.714)
p	-	0.5387	0.0018	<0.0001	-	0.0562	0.0005	0.0001	-	0.0733	<0.0001	<0.0001
AOR	1	0.934	0.706	0.631	1	0.884	0.731	0.880	1	0.906	0.733	0.770
(95% CI) [2]	-	(0.685–1.274)	(0.511–0.976)	(0.452–0.882)	-	(0.675–1.157)	(0.554–0.965)	(0.663–1.167)	-	(0.740–1.111)	(0.595–0.904)	(0.621–0.955)
p	-	0.6654	0.0353	0.0071	-	0.3692	0.0268	0.3744	-	0.3432	0.0037	0.0176
AOR	1	0.916	0.698	0.610	1	0.882	0.726	0.877	1	0.904	0.726	0.752
(95% CI) [3]	-	(0.671–1.251)	(0.504–0.967)	(0.436–0.855)	-	(0.673–1.155)	(0.550–0.958)	(0.660–1.165)	-	(0.737–1.108)	(0.588–0.895)	(0.605–0.935)
p	-	0.5830	0.0306	0.0041	-	0.3604	0.0237	0.3636	-	0.3304	0.0028	0.0102

n—Number; OR—odds ratio; AOR—adjusted odds ratio; CI—confidence interval; OR—odds ratio; CVD—cardiovascular disease; DTAC—dietary total antioxidant capacity; [A]—adjusted for age in men and women and for sex and age in the total group; [1]—crude OR in men and women or adjusted for sex in the total group; [2]—adjusted for age, BMI, HDI, smoking status, and alcohol intake in men and women and additionally for sex in the total group; [3]—adjusted for age, BMI, HDI, smoking status, alcohol intake, educational level, and physical activity in men and women and additionally for sex in the total group; *—p-value for comparisons between adjusted means. Difference in CVD prevalence over quartiles was tested using a general linear model, while logistic regression analysis was performed to estimate the odds of CVD between pairs of quartiles: Q2 vs. Q1, Q3 vs. Q1, and Q4 vs. Q1 for each model.

When the analysis was adjusted for multiple variables (model 3), the OR of CVDs in men was 30.2% lower in Q3 (OR = 0.698, 95% CI = 0.504–0.967, $p = 0.0306$) and 39.0% lower in Q4 (OR = 0.610, 95% CI = 0.436–0.855, $p = 0.0041$) in comparison to Q1. In women, the OR of CVDs was 27.4% lower in Q3 (OR = 0.726, 95% CI = 0.550–0.958, $p = 0.0237$) relative to Q1. In the total group, the OR of CVDs was 27.4% lower in Q3 (OR = 0.726, 95% CI = 0.588–0.895, $p = 0.0028$) and 24.8% lower in Q4 (OR = 0.752, 95% CI = 0.605–0.935, $p = 0.0102$) in comparison to Q1.

The mean values of HDI, after adjustment for age in men and women and additionally for sex in the total group, are presented in Figure 1. The average HDI increased significantly across quartiles of DTAC from 2.83 (95% CI = 2.72–2.92) to 3.63 (95% CI = 3.53–3.73) in men, from 2.95 (95% CI = 2.87–3.04) to 3.64 (95% CI = 3.55–3.72) in women, and from 2.89 (95% CI = 2.83–2.96) to 3.63 (95% CI = 3.57–3.70) overall.

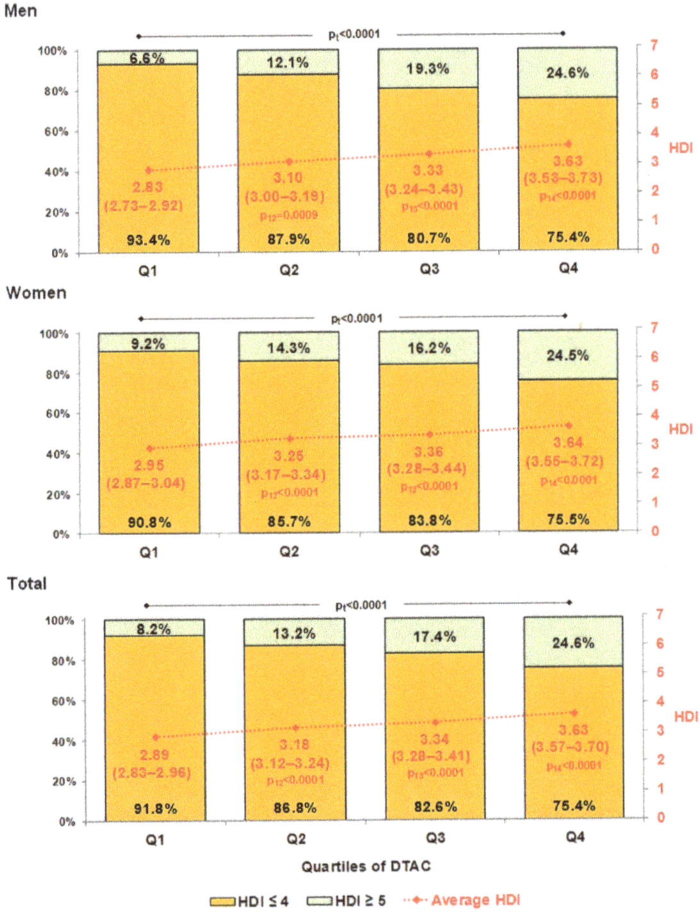

Figure 1. Distribution of HDI score (HDI ≥ 5 vs. HDI ≤ 4) and average HDI score (95% CI) by quartiles of DTAC. Mean HDI adjusted for age in men and women and additionally for sex in the total group. p_t—p-value for trend test; p_{12}, p_{13}, and p_{14}—p-values for comparisons with reference quartile Q1. DTAC—dietary total antioxidant capacity; HDI—Healthy Diet Indicator. Difference between average HDI was tested using a general linear model with Tukey–Kramer adjustment for multiple comparisons, while the distribution of HDI was tested by Cochran–Armitage test for trend.

Furthermore, the percentage of people who followed a healthy diet was also associated with an increase in DTAC. In our study, a proper diet (HDI $\geq$ 5) was followed by 6.6% and 24.6% of men, 9.2% and 24.5% of women, and 8.2% and 24.6% of the total in the first (Q1) and fourth (Q4) quartiles of DTAC, respectively (Figure 1).

4. Discussion

In the present study, we evaluated for the first time the relationship between DTAC, measured by the FRAP method, and the prevalence of CVDs in a representative sample of the Polish adult population (WOBASZ II study). It was found that higher DTAC was associated with a lower prevalence of CVDs. The odds ratio of developing CVD, after adjustment for confounding variables (model 3), was 24.8–39.0% lower in the third and fourth quartiles in comparison to the first quartile of DTAC.

CVDs are the leading cause of death and disability around the world [1]. While genetic and environmental contributors to developing CVD are important, modifiable risk factors associated with lifestyle also play a large role. Food choices influence the development of obesity, hypertension, and dyslipidemia, which increase the risk of CVD [31]. Therefore, guidelines for the prevention of CVDs also include healthy diet strategies [20]. In our study, people with CVDs were more likely to have obesity, hypertension, diabetes mellitus, hypercholesterolemia, hypertriglyceridemia, and metabolic syndrome.

A large body of evidence supports the intake of natural nutrients such as polyphenols [32] and antioxidant vitamins [33], as well as dietary patterns such as the Mediterranean diet [34], Dietary Approaches to Stop Hypertension [35], the Nordic Diet [36], and Traditional Asian Diets [37–39], can prevent CVDs. The findings of three large prospective cohorts with up to 32 years of follow-up showed that greater adherence to various healthy eating patterns was associated with a lower risk of CVD [40]. Healthy dietary patterns share common characteristics based on higher consumption of vegetables, fruits, legumes, nuts and seeds, and whole grains, which are the most important sources of antioxidants [41]. In contrast to natural antioxidants, exogenous supplementation of antioxidant vitamins and minerals in people without nutritional deficiencies has no beneficial effects on the prevention of CVDs [42]. The findings of the NHANES study showed that US adults with high DTAC showed higher adherence to popular diet quality index scores: HEI-2015 (Healthy Eating Index-2015), AHEI-2010 (Alternative Healthy Eating Index-2010), aMED (alternate Mediterranean Diet), and DASH (Dietary Approaches to Stop Hypertension). However, in this study, dietary total antioxidant capacity was measured using the ABTS (2,2′-azino-bis-3-ethylbenzthiazoline-6-sulphonic acid) assay in individual dietary antioxidant vitamins and polyphenols and expressed as the sum of individual antioxidant capacities [43]. In our study, DTAC measured by the FRAP method was positively associated with the intake of individual dietary antioxidants (from foods and supplements), such as total polyphenols, antioxidant vitamins (C, E, and β-carotene), and minerals (Fe, Zn, Cu, and Mn (as antioxidant enzyme cofactors)). However, after recalculation of the results per 1000 kcal of the diet, significant positive associations were found only for DTPI, vitamin C, and Cu in men and women and for DTPI, vitamin C, Fe, Mn, and Cu in the total group. Surprisingly, inverse associations were found for Zn in men and women and for β-carotene and Zn in the total group. The measurement of the total antioxidant potential of the diet includes different individual antioxidants with synergistic, additive, or inverse effects. The consumption of individual antioxidants was measured in both the diet and supplements, while excessive consumption of antioxidants with supplements may have a pro-oxidative effect.

In this study, DTAC was positively associated with healthy diet quality measured by the HDI score, although the quality of the diet in the Polish adult population was relatively low (mean HDI less than 4 in men, women, and total).

Dietary antioxidants can support the action of endogenous antioxidants (enzymatic, e.g., superoxide dismutase, glutathione peroxidase, and catalase, and non-enzymatic, e.g., glutathione, metal-binding proteins, uric acid, melatonin, and bilirubin) in alleviating the destructive effects of oxidative stress, e.g., can inhibit LDL-C oxidation and prevent endothelial dysfunction. The overproduction of free radicals (ROS—reactive oxygen species; RNS—reactive nitrogen species) in oxidative stress can damage the body's DNA, as well as proteins and lipids, and lead to non-communicable diseases, including CVDs. There is emerging evidence that inflammatory mechanisms and oxidative stress lead to atherosclerosis, arterial hypertension, arterial fibrillation, and heart failure. However, increasing oxidative stress can be considered a contributing factor, not the primary pathophysiologic mechanism, because CVDs are very complex in their pathogenesis [44].

In previous population studies, DTAC, measured by different methods, was inversely associated with prediabetes, insulin resistance, and diabetes [14,45,46], cancer [47], myocardial infarction [48], heart failure [49], and stroke [50]. In a large cross-sectional study, DTAC, measured by the FRAP method, was inversely associated with the hypertriglyceridemic waist phenotype and amputation due to arterial disease in people undergoing secondary care for CVDs [12]. In another cross-sectional study, high DTAC, measured by the FRAP method, had a protective effect against oxidative DNA damage in individuals at cardiovascular risk [51]. Findings from a meta-analysis of prospective cohort studies have shown that high DTAC was associated with a lower risk of mortality from all causes, cancer, and CVDs. In addition, a 5 mmol/day increase in DTAC, based on the FRAP method, was associated with a 7% lower risk of all-cause mortality [11].

Population studies on the relationship between DTAC, measured by the FRAP method, and CVDs are quite limited. Therefore, further research is needed to investigate these associations and mechanisms of action.

Limitations

The present study has some strengths and limitations. The strength of this study is that it was a cross-sectional study conducted on a large, representative sample of the adult Polish population. Moreover, dietary antioxidants were calculated from foods and supplements. The first limitation of this study is the use of a single 24 h recall, which does not take into account habitual food intake, although this method is commonly used in large population studies [43]. Moreover, participants who indicated that their 24 h recall differed from their typical diet (usual eating habits, typical of most days of the year) were excluded from the study. The second limitation is that the FRAP database, although very extensive, did not contain all foods, beverages, herbs, and supplements. Thirdly, this study did not take into account the consumption of selenium, which is also a component of antioxidant enzymes, because Polish Food Composition Tables do not contain this micronutrient.

5. Conclusions

This study demonstrated that a higher quartile of DTAC was significantly associated with a reduced odds ratio for cardiovascular diseases in the Polish adult population, as well as with a higher Healthy Diet Indicator. Therefore, dietary recommendations for the prevention and therapy of cardiovascular diseases should take into account a high DTAC. DTAC, measured by the FRAP method, can be considered an indicator of healthy diet quality.

Author Contributions: Conceptualization, A.W., M.E.Z. and A.C.-M.; methodology, A.W., M.E.Z. and A.M.W.; software, A.W., M.E.Z. and A.C.-M.; validation, A.W., M.E.Z. and A.C.-M.; formal analysis, A.C.-M. and A.W.; investigation, A.W. and M.E.Z.; resources, M.E.Z., A.W., A.M.W., K.Z. and W.D.; data curation, A.W. and M.E.Z.; writing—original draft preparation, M.E.Z.; writing—review and editing, A.W., A.M.W., K.Z., A.C.-M. and W.D.; visualization, A.W., M.E.Z. and A.C.-M.; supervision, A.W., M.E.Z. and W.D.; project administration, M.E.Z. and A.W.; funding acquisition, A.W. and M.E.Z. All authors have read and agreed to the published version of the manuscript.

Funding: This research was funded by the National Institute of Cardiology (Grant No. 2.20/I/20) and Medical University of Bialystok (Grant No. SUB/3/DN/22/004/3317).

Institutional Review Board Statement: The study was conducted according to the guidelines of the Declaration of Helsinki and approved by the Bioethics Committee of the National Institute of Cardiology (protocol code 1344, date of approval 5 November 2012, and protocol code 1837, date of approval 14 January 2020).

Informed Consent Statement: Informed consent was obtained from all subjects involved in the study.

Data Availability Statement: All data in this study are available upon request to the authors at the following e-mail address: malgorzata.zujko@umb.edu.pl or awaskiewicz@ikard.pl.

Conflicts of Interest: The authors declare no conflict of interest.

References

1. World Health Organization. Cardiovascular Diseases (CVDs). Available online: https://www.who.int/news-room/fact-sheets/detail/cardiovascular-diseases-(cvds) (accessed on 13 March 2022).
2. Van Horn, L.; Carson, J.A.S.; Appel, L.J.; Burke, L.E.; Economos, C.; Karmally, W.; Lancaster, K.; Lichtenstein, A.H.; Johnson, R.K.; Thomas, R.J.; et al. Recommended dietary pattern to achieve adherence to the American Heart Association/American College of Cardiology (AHA/ACC) Guidelines: A Scientific Statement from the American Heart Association. *Circulation* **2016**, *134*, e505–e529. [CrossRef] [PubMed]
3. Nestel, P.J.; Mori, T.A. Dietary patterns, dietary nutrients and cardiovascular disease. *Rev. Cardiovasc. Med.* **2022**, *23*, 17. [CrossRef] [PubMed]
4. Micek, A.; Godos, J.; Del Rio, D.; Galvano, F.; Grosso, G. Dietary Flavonoids and Cardiovascular Disease: A Comprehensive dose-response meta-analysis. *Mol. Nutr. Food Res.* **2021**, *65*, e2001019. [CrossRef] [PubMed]
5. Gan, Z.H.; Cheong, H.C.; Tu, Y.K.; Kuo, P.H. Association between plant-based dietary patterns and risk of cardiovascular disease: A systematic review and meta-analysis of prospective cohort studies. *Nutrients* **2021**, *13*, 3952. [CrossRef] [PubMed]
6. Zujko, M.E.; Rożniata, M.; Zujko, K. Individual Diet Modification Reduces the Metabolic Syndrome in Patients Before Pharmacological Treatment. *Nutrients* **2021**, *13*, 2102. [CrossRef]
7. Witkowska, A.M.; Waśkiewicz, A.; Zujko, M.E.; Szcześniewska, D.; Stepaniak, U.; Pająk, A.; Drygas, W. Are total and individual dietary lignans related to cardiovascular disease and its risk factors in postmenopausal women? A nationwide study. *Nutrients* **2018**, *10*, 865. [CrossRef]
8. Nascimento-Souza, M.A.; Paiva, P.G.; Martino, H.S.D.; Ribeiro, A.Q. Dietary total antioxidant capacity as a tool in health outcomes in middle-aged and older adults: A systematic review. *Crit. Rev. Food Sci. Nutr.* **2018**, *58*, 905–912. [CrossRef]
9. Pellegrini, N.; Vitaglione, P.; Granato, D.; Fogliano, V. Twenty-five years of total antioxidant capacity measurement of foods and biological fluids: Merits and limitations. *J. Sci. Food Agric.* **2020**, *100*, 5064–5078. [CrossRef]
10. Salari-Moghaddam, A.; Nouri-Majd, S.; Keshteli, A.H.; Emami, F.; Esmaillzadeh, A.; Adibi, P. Association between Dietary Total Antioxidant Capacity and Diet Quality in Adults. *Front. Nutr.* **2022**, *9*, 838752. [CrossRef]
11. Parohan, M.; Anjom-Shoae, J.; Nasiri, M.; Khodadost, M.; Khatibi, S.R.; Sadeghi, O. Dietary total antioxidant capacity and mortality from all causes, cardiovascular disease and cancer: A systematic review and dose-response meta-analysis of prospective cohort studies. *Eur. J. Nutr.* **2019**, *58*, 2175–2189. [CrossRef]
12. da Silva, A.; Caldas, A.P.S.; Pinto, S.L.; Hermsdorff, H.H.M.; Marcadenti, A.; Bersch-Ferreira, A.C.; Torreglosa, C.R.; Weber, B.; Bressan, J. Dietary total antioxidant capacity is inversely associated with cardiovascular events and cardiometabolic risk factors: A cross-sectional study. *Nutrition* **2021**, *89*, 111140. [CrossRef] [PubMed]
13. Zujko, M.E.; Waśkiewicz, A.; Witkowska, A.M.; Szcześniewska, D.; Zdrojewski, T.; Kozakiewicz, K.; Drygas, W. Dietary total antioxidant capacity and dietary polyphenol intake and prevalence of metabolic syndrome in Polish adults: A nationwide study. *Oxid. Med. Cell. Longev.* **2018**, *2018*, 7487816. [CrossRef] [PubMed]
14. Cyuńczyk, M.; Zujko, M.E.; Jamiołkowski, J.; Zujko, K.; Łapińska, M.; Zalewska, M.; Kondraciuk, M.; Witkowska, A.M.; Kamiński, K.A. Dietary total antioxidant capacity is inversely associated with prediabetes and insulin resistance in Bialystok PLUS population. *Antioxidants* **2022**, *11*, 283. [CrossRef]
15. Hermsdorff, H.H.; Puchau, B.; Volp, A.C.; Barbosa, K.B.; Bressan, J.; Zulet, M.Á.; Martínez, J.A. Dietary total antioxidant capacity is inversely related to central adiposity as well as to metabolic and oxidative stress markers in healthy young adults. *Nutr. Metab.* **2011**, *8*, 59. [CrossRef]
16. Georgoulis, M.; Fragopoulou, E.; Kontogianni, M.D.; Margariti, A.; Boulamatsi, O.; Detopoulou, P.; Tiniakos, D.; Zafiropoulou, R.; Papatheodoridis, G. Blood redox status is associated with the likelihood of nonalcoholic fatty liver disease irrespectively of diet's total antioxidant capacity. *Nutr. Res.* **2015**, *35*, 41–48. [CrossRef] [PubMed]
17. Drygas, W.; Niklas, A.A.; Piwońska, A.; Piotrowski, W.; Flotyńska, A.; Kwaśniewska, M.; Nadrowski, P.; Puch-Walczak, A.; Szafraniec, K.; Bielecki, W.; et al. Multi-center National Population Health Examination Survey (WOBASZ II study): Assumptions, methods and implementation. *Kardiol. Pol.* **2016**, *74*, 681–690. [CrossRef]
18. Witkowska, A.M.; Waśkiewicz, A.; Zujko, M.E.; Cicha-Mikołajczyk, A.; Mirończuk-Chodakowska, I.; Drygas, W. Dietary intake of plant sterols and phytosterol-enriched margarines and their relationship with cardiovascular disease among Polish men and women: Results of the WOBASZ II cross-sectional study. *Nutrients* **2022**, *14*, 2665. [CrossRef] [PubMed]

19. WHO. Body Mass Index—BMI. Available online: https://www.euro.who.int/en/health-topics/disease-prevention/nutrition/ahealthy-lifestyle/body-mass-index-bmi (accessed on 10 April 2022).
20. Visseren, F.L.J.; Mach, F.; Smulders, Y.M.; Carballo, D.; Koskinas, K.C.; Bäck, M.; Benetos, A.; Biffi, A.; Boavida, J.M.; Capodanno, D.; et al. 2021 ESC Guidelines on cardiovascular disease prevention in clinical practice with representatives of the European Society of Cardiology and 12 medical societies with the special contribution of the European Association of Preventive Cardiology (EAPC). *Eur. Heart. J.* **2021**, *42*, 3227–3337. [CrossRef]
21. American Diabetes Association. Standards of Medical Care in Diabetes-2022, Abridged for Primary Care Providers. *Clin. Diabetes* **2022**, *40*, 10–38. [CrossRef]
22. Alberti, K.G.; Eckel, R.H.; Grundy, S.M.; Zimmet, P.Z.; Cleeman, J.I.; Donato, K.A.; Fruchart, J.C.; James, W.P.; Loria, C.M.; Smith, S.C., Jr. Harmonizing the metabolic syndrome. A joint interim statement of the International Diabetes Federation Task Force on Epidemiology and Prevention; National Heart, Lung, and Blood Institute; American Heart Association; World Heart Federation; International Atherosclerosis Society; and International Association for the Study of Obesity. *Circulation* **2009**, *120*, 1640–1645.
23. Szponar, L.; Wolnicka, K.; Rychlik, E. *Album of Photographs of Food Products and Dishes*; National Food and Nutrition Institute Press: Warsaw, Poland, 2000.
24. Kunachowicz, H.; Nadolna, I.; Przygoda, B.; Iwanow, K. *Food Composition Tables*; PZWL: Warsaw, Poland, 2005.
25. Neveu, V.; Perez-Jiménez, J.; Vos, F. Phenol-Explorer: An online comprehensive database on polyphenol contents in foods. *J. Biol. Databases Curation* **2010**, *2010*, bap024. [CrossRef] [PubMed]
26. Zujko, M.E.; Witkowska, A.M. Antioxidant potential and polyphenol content of selected food. *Int. J. Food Prop.* **2011**, *14*, 300–308. [CrossRef]
27. Zujko, M.E.; Witkowska, A.M. Antioxidant potential and polyphenol content of beverages, chocolates, nuts, and seeds. *Int. J. Food Prop.* **2014**, *17*, 86–92. [CrossRef]
28. Carlsen, M.H.; Halvorsen, B.L.; Holte, K.; Bøhn, S.K.; Dragland, S.; Sampson, L.; Willey, C.; Senoo, H.; Umezono, Y.; Sanada, C.; et al. The total antioxidant content of more than 3100 foods, beverages, spices, herbs and supplements used worldwide. *Nutr. J.* **2010**, *9*, 3. [CrossRef] [PubMed]
29. World Health Organisation. Diet, nutrition, and the prevention of chronic diseases. Report of a WHO Study Group. *World Health Organ. Tech. Rep. Ser.* **1990**, *797*, 1–204.
30. Fransen, H.P.; Beulens, J.W.; May, A.M.; Struijk, E.A.; Boer, J.M.; de Wit, G.A.; Onland-Moret, N.C.; van der Schouw, Y.T.; Bueno-de-Mesquita, H.B.; Hoekstra, J.; et al. Dietary patterns in relation to quality-adjusted life years in the EPIC-NL cohort. *Prev. Med.* **2015**, *77*, 119–124. [CrossRef] [PubMed]
31. Koene, R.J.; Prizment, A.E.; Blaes, A.; Konety, S.H. Shared risk factors in cardiovascular disease and cancer. *Circulation* **2016**, *133*, 1104–1114. [CrossRef] [PubMed]
32. Grosso, G.; Godos, J.; Currenti, W.; Micek, A.; Falzone, L.; Libra, M.; Giampieri, F.; Forbes-Hernández, T.Y.; Quiles, J.L.; Battino, M.; et al. The effect of dietary polyphenols on vascular health and hypertension: Current evidence and mechanisms of action. *Nutrients* **2022**, *14*, 545. [CrossRef]
33. Núñez-Córdoba, J.M.; Martínez-González, M.A. Antioxidant vitamins and cardiovascular disease. *Curr. Top. Med. Chem.* **2011**, *11*, 1861–1869. [CrossRef]
34. Grosso, G.; Marventano, S.; Yang, J.; Micek, A.; Pajak, A.; Scalfi, L.; Galvano, F.; Kales, S.N. A comprehensive meta-analysis on evidence of Mediterranean diet and cardiovascular disease: Are individual components equal? *Crit. Rev. Food Sci. Nutr.* **2017**, *57*, 3218–3232. [CrossRef]
35. Siervo, M.; Lara, J.; Chowdhury, S.; Ashor, A.; Oggioni, C.; Mathers, J.C. Effects of the Dietary Approach to Stop Hypertension (DASH) diet on cardiovascular risk factors: A systematic review and meta-analysis. *Br. J. Nutr.* **2015**, *113*, 1–15. [CrossRef] [PubMed]
36. Adamsson, V.; Reumark, A.; Fredriksson, I.B.; Hammarstrom, E.; Vessby, B.; Johansson, G.; Riserus, U. Effects of a healthy Nordic diet on cardiovascular risk factors in hypercholesterolaemic subjects: A randomized controlled trial (NORDIET). *J. Intern. Med.* **2011**, *269*, 150–159. [CrossRef] [PubMed]
37. Lee, K.W.; Cho, M.S. The traditional Korean dietary pattern is associated with decreased risk of metabolic syndrome: Findings from the Korean National Health and Nutrition Examination Survey, 1998–2009. *J. Med. Food* **2014**, *17*, 43–56. [CrossRef] [PubMed]
38. Niu, K.; Momma, H.; Kobayashi, Y.; Guan, L.; Chujo, M.; Otomo, A.; Ouchi, E.; Nagatomi, R. The traditional Japanese dietary pattern and longitudinal changes in cardiovascular disease risk factors in apparently healthy Japanese adults. *Eur. J. Nutr.* **2016**, *55*, 267–279. [CrossRef]
39. Leonetti, F.; Liguori, A.; Petti, F.; Rughini, S.; Silli, L.; Liguori, S.; Bangrazi, S. Effects of basic traditional Chinese diet on body mass index, lean body mass, and eating and hunger behaviours in overweight or obese individuals. *J. Tradit. Chin. Med.* **2016**, *36*, 456–463. [CrossRef]
40. Shan, Z.; Li, Y.; Baden, M.Y.; Bhupathiraju, S.N.; Wang, D.D.; Sun, Q.; Rexrode, K.M.; Rimm, E.B.; Qi, L.; Willett, W.C.; et al. Association between healthy eating patterns and risk of cardiovascular disease. *JAMA Intern. Med.* **2020**, *180*, 1090–1100. [CrossRef]
41. Cena, H.; Calder, P.C. Defining a healthy diet: Evidence for the role of contemporary dietary patterns in health and disease. *Nutrients* **2020**, *12*, 334. [CrossRef]
42. Sunkara, A.; Raizner, A. Supplemental vitamins and minerals for cardiovascular disease prevention and treatment. *Methodist Debakey Cardiovasc. J.* **2019**, *15*, 179–184. [CrossRef]

43. Ha, K.; Kim, K.; Sakaki, J.R.; Chun, O.K. Relative validity of dietary total antioxidant capacity for predicting all-cause mortality in comparison to diet quality indexes in US adults. *Nutrients* **2020**, *12*, 1210. [CrossRef]
44. Senoner, T.; Dichtl, W. Oxidative stress in cardiovascular diseases: Still a therapeutic target? *Nutrients* **2019**, *11*, 2090. [CrossRef]
45. Mancini, F.R.; Aurélie Affret, A.; Dow, C.; Balkau, B.; Bonnet, F.; Boutron-Ruault, M.C.; Fagherazzi, G. Dietary antioxidant capacity and risk of type 2 diabetes in the large prospective E3N-EPIC cohort. *Diabetologia* **2018**, *61*, 308–316. [CrossRef] [PubMed]
46. van der Schaft, N.; Schoufour, J.D.; Nano, J.; Kiefte-de Jong, J.C.; Muka, T.; Sijbrands, E.J.G.; Ikram, M.A.; Franco, O.H.; Voortman, T. Dietary antioxidant capacity and risk of type 2 diabetes mellitus, prediabetes and insulin resistance: The Rotterdam Study. *Eur. J. Epidemiol.* **2019**, *34*, 853–861. [CrossRef] [PubMed]
47. Abbasalizad Farhangi, M.; Vajdi, M. Dietary total antioxidant capacity (TAC) significantly reduces the risk of site-specific cancers: An updated systematic review and meta-analysis. *Nutr. Cancer* **2021**, *73*, 721–739. [CrossRef]
48. Rautiainen, S.; Levitan, E.B.; Orsini, N.; Åkesson, A.; Morgenstern, R.; Mittleman, M.A.; Wolk, A. Total antioxidant capacity from diet and risk of myocardial infarction: A prospective cohort of women. *Am. J. Med.* **2012**, *125*, 974–980. [CrossRef] [PubMed]
49. Rautiainen, S.; Levitan, E.B.; Mittleman, M.A.; Wolk, A. Total antioxidant capacity of diet and risk of heart failure: A population-based prospective cohort of women. *Am. J. Med.* **2013**, *126*, 494–500. [CrossRef]
50. Colarusso, L.; Serafini, M.; Lagerros, Y.T.; Nyren, O.; La Vecchia, C.; Rossi, M.; Ye, W.; Tavani, A.; Adami, H.O.; Grotta, A.; et al. Dietary antioxidant capacity and risk for stroke in a prospective cohort study of Swedish men and women. *Nutrition* **2017**, *33*, 234–239. [CrossRef]
51. de Lima-Reis, S.R.; Silva, T.A.; Costa, L.S.A.; Volp, A.C.P.; Rios-Santos, F.; Reis, É.M.; Bassi-Branco, C.L. Serum levels of vitamin A, selenium and better dietary total antioxidant capacity are related to lower oxidative DNA damage: A cross-sectional study of individuals at cardiovascular risk. *J. Nutr. Biochem.* **2022**, *107*, 109070. [CrossRef]

nutrients

MDPI

Article

Relation of Dietary Patterns and Nutritional Profile to Hepatic Fibrosis in a Sample of Lebanese Non-Alcoholic Fatty Liver Disease Patients

Nicole Fakhoury Sayegh [1,*], Gessica N. H. A. Heraoui [1], Hassan Younes [2], Lea Nicole Sayegh [3], Christa Boulos [1] and Raymond Sayegh [4]

1 Department of Nutrition, Faculty of Pharmacy, Saint Joseph University, Damascus Road, Riad el Solh, Beirut P.O. Box 11-5076, Lebanon; h.gessica10@gmail.com (G.N.H.A.H.); christa.boulos@gmail.com (C.B.)
2 College Health, équipe PANASH-ULR 7519, Institut Polytechnique UniLaSalle, 19, Rue Pierre Waguet, CEDEX, 60026 Beauvais, France; hassan.younes@unilasalle.fr
3 Faculty of Medicine, American University of Beirut, Beirut P.O. Box 11-0236, Lebanon; lea.n.sayegh@gmail.com
4 Department of Gastroenterology and Hepatology, Faculty of Medicine, Saint Joseph University, Damascus Road, Riad el Solh, Beirut P.O. Box 11-5076, Lebanon; drrsayegh@gmail.com
* Correspondence: nicolsayegh@gmail.com or nicole.fakhourysayegh@usj.edu.lb

Abstract: Non-alcoholic fatty liver disease (NAFLD) is considered the most common liver injury worldwide. NAFLD can evolve into non-alcoholic steatohepatitis (NASH) with or without fibrosis. The objectives of this study were to determine the nutritional profile and dietary patterns of NAFLD Lebanese patients and to report the type of diet-related to the presence of hepatic fibrosis. We hypothesized that the traditional pattern was related to a low risk of fibrosis. This cross-sectional study included 320 eligible Lebanese NAFLD patients. Three dietary patterns were identified: the Traditional diet, the High Fruit diet, and the Westernized diet. Multivariate analysis showed a significant relationship between high adherence to the traditional diet and absence of hepatic fibrosis with a decreased risk of 82%, $p = 0.031$ after adjusting for its covariables. Fruits were absent from this dietary pattern. Although our results pointed to a possible relationship between fibrosis in NAFLD patients and fruit intake, experimental studies are needed to show whether this is a causal relationship. However, the results obtained in this study may contribute to the planning of dietary interventions and recommendations and enable a better follow-up for NAFLD patients with fibrosis.

Keywords: hepatic fibrosis; dietary patterns; the traditional diet; the high fruit diet

Citation: Sayegh, N.F.; Heraoui, G.N.H.A.; Younes, H.; Sayegh, L.N.; Boulos, C.; Sayegh, R. Relation of Dietary Patterns and Nutritional Profile to Hepatic Fibrosis in a Sample of Lebanese Non-Alcoholic Fatty Liver Disease Patients. *Nutrients* **2022**, *14*, 2554. https://doi.org/10.3390/nu14122554

Academic Editors: Andriana Kaliora, Chara Tzavara and Charalampia Amerikanou

Received: 25 May 2022
Accepted: 17 June 2022
Published: 20 June 2022

1. Introduction

Non-alcoholic fatty liver disease (NAFLD) is considered the most known liver injury in the USA and probably worldwide [1]. The simplest form of this disease is the pure fatty liver, which is characterized by the accumulation of triglycerides in hepatocytes at a percentage greater than 5% of the total number of cells [2]. The prevalence of the disease varies remarkably according to the diagnostic tools (liver biopsy, radiological tests, and blood tests) and the chosen population [2]. It is approximately 20 to 30% in the Middle East and in obese patients, the prevalence has ranged from 50 to 90% [3]. Concerning gender, some studies have shown that the prevalence of the disease is two to three times higher in men than in women [4].

NAFLD can progress into non-alcoholic steatohepatitis (NASH), which is defined as steatosis associated with inflammatory lesions and hepatocyte ballooning with or without fibrosis [5]. NASH can progress into cirrhosis and ultimately into hepatic cellular carcinoma [5]. Recently, studies have shown a positive association between the characteristics of the metabolic syndrome and NAFLD/NASH, especially abdominal obesity, hyperglycemia, hypertriglyceridemia, dyslipidemia, and arterial hypertension [6]. Non-alcoholic fatty liver disease is recognized as the liver manifestation of metabolic syndrome [6].

A biopsy is considered the most reliable method for detecting the presence of NAFLD and for assessing the inflammatory state of the liver as well as the presence or absence of fibrosis [7]. Nevertheless, it is rarely used in clinical practice because of its invasive nature and potential complications. Other non-invasive methods have emerged, such as liver biological tests and especially the NAFLD Fibrosis Score (NFS), which is defined as follows:

$$\text{NFS score} = -1.675 + 0.037 \times \mathit{age}\ (\mathit{years}) + 0.094 \times \mathit{BMI}\ (\text{kg/m}^2) + 1.13 \times \mathit{diabetes}\ (\mathit{yes/no}) + 0.99 \times \mathit{ASAT/ALAT} - 0.013 \times \mathit{platelet\ count}\ (109\ \text{Giga/L}) \times (-0.66)\ \mathit{albumin}\ (\text{g/dL}).$$

Values less than -1.45 indicate a state of non-fibrosis, values between -1.45 and 0.67 indicate a possible presence of fibrosis, and values above 0.67 indicate advanced fibrosis [8].

Dietary patterns have a significant impact on the biological and physical profile of patients [5]. The use of dietary patterns in establishing an association between dietary intake and NAFLD, mainly in its advanced stage, is of importance. This will enable the investigation and measure of the overall diet, which is more likely to be associated with NAFLD than each studied nutrient or single food [9].

Currently, there is no data on the prevalence of NAFLD in Lebanon and no studies describe the nutritional profile of NAFLD Lebanese patients with fibrosis. This observational cross-sectional study could be a preliminary study for more powerful ones such as cohorts and experimental studies. The main objectives were to analyze the dietary patterns and nutritional profile of Lebanese NAFLD patients in relation to the absence or presence of hepatic fibrosis. The main hypothesis was that the traditional diet was inversely associated with the presence of fibrosis.

2. Materials and Methods

2.1. Study Design

From November 2014 to June 2019, 500 Lebanese participants aged between 18 and 70 years, visiting an outpatient clinic of the department of gastroenterology in an academic hospital in Beirut, were invited to participate in the study after providing their informed consent. Four hundred patients diagnosed with NAFLD were recruited according to the inclusion criteria. Three hundred and twenty Lebanese patients completed the clinical, anthropometric, and dietary data. The response rate was 80%, which is statistically satisfactory. Figure 1 summarizes the course of the study. This sample size corresponded to the expected proportion (30%) of NAFLD patients with a confidence interval of 95% and an inaccuracy gap of 5% [10].

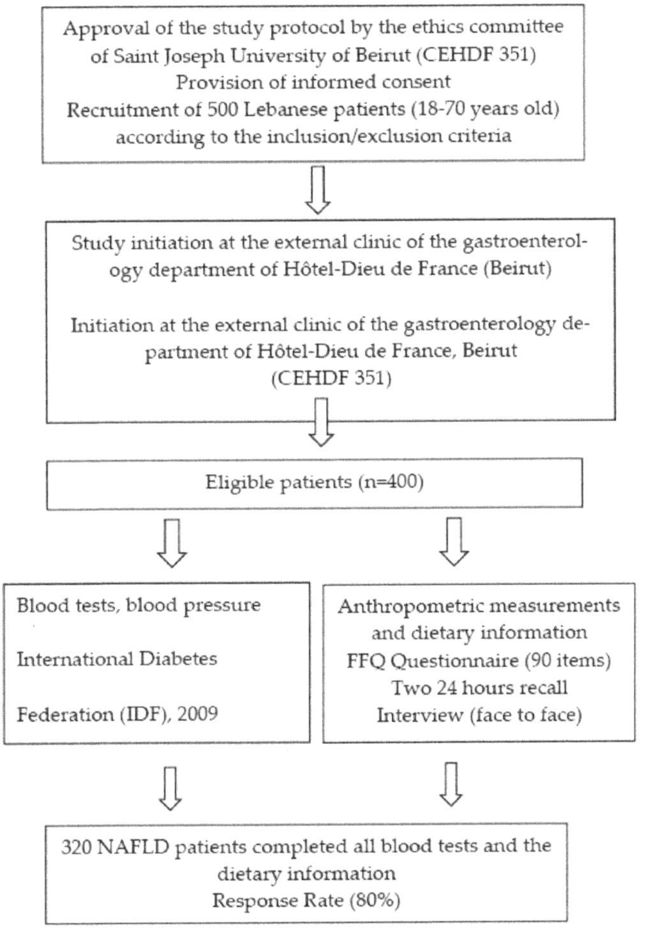

Figure 1. Flow chart for selection and enrolment of the study participants. NAFLD: Non -Alcoholic Fatty Liver Disease, FFQ Food Frequency Questionnaire [11], International Diabetes Federation (IDF) [12].

2.2. Sample Selection, Inclusion Criteria

Eligible patients were Lebanese men and women without: (1) biliary diseases or recognized cirrhosis; (2) infection with hepatitis A, B, or C virus; (3) genetic metabolic disease; (4) auto-immune liver diseases; (5) type 1 diabetes. Other inclusion criteria were (6) non-pregnancy among women; (7) less than or equal to two servings/day of alcohol consumption; and (8) the absence of drugs inducing hepatotoxicity (tamoxifen, steroids, amiodarone).

2.3. Study Protocol

The 320 patients were recruited and subsequently solicited for a face-to-face interview. These patients were recognized as having NAFLD following abdominal ultrasound by the same radiologist and with the same equipment (Hitachi-Aloka ProSound F75, Tokyo, Japan). The liver steatosis was estimated with the evaluation of the image brightness of the echo pattern. Abdominal ultrasound cannot identify hepatic fat deposition if it is less than 33% of the total liver and, accordingly, all patients with a lower percentage were categorized as free of the disease. A blood test, after a 10-h overnight fast, was requested during the interview with the patients after obtaining their informed consent. Serum samples were obtained from the coagulated blood after centrifugation and were immediately stored at −20 °C. Serum fasting blood glucose (mmol/L), triglycerides (mmol/L), ALAT (U/L), ASAT (U/L) HDL-C (mmol/L) and Albumin (g/dL) were assayed spectrophotometrically on Cobas c 501 (Roche, Germany). The serum insulin level (U/L) was measured by Electrochemiluminescence immunoassay (ECLIA) on a Cobas c 501 (Roche, Germany) and serum platelet count (Giga/L) was determined by Yumizen H2500/H1500 (Horiba, Japan). The reference values of the different biological parameters such as transaminases, albumin, platelets, and insulin corresponded to the reference values of the hospital laboratory. The blood pressure (mmHg) was taken using a manual blood pressure monitor (Bokang, China, 2009).

2.4. Measuring Tools

Other measuring tools included a face-to-face interview and anthropometric measurements. Dietary information was obtained using two questionnaires: a food frequency questionnaire (Harvard, Nurses' Health Study, 2016) [11] translated into Arabic and adapted to Lebanese food, and two 24-hr recalls which summarize typical weekdays and weekend days (Supplementary Materials) It was administered by the same dietician throughout the study, thereby limiting inter-investigator errors. It included the quantitative variables such as age (years), body mass index (BMI) (kg/m^2) and socio-demographic qualitative variables such as gender, marital status, place of residence and birth, and occupation. Other qualitative questions referred to family history of obesity (limited to first-degree relatives), the presence of diabetes mellitus and dyslipidemia, and drug and supplement intake during the six months prior to diagnosis.

Questions about practicing physical activity, smoking status, and the number of cigarettes/day (qualitative variables) were asked. Physical activities were classified into two categories: mild/moderate or vigorous, according to the Centre for Disease Control Guidelines (CDC, 1993) [13]. Activities such as such as walking, housework, and gardening were considered mild or moderate (3.5 Kcal/min), while tennis, football, or fast swimming were considered vigorous (7 Kcal/min). Patients were vigorously active if the activity was vigorous and exceeded 20 min/day and was performed at least three times/week. It was considered moderate if the activity was moderate and exceeded 30 min/day and was performed at least five times/week and was during the last three months before the start of the study [14].

Photos with portion measurements (Numed, s. a. r. l) were used to determine the daily portion of different foods [15]. Certified software (Nutrilog, France, version 2.33) was used to analyze and determine the composition of different foods in micro and macronutrients. The frequency category selected for food (monthly, weekly, or daily) has been converted into a daily intake. The software analyzed the total daily energy intake of patients and the dietary composition of micro and macro-elements of different foods consumed. Fructose, ω-3, ω-6, Eicosapentaenoic acid (EPA) and Docosahexaenoic acid (DHA) intake was calculated according to the US Department of Agriculture (USDA) [16]. The database (1970) of the American University of Beirut was used for some national ingredients or recipes. Food fructose was analyzed as free fructose and/or as fructose from sucrose. Simple sugar has been defined as disaccharides and/or monosaccharides naturally present in foods or added sugar in commercial foods.

Questions such as the type and frequency of dietary supplements and vitamins, the frequency of fruit portions, soft drinks, and others (weekly, monthly, or daily), and the type of oil used in cooking were asked. To minimize response biases, the questionnaire included duplicate questions and patients were interviewed prior to their knowledge of the presence of fatty liver (Supplementary Materials).

Anthropometric measures such as weight (kg) and height (cm), BMI (kg/m^2) were taken by the same nutritionist and the same mechanical balance and stadiometer. The waist circumference (cm) and the hip circumference (cm) were taken by the same calibrated band throughout the study (average of three consecutive measures for each variable). Patients were classified as obese, overweight, or normal weight according to the WHO obesity classification, 2004: normal weight (BMI between 18 and 24.9 kg/m^2), overweight (BMI between 25 and 29.9 kg)/m^2) and obese (BMI > 30 kg/m^2) [17].

Insulin resistance was studied by an evaluation index, the Homeostasis Model Assessment of Insulin Resistance (HOMA-IR). This index is calculated using the following formula: HOMA-IR = $fasting\ plasma\ glucose$ (mmol/L) × $fasting\ serum\ insulin$ (U/l)/22.5. A value greater than 3 indicates a state of insulin resistance [18]. Parameters that met the criteria of the International Diabetes Federation (IDF 2009) were identified to categorize patients with or without metabolic syndrome [12]. All clinical parameters were analyzed at the hospital laboratory using standard methods. Patients who were on drugs for hypertension, hyperglycemia or dyslipidemia were classified with metabolic syndrome regardless of laboratory findings.

2.5. Validity and Reproducibility of the Food Frequency Questionnaire

The reliability and validity of the Food Frequency Questionnaire (FFQ) were tested in a previous study [19]. The FFQ questionnaire was adapted from the Harvard Nurses' Health Study, 2016, taking the general format (once/month, once/week, two to three times/week etc.). It was translated into Arabic and modified to Lebanese food items. It was administered into the native language, Arabic, and in a face-to-face method, by the same nutritionist throughout the study. The reproducibility of the FFQ was confirmed by administering it to 50 patients at the start of the study, prior to diagnosis. The re-interview of this subsample by the same nutritionist after one month yielded an interclass correlation co-efficient (ICC) = 0.957 (0.917–0.978), $p = 0.0001$ for energy intake/day of all participants. According to macronutrient intake/day, such as the percentage of carbohydrates and proteins of total energy intake/day, ICC varied between 0.969 (0.939–0.984) and 0.961 (0.924–0.980), respectively ($p = 0.0001$).

This coefficient corresponds to the agreement in energy intake/day (kcal/day) and macronutrients intake/day (g) or in their percentage of the total energy intake at two time points for each participant.

The estimate of validity was performed, using Bland-Altman analyses, on 100 patients, prior to diagnosis, who fulfilled both the FFQ questionnaire and the two 24-h recalls. The difference in dietary intake between the FFQ and the mean of estimated nutrients of both 24-h recalls was plotted on the Y axis, and the mean intake of both tools on the X axis. Most data points were clustered around the mean difference line between the two limits of agreement [19].

2.6. Ethical Considerations

The present study has complied with all ethical principles according to the revised Helsinki Declaration, 1975. The study has guaranteed the confidentiality and anonymity of the data. An informed consent letter was distributed and signed by the patients. The study started after obtaining the consent of the Ethics Committee of Saint Joseph University of Beirut (CEHDF 351) and was supported with grants by the research council of Saint Joseph University, Lebanon (FPH 34).

2.7. Analysis Plan and Statistical Tests

Descriptive data such as age, BMI, and waist circumference were presented as means and standard deviation. Categorical variables, such as gender, family history of obesity, cardiovascular disease (CVD), or type 2 diabetes were presented as frequency and percentage. Correlational studies were done by the chi-square test (bivariate study). Log10 of all quantitative variables with non-normal distribution was used. Logistic regression identified the main independent predictors of the absence or presence of fibrosis after adjustment for covariates. The score of each dietary pattern was entered as an independent variable with other co-variables. The threshold of significance was set at 5%. A statistical analysis was performed on SPSS 20 for Windows (IBM Corp., Released 2011, IBM SPSS Statistics for Windows, version 20.0. Armonk, NY, USA, IBM corp.)

Factor analysis was applied to extract dietary patterns from the FFQ. Food items were grouped into 25 groups according to food family and nutrient profile (Supplementary Materials). The total consumption for each food group was set by determining the daily intake of servings from each item in this group. by The Kaiser Meyer-Olkin measure of sampling adequacy value was 0.788. The Bartlett's test of sphericity value was significant ($p < 0.0001$). The number of components to extract was based on the Kaiser criterion (eigenvalues > 1), the change in the shape of the scree plot, and the loading of the items in the components generated (component matrix). Varimax rotation was conducted, and dietary patterns were named according to food groups with a factor loading greater than 0.2. Each participant had a factor score for each dietary pattern. The factor score for each participant was calculated by the Anderson and Rubin (1956) method [20]. The Traditional Lebanese pattern represents a group of related variables such as vegetables, chickpeas, red beans, olive oil, almonds, walnuts, and fish with a correlation matrix of $r > 0.3$. The High Fruit Pattern represents a group of related variables, all belonging to fruit families, such as plums, peaches, apricots, and apples. The variables of The Westernized pattern which represent an $r > 0.3$ were mainly beef meat, chicken, chips, carbonated beverages, pizza, and hot dogs. We also categorized patients into low, medium, and high adherence to each nutrient and dietary pattern based on tertiles of nutrients and dietary pattern scores. The association of dietary patterns and nutrients with hepatic fibrosis status (absence/presence or at risk) was determined using binary logistic regression. We defined three models with the crude one to compute multivariable–adjusted odds ratios (ORs).

3. Results

3.1. Variation of Age, Sociodemographic, Clinical, and Biochemical Parameters of the NAFLD Patients between the NFS Group ($<-1.45/\geq-1.45$)

Values of the NFS score which are less than -1.45 indicate a state of non-fibrosis, while values ≥ -1.45 indicate a possible fibrosis [8]. The mean age (years) of the sample was 43.34 ± 0.12 (Table 1), and the risk of fibrosis was associated with increasing age ($p = 0.0001$). However, no significance was found in the risk of fibrosis according to sex, environmental, social status, and education. In total, 34.7% of participants lived in the Mount Lebanon district and 24.1% lived in Beirut. The remaining participants (35.9%) lived in the various Lebanese regions such as the North, South, Nabatiyeh, and Beqaa, with 5.3% living abroad. Regarding social status and education, 75.9% of the participants were married, 52.8% were university students and 0.9% were illiterate. The risk of fibrosis was significant according to the profession; 43.1% were self-employed, 31.2% were employees, and 21.2% were retired or unemployed ($p = 0.001$). There was no association between high economic status and risk of fibrosis, and 77.1% of participants had a crowding index (amount of residents/number of rooms) $\leq$1 (Table 1).

In total, 63.7% of the sample were obese (Table 1) and 83.9% had metabolic syndrome parameters. In addition, 21.2% had type 2 diabetes and 49.4% had Homa-IR > 3. As for family history, 84.7% of patients had a family history of cardiovascular diseases and/or diabetes, and/or obesity (Table 1). The presence of obesity, CVD, diabetes type 2 as well as the ratio of ASAT/ALAT were significantly different according to the NFS scores ($p < 0.05$).

Table 1. Sociodemographic, clinical, and environmental parameters according to the NFS scores (n = 320).

	NFS Scores		*p*-Value	Total (n = 320)
	<−1.45 n **= 191**	**≥−1.45** n **= 129**		
Age (years), mean ± SD	40.04 ± 0.11	47.65 ± 0.13	0.0001	43.34 ± 0.12
Sex				
Male	125 (65.5%)	88 (67.9%)	0.778	213 (66.5%)
Female	66 (34.5%)	41 (32.1%)		107 (33.6%)
Place of residence, n (%)				
Mount Lebanon	57 (29.8%)	54 (41.9%)		111 (34.7%)
North	23 (11.9%)	6 (4.7%)		29 (9.1%)
South	14 (7.4%)	9 (6.9%)		23 (7.2%)
Beirut	47 (24.4%)	30 (23.6%)	0.314	77 (24.1%)
Bekaa	22 (11.4%)	11 (8.5%)		33 (10.3%)
Nabatieh	17 (9.1%)	13 (10.1%)		30 (9.3%)
Abroad	11 (5.7%)	6 (4.7%)		17 (5.3%)
Marital Status, n (%)				
Single	42 (22.0%)	24 (18.9%)		66 (20.6%)
Married	144 (75.4%)	100 (77.5%)	0.853	244 (75.9%)
Divorced	3 (1.6%)	4 (2.8%)		7 (2.10%)
Widow/er	2 (1.04%)	1 (0.8%)		3 (1.10%)
Academic level, n (%)				
Illiterate	1 (0.6%)	2 (1.5%)		3 (0.9%)
Elementary	15 (8.0%)	18 (13.9%)	0.329	33 (10.3%)
Intermediate, Secondary	72 (37.4%)	43 (33.3%)		115 (35.9%)
University	103 (54.0%)	66 (51.2%)		169 (52.8%)
Occupation, n (%)				
Self-employed	85 (44.6%)	53 (41%)		138 (43.1%)
Employee	60 (31.4%)	40 (31.4%)	0.001	100 (31.2%)
Retired/unemployment	36 (18.9%)	32 (24.7%)		68 (21.2%)
Others	10 (5.2%)	4 (3.1%)		14 (4.4%)
Crowding index †, n (%)				
≤1	144 (75.4%)	103 (79.8%)	0.53	247 (77.1%)
>1	47 (24.6%)	26 (20.2%)		73 (22.9%)
Presence of metabolic syndrome, n (%)	155 (81.2%)	114 (88.5%)	0.156	269 (83.9%)
Obesity (yes), n (%)	103 (53.9%)	101 (78.2%)	0.001	204 (63.7%)
CVD (yes), n (%)	8 (4.2%)	21(16%)	0.0001	29 (9.1%)
Family medical history (yes), n (%)	167 (87.6%)	104 (81.0%)	0.182	271 (84.7%)
Smoking (yes), n (%)	78 (40.8%)	55(42.6%)	0.187	133 (41.5%)
Physical Activities (Kcals)(yes)	41 (21.46%)	23 (17.82%)	0.469	64 (20%)
Energy intake (kcal), (M), mean ± SD	4525.8 ± 0.2	4127.6 ± 0.2	0.110	4162.9 ± 0.2
Energy intake (kcal), (F), mean ± SD	2829.4 ± 0.2	2731.5 ± 0.3	0.737	2747.3 ± 0.3
Waist circumference (cm) (M)≥ 94 §	130 (68.2%)	107 (83.0%)	0.006	237 (74.2%)
(F) ≥80 §	180 (94.5%)	127 (98.7%)	0.129	307 (96.0%)
Waist/hip ratio				
(M) > 0.90 §	144 (75.6%)	103 (79.6%)	0.438	247 (77.3%)
(F) > 0.85 §	171 (89.8%)	124 (96.1%)	0.058	295 (92.2%)
Diabetes type 2, n (%)	25 (13.0%)	43 (33.0%)	0.0001	68 (21.2%)
Homa > 3, n (%)	92 (48.3%)	66 (51.5%)	0.697	158 (49.4%)
ASAT/ALAT ≥ 1	29 (15.3%)	34 (26.4%)	0.032	63 (19.7%)
Current dietary ** supplementation use, n (%)	81 (42.4%)	70 (54.3%)	0.053	151 (47.2%)

Continuous variables were reported as geometric means± standard deviations. Categorical variables were reported as numbers and percentage. Statistical tests used: independent, *t*-test (continuous variables), test $\chi 2$-test (categorical variables). SD, standard deviation, $p < 0.05$. † crowding index: amount of residents/number of rooms. § waist circumference and waist to hip ration: values according to the IDF, 2009 (M/F). ** Current dietary supplementation use; 75% of supplements were Vit D.

3.2. Dietary Patterns

In this study, three dietary patterns were generated: The Traditional Lebanese, the High Fruit, and the Westernized diet. In total, 33.70% of NAFLD patients followed the Traditional diet (40.60% females, 30.20% males), (Figure 2). The adherence was almost the same for the High Fruits diet group (total 33.70%; 33.90% males, 33.30% females) and the Westernized diet group (total 32.60%; 35.90% males, 26.10% females). The Traditional diet contributed to 18.22% of the total variance. The High Fruit diet accounted for 9.84% of the total variance and the Westernized diet explained 7.22% of the total variance (Table 2).

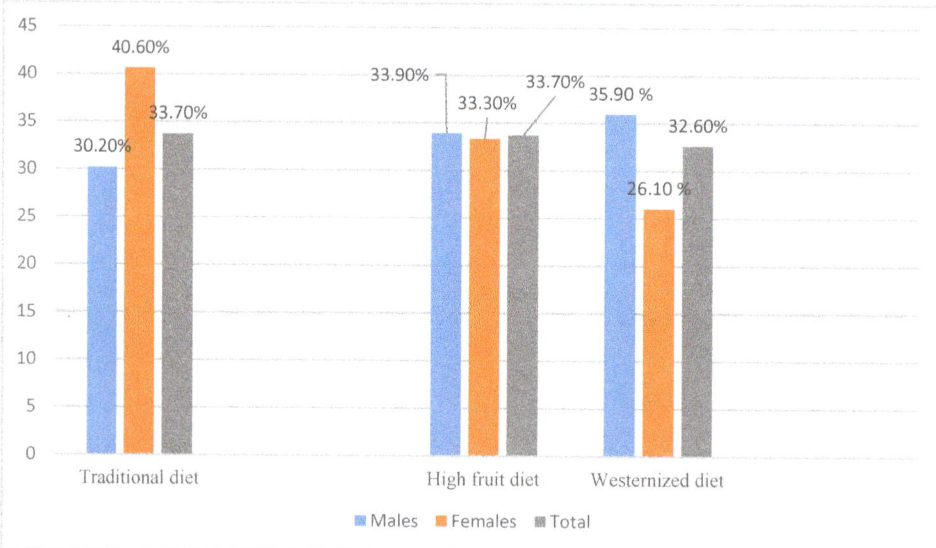

Figure 2. Dietary patterns distribution between males and females ($n = 320$).

Table 2. Factor loading matrix for the three identified dietary patterns in the study population.

Food Group	Pattern Traditional Lebanese	 High Fruits	 Westernized
Vegetables	0.85		
Chickpeas, red beans, lentils, peas	0.50		
Vegetable oil/olives	0.33		0.20
Almonds, walnuts, hazelnuts, sesames	0.27		
Fish	0.21		
Sea Food			0.36
Red Wine	0.21		
Fruits and fruit juices		0.73	
Hamburger and fries			0.63
Beef meat			0.53
Chicken			0.53
Carbonated beverages			0.52
Pizza			0.52
Chips			0.50
Pork			0.41
Hot Dog			0.45

Table 2. *Cont.*

Food Group	Traditional Lebanese	Pattern High Fruits	Westernized
Ketchup			0.50
Mayonnaise or mustard			0.41
1 chicken egg			0.44
Spaghetti or noodles			0.41
Cooked rice	0.24		0.39
Pies or fatayer			0.39
Bread			0.27
Desserts, Arabic pastries			0.30
Milk chocolate			0.39
Laban/Lebanese yogurt			0.25
Energy drink			0.22
Beer			0.21
Pop Corn			0.20
Percent variance explained by each pattern	18.22%	9.84%	7.22%

Extraction method: principal component analysis; Rotation method: Varimax with Kaiser normalization; Absolute values ≤ 0.2 were excluded from the table.

3.3. Association between Dietary Patterns and Sociodemographic, Environmental, and Clinical Characteristics in the Study Population

The traditional pattern was associated with aging (years) (OR: 1.60; 95% CI: 1.03–1.09) which was inversely associated with a westernized one (OR: 0.92; 95% CI: 0.89–0.95), (Table 3). The latter was positively associated with an increase in BMI, $p < 0.05$. The traditional pattern was highly associated with the university level (OR: 2.90; 95% CI: 1.50–5.76), which was inversely associated with a High Fruit diet ($p < 0.05$).

Table 3. Association between dietary patterns, sociodemographic and clinical characteristics in the study population.

	Traditional Lebanese		High Fruits		Westernized	
	OR	CI	OR	CI	OR	CI
Age (years)	1.60	1.03–1.09 *	1.01	0.98–1.03	0.92	0.89–0.95 *
BMI (kg/m^2)	0.90	0.91–1.02	0.95	0.89–1.01	1.11	1.04–1.18 *
Education ** (university level/others)	2.90	1.50–5.76 *	0.50	0.27–0.91 *	0.76	0.39–1.48

* Test statistic; Multivariate -adjusted OR (95% CI) using binary logistic regression, $p < 0.05$. The Model was adjusted for gender, crowding index, presence of metabolic syndrome (no/yes), physical activity (yes/no), family history (no/yes), marital status (married, single, widow), smoking (no/yes) and profession (freelance, employee, unemployed, retirement and others). ** Education (university level versus illiterate, primary, secondary, and high school level).

3.4. Hepatic Fibrosis across Tertiles (T) of Dietary Pattern Scores

The risk of hepatic fibrosis across tertiles of the three dietary pattern scores is presented in Table 4. The association between the high adherence versus low adherence to the traditional pattern became statistically significant after adjustment for confounding variables. In the crude model, high adherence versus low adherence to the traditional pattern was associated with low odds of fibrosis (OR: 0.36; 95% CI: 0.13–1.03). In the third model, after further adjustment with the confounding variables, a significant decrease in the trend of odds of fibrosis was reported with high adherence versus low adherence to the traditional pattern (OR:0.18; 95% CI: 0.04–0.85). The other dietary patterns (High Fruits and Westernized patterns) showed no significant association with hepatic fibrosis, $p > 0.05$.

Table 4. Hepatic fibrosis across tertiles (T) of dietary pattern scores.

Dietary Pattern	T1 (Low Adherence)	T2 (Medium Adherence)	T3 (High Adherence)	*p*-Trend
Traditional Lebanese				
Crude Model	Ref *	0.41 (0.15–1.12)	0.36 (0.13–1.03)	0.057
Model 1	Ref	0.31 (0.99–1.18)	0.21 (0.07–0.82) ‡	0.024
Model 2	Ref	0.32 (0.07–1.21)	0.18 (0.04–0.79) ‡	0.023
Model 3	Ref	0.42 (0.11–1.96)	0.18 (0.04–0.85) ‡	0.031
High Fruits				
Crude Model	Ref	1.22 (0.42–3.53)	0.98 (0.32–3.02)	0.969
Model 1	Ref	1.68 (0.39–7.11)	1.05 (0.24–4.69)	0.943
Model 2	Ref	1.59 (0.32–7.93)	1.06 (0.21–5.35)	0.946
Model 3	Ref	2.53 (0.43–14.97)	1.78 (0.28–11.17)	0.537
Westernized				
Crude Model	Ref	1.51 (0.47–4.82)	1.03 (0.31–3.39)	0.956
Model 1	Ref	1.78 (0.53–6.01)	1.41 (0.38–5.28)	0.606
Model 2	Ref	2.43 (0.59–9.99)	1.19 (0.28–5.02)	0.813
Model 3	Ref	2.22 (0.51–9.61)	1.04 (0.24–4.51)	0.959

‡ Test statistic; Multivariate -adjusted OR (95%CI) using binary logistic regression, $p < 0.05$. The Model 1 was adjusted for age, gender, and the crowding index. Model 2; Model 1 + presence of metabolic syndrome (no/yes), physical activity (yes/no), obesity (no, yes), diabetes type 2 (no, yes), family history (no/yes) and smoking (no/yes). Model 3; Model 2 adjusted for marital status (married, single, widow), education (illiterate, primary, high school, and university) and profession (freelance, employee, unemployed, retirement and others). * Ref referred to the first tertile of dietary pattern (low adherence to the corresponding dietary pattern).

3.5. *Hepatic Fibrosis across Tertiles of Food Groups*

A significantly lower risk of fibrosis was observed with medium intake of EPA and DHA (tertile 2) versus low intake respectively (OR: 0.35; 95% CI: 0.15–0.82), (OR: 0.26; 95% CI: 0.09–0.73). A significantly higher risk of fibrosis was observed with medium intake of ω-6 (g) and simple carbohydrates (g) as compared to the low intake of these two nutrients (OR: 2.24; 95% CI: 1.06–4.74), (OR: 2.43; 95% CI: 1.12–5.26). The other food groups showed no significant association with the absence or presence of hepatic fibrosis (Table 5).

Table 5. Hepatic fibrosis across tertiles of food groups.

Food Groups	Tertile 1	Tertile 2	Tertile 3	*p*-Trend
Fructose (g/day)	Ref *	1.07(0.52–2.20)	1.15(0.54–2.45)	0.718
Fibres (g/day)	Ref	0.67(0.32–1.39)	1.24(0.61–2.52)	0.554
Monounsaturated Fatty acids (g/day)	Ref	0.91(0.45–1.84)	0.69(0.33–1.46)	0.342
Polyunsaturated Fatty acids (g/day)	Ref	0.77 (0.39–1.52)	0.65(0.31–1.36)	0.253
Saturated Fatty acids (g/day)	Ref	1.50 (0.74–3.06)	0.83(0.38–1.83)	0.653
ω-3 (g/day)	Ref	1.12 (0.55–2.29)	1.03(0.49–2.16)	0.944
ω-6 (g/day)	(1.20–13.70) Ref	(13.71–19.01) 2.24 (1.06–4.74) §	(19.02–52.70) 1.04(0.50–2.14)	0.922
Cholesterol (mg/day)	Ref	0.79 (0.39–1.61)	1.10(0.52–2.30)	0.792
Protein (g/day)	Ref	0.91 (0.43–1.91)	0.65(0.28–1.50)	0.311
Fat (g/day)	Ref	1.06 (0.50–2.23)	0.55(0.24–1.26)	0.159
Simple Carbohydrates (g/day)	(8.15–101.4) Ref	(101.50–164.69) 2.43 (1.12–5.26) §	(164.70–469.64) 2.06(0.87–4.88)	0.099
EPA (mg/day)	(0.1–2.0) Ref	(2.10–5.0) 0.35 (0.15–0.82) §	(5.1–160) 0.65(0.26–1.63)	0.357
DHA (mg/day)	(0.1–1.0) Ref	(1.1–9.0) 0.26 (0.09–0.73) §	(9.1–230) 0.53 (0.19–1.44)	0.212
Energy/day (Kcals), males	Ref	0.63 (0.26–1.57)	1.01 (0.38–2.64)	0.994
Energy/day (Kcals), females	Ref	2.27 (0.48–10.80)	1.53 (0.33–7.09)	0.587

Test statistic; § Multivariate -adjusted OR (95%CI) using binary logistic regression, $p < 0.05$. The Model is adjusted for age, gender, crowding index, presence of metabolic syndrome(no/yes), obesity (no, yes), diabetes type 2 (no, yes) physical activity (yes/no), family history (no/yes), marital status (married, single, widow), smoking (no/yes), education (illiterate, primary, high school, and university) and profession (freelance, employee, unemployed, retirement and others). * Ref referred to the first tertile of nutrients consumed (g/day).

4. Discussion

Several research studies have reported an increase in the prevalence of NAFLD with aging [21,22]. However, the mean age of our sample (43.3 ± 0.1 years) was below the average found in other studies [21,22]. In addition, this study highlighted a high prevalence of metabolic syndrome (more than 80%) independently of the NFS score. This value is higher than that determined by Sibai et al., who found that the prevalence of metabolic syndrome among the general population in Lebanon was 31.2% [23]. This feature has been studied by Marchesini et al., who reported hypertriglyceridemia and hyperinsulinemia in patients diagnosed with NAFLD [6]. The latter is usually associated with obesity [24], high waist circumference (cm), and high waist circumference/hip circumference [25]. In this study, about 80% of male NAFLD patients had a waist circumference or waist-to-hip ratio exceeding the WHO recommendations [26]. This reflects an increase in visceral adiposity, which is generally related to a state of insulin resistance responsible for the onset of metabolic syndrome and fatty liver disease [27].

More than eighty-five percent of observed patients have a positive family history of metabolic disorders. Similarly, Chehreh et al. reported a high prevalence of a family history of type 2 diabetes and hypertension in NAFLD patients in their study. This could be of ethnic, environmental, or hereditary origin [28].

As for the environmental characteristics, the rate of sedentary behavior is high in our sample, with only 20% of the sample exercising regularly. It has been shown that low physical activity had been correlated with NAFLD complications [29], while regular exercise has been shown to improve liver enzymes and fatty liver, regardless of the type or frequency of activity [30]. This is also accompanied by an improvement in insulin resistance, blood triglycerides, and liver histology in patients with NASH [30]. The smoking rate was also high in the sample, in concordance with the study made by Suzuki et al., which correlated smoking with increased oxidative stress and liver injury [31]. This smoking rate is higher than the national average of 41.5% that was reported by Sibai et al. following a national survey done in Lebanon in 2010 [32].

Regarding liver fibrosis, no significant differences in clinical and environmental parameters were found between the two NFS score groups, except for the age and the presence of obesity, diabetes type 2, cardiovascular diseases (CVD), and the ratio of ASAT to ALAT. These parameters were found to be associated with the presence of hepatic fibrosis [33], and the ASAT/ALAT ratio could be used as a screening tool for liver evaluation and detection of advanced liver fibrosis [34]. Their increase is related to the presence of free inflammatory fatty acids, which contribute to a state of insulin resistance, hence increasing the degree of severity of non-alcoholic steatosis (Table 3) [35].

Three major dietary patterns were derived from our sample in the current study. The Traditional, the High Fruit pattern, and the Westernized pattern. The presence of the High Fruit pattern among NAFLD Lebanese patients had been discussed in a previous study [19]. This is in accordance with results obtained in other Mediterranean populations, especially that fruits such as apples, plums, and raisins are of low cost and are available in the Lebanese market. Fructose as a main monosaccharide present in fruits had been linked to NAFLD and, more specifically, to fibrosis [36]. An increase in its consumption increases endoplasmic reticulum stress, promotes activation of stress-related kinases, and induces mitochondrial dysfunctions [36].

An interesting result obtained in this study is that NAFLD Lebanese patients were more inclined to consume a traditional diet while aging with an increase in odds by 1.6, $p < 0.05$ (Table 3). This feature is also present in educated people, with an increase in odds of 2.90 towards a traditional diet, versus a decrease by 50% in adherence to a high fruit diet. A study done by Hiza et al. reported that children and older adults had better-quality diets than younger and middle-aged adults with a clearer tendency toward vegetables, whole grains, and legumes [37]. Moreover, adults with a college diploma had a higher adherence to vegetables and whole grains. This indicates an ability to translate nutritional knowledge into better dietary practices [37,38]. Another finding was an increase in BMI following a Westernized diet. Pliego et al. found that the higher the score of unhealthy patterns, the higher the BMI [39]. These results suggest that weight gain is determined by both qualitative and quantitative aspects of food consumption [39].

The multivariate analysis confirmed the hypothesis that a traditional diet was correlated with a low risk of fibrosis. The obtained results showed a significant relationship between high adherence to the traditional diet and the absence of fibrosis with a decreased risk of getting fibrosis by 82%, $p = 0.031$, after adjusting for its covariables (Table 4). Diets enriched in vegetables, legumes, olive oil, seeds, and red wine have proven their beneficial effects on all the risk factors associated with metabolic syndrome and NAFLD [40]. This can be explained through several mechanisms that can vary from an effective dietary approach for weight loss to a model diet that is plentiful in some beneficial nutrients such as antioxidants, vitamins, and monounsaturated fatty acid (MUFA) through the presence of olive oil as the main contributor of fat [41]. A westernized diet characterized by a high intake of pasta, red meat, desserts, and pizza, rich in simple sugar as well as in saturated and trans fatty acids is well known to trigger an increase in weight, higher postprandial insulin secretion and ultimately an increase in liver fat storage [42].

Concerning nutrients, simple carbohydrates were identified as risk factors for NAFLD, increasing the risk of fibrosis twofold, $p = 0.09$ (Table 5). An increase in the intake of simple carbohydrates was also associated with hepatic de novo lipogenesis (DNL) and hepatic inflammation [43,44]. According to Zykovic et al., a reduction in the amount of total carbohydrates, especially simple sugars, would reduce the total pool of acetyl CoA in the liver and, therefore, reduce the flux through the DNL pathway [45]. The reduction in triacylglycerol synthesis would also prevent the excess accumulation of total fat in the liver [46].

The consumption of long-chain polyunsaturated fatty acids was low in the sample. This is realistic, since Lebanese people rarely consume fatty fish. Supplements enriched with EPA, DHA, and ω-3 are rarely used among Lebanese NAFLD patients. Although Lebanon is a coastal country, the Lebanese population avoids eating seafood for cultural, economic, and public health reasons. The general belief is that the coastline is polluted, and consumable fish are unavailable or very expensive [19]. A study done by Nassreddine et al. reported this low consumption of seafood in Lebanese subjects, with 73.6% of Lebanese adult participants consuming less than two servings of fish per week [47]. A medium adherence to EPA and DHA was identified as decreasing fibrosis by approximately 65% and 74%, respectively, $p < 0.05$, as compared to a low adherence after adjustment for the covariables (Table 5). sThese two nutrients are well known for their anti-inflammatory and anti-oxidative properties and their beneficial effect in the treatment of NASH. EPA and DHA converted from ω-3 polyunsaturated fatty acids (PUFAs) might inhibit the accumulation of triglycerides by modifying hepatic lipid metabolism leading to an increase in triglyceride transportation from hepatocytes [46], and could reduce inflammatory and oxidative status [44,46]. In addition, long-chain polyunsaturated fatty acids may induce transcription of genes encoding enzymes for fatty acid oxidation through their ability to act as ligand activators of PPAR-α (peroxisome proliferator-activated receptor-α) [48]. Dietary EPA and DHA were identified as linear, independent, and preventive parameters for NAFLD in Japanese men who generally consume more fish than Western men [49].

Concerning ω-6, a medium intake/day versus a low intake/day was found to increase the risk of fibrosis by 2.24-fold, $p < 0.05$. This is in concordance with other studies which showed an association between ω-6 intake and the development of steatosis and fibrosis in animal models [50–54]. It had been reported that ω-6 metabolism causes oxidative stress and mitochondrial dysfunction. It enhances liver Kupffer cell production of inflammatory cytokines, exacerbates systemic and hepatic insulin resistance, and worsens inflammation and fibrosis [55]. This contributes to the effect of ω-6 intake on the risk of fibrosis as reported in our results after adjustment with the covariables.

Finally, it should be noted that the study had some limitations, such as selection and sampling bias. The sampling bias was mainly due to the difficulty of getting a significant sample of the Lebanese population because the study was done in one outpatient clinic in an academic hospital in Beirut. The second limitation was the ultrasound used to assess the presence or not of fatty liver (sensitivity of ultrasound varies between 60 and 65%, almost no detection for a degree of steatosis <30%) [56]. The third limitation came from recall bias. Patients may overestimate the portion sizes, have memory loss, and over-report their physical activity level. To overcome these biases, patients were interviewed to report their dietary intake prior to the disease diagnosis or prior to any diet change due to medical advice or used medications. Another bias is related to the use of factor analysis, which requires subjective decisions for grouping food, choosing the method of rotation or determining dietary patterns according to their loading factors. However, the results obtained were in line with those obtained in other studies [6,19,57], and highlighted the nutritional profile and dietary patterns of Lebanese NAFLD patients in relation to the absence or presence of hepatic fibrosis.

5. Conclusions

Three dietary patterns, the Traditional, the High Fruit, and the Westernized patterns, characterized the nutritional profile of Lebanese NAFLD patients. Fruits were absent from the Traditional dietary pattern and constituted a new High Fruit pattern. The Traditional diet was composed mainly of vegetables, nuts, and legumes, which are high in fibers and antioxidants, fish known to be rich in long-chain polyunsaturated fatty acids, and olive oil, which is high in polyphenols and monounsaturated fatty acids. It was found to be inversely associated with the risk or presence of fibrosis. This could be an effective dietary approach for NAFLD patients. The role of fruits in the progression or prevention of the disease has yet to be determined. Further experimental studies are needed to establish a possible relationship between fruit intake/day and the presence of fibrosis in NAFLD patients. This is important to generate dietary NAFLD guidelines and reach a conclusion on the quantity and type of fruits to be consumed by these patients.

Supplementary Materials: The following are available online at https://www.mdpi.com/article/10.3390/nu14122554/s1, File S1: Questionnaire of the study, File S2: Food group. File S3: Abbreviations Page.

Author Contributions: N.F.S., as the principal investigator, developed the idea, performed literature review, and wrote and edited the manuscript. G.N.H.A.H., H.Y., L.N.S., C.B. and R.S. also undertook literature review, wrote, and edited the manuscript and acted as lead reviewers. G.N.H.A.H. was responsible of reviewing and editing tables and the figure. All authors have read and agreed to the published version of the manuscript.

Funding: This research (FPH 34) received no external funding.

Institutional Review Board Statement: The study protocol got the approval of the ethics committee of Saint Joseph University of Beirut (CEHDF 351).

Informed Consent Statement: Informed consent was obtained from all patients involved in this study.

Data Availability Statement: Details on FFQ questionnaire, food groups can be found in Supplementary Materials, Files S1 and S2. Data sets can be provided upon requests.

Acknowledgments: This research (FPH 34) was accepted by the research council of Saint Joseph University. We thank them for that.

Conflicts of Interest: The authors declare that they have no conflict of interest.

References

1. Ong, J.P.; Younossi, Z.M. Epidemiology and Natural History of NAFLD and NASH. *Clin. Liver Dis.* **2007**, *11*, 1–16. [CrossRef] [PubMed]
2. Vernon, G.; Baranova, A.; Younossi, Z.M. Systematic review: The epidemiology and natural history of non-alcoholic fatty liver disease and non-alcoholic steatohepatitis in adults. *Aliment. Pharmacol. Ther.* **2011**, *34*, 274–285. [CrossRef] [PubMed]

3. Ong, J.P.; Pitts, A.; Younossi, Z.M. Increased overall mortality and liver-related mortality in non-alcoholic fatty liver disease. *J Hepatol.* **2008**, *49*, 608–612. [CrossRef] [PubMed]
4. Hashimoto, E.; Tokushige, K. Prevalence, gender, ethnic variations, and prognosis of NASH. *J. Gastroenterol.* **2011**, *46*, 63–69. [CrossRef]
5. Rikhi, R.; Singh, T.; Esfeh, J.M. Work up of fatty liver by primary care physicians, review. *Ann. Med. Surg.* **2020**, *50*, 41–48. [CrossRef]
6. Marchesini, G.; Brizi, M.; Morselli/labate, A.M.; Bianchi, G.; Bugianesi, E.; McCullough, A.J.; Forlani, G.; Melchionda, N. Association of nonalcoholic fatty liver disease with insulin resistance. *Am. J. Med.* **1999**, *107*, 450–455. [CrossRef]
7. Adams, L.A.; Angulo, P. Role of Liver Biopsy and Serum Markers of Liver Fibrosis in Non-alcoholic Fatty Liver Disease. *Clin. Liver Dis.* **2007**, *11*, 25–35. [CrossRef]
8. Nakano, M.; Murohisa, T.; Imai, Y.; Hiraishi, H. Validity of the NAFLD fibrosis score in a Japanese population. *Nihon Shokakibyo Gakkai Zasshi* **2012**, *109*, 751–759.
9. Koch, M.; Nöthlings, U.; Lieb, W. Dietary patterns and fatty liver disease. *Curr. Opin. Lipidol.* **2015**, *26*, 35–41. [CrossRef]
10. Landrivon, G.; Delahaye, F. *La Recherche Clinique de l'idée à la Publication*; Elsevier-Masson: Paris, France, 1997; p. 296.
11. Hu, F.B.; Satija, A.; Rimm, E.B.; Spiegelman, D.; Sampson, L.; Rosner, B.; Camargo, C.A.; Stampfer, M.; Willett, W.C. Diet Assessment Methods in the Nurses' Health Studies and Contribution to Evidence-Based Nutritional Policies and Guidelines. *Am. J. Public Health* **2016**, *106*, 1567–1572. [CrossRef]
12. Alberti, K.G.M.M.; Eckel, R.H.; Grundy, S.M.; Zimmet, P.Z.; Cleeman, J.I.; Donato, K.A.; Fruchart, J.C.; James, W.P.T.; Loria, C.M.; Smith, S.C., Jr. Harmonizing the metabolic syndrome: A joint interim statement of the international diabetes federation task force on epidemiology and prevention; National heart, lung, and blood institute; American heart association; World heart federation; International Atherosclerosis Society; and International Association for the Study of Obesity. *Circulation* **2009**, *120*, 1640–1645. [CrossRef] [PubMed]
13. Ainsworth, B.E.; Haskell, W.L.; Leon, A.S.; Jacobs, D.R., Jr.; Montoye, H.J.; Sallis, J.F.; Paffenbarger, R.S., Jr. Compendium of Physical Activities: Classification of energy costs of human physical activities. *Med. Sci. Sports Exerc.* **1993**, *25*, 71–80. [CrossRef] [PubMed]
14. International Physical Activity Questionnaires. (May 2001); IPAQ: Short Last 7 Days Self-Administered Format. Available online: www.ipaq.ki.se (accessed on 16 December 2021).
15. Nelson, M.; Atkinson, M.; Darbyshire, S. Food photography II: Use of food photographs for estimating portion size and the nutrient content of meals. *Br. J. Nutr.* **1996**, *76*, 31–49. [CrossRef] [PubMed]
16. Food Composition | Food and Nutrition Information Center | NAL | USDA. 2020. Available online: https://www.nal.usda.gov/fnic/food-composition (accessed on 10 December 2021).
17. Weir, C.B.; Jan, A. BMI Classification Percentile and Cut off Points. In *StatPearls*; StatPearls Publishing: Treasure Island, FL, USA, 2022.
18. Stern, S.E.; Williams, K.; Ferrannini, E.; DeFronzo, R.A.; Bogardus, C.; Stern, M.P. Identification of Individuals With Insulin Resistance Using Routine Clinical Measurements. *Diabetes* **2005**, *54*, 333–339. [CrossRef] [PubMed]
19. Fakhoury-Sayegh, N.; Younes, H.; Heraoui, G.N.H.A.; Sayegh, R. Nutritional Profile and Dietary Patterns of Lebanese Non-Alcoholic Fatty Liver Disease Patients: A Case-Control Study. *Nutrients* **2017**, *14*, 1245. [CrossRef] [PubMed]
20. Distefano, C.; Zhu, M.; Mindrilla, D. Understanding and using factor scores: Considerations for the applied researcher. *Pract. Assess. Res. Eval.* **2009**, *14*, 1–11. [CrossRef]
21. Bedogni, G.; Miglioli, L.; Masutti, F.; Tiribelli, C.; Marchesini, G.; Bellentani, S. Prevalence of and risk factors for nonalcoholic fatty liver disease: The Dionysos nutrition and liver study. *Hepatology* **2005**, *42*, 44–52. [CrossRef]
22. Shen, L.; Fan, J.-G.; Shao, Y.; Zeng, M.-D.; Wang, J.-R.; Luo, G.-H.; Li, J.-Q.; Chen, S.-Y. Prevalence of nonalcoholic fatty liver among administrative officers in Shanghai: An epidemiological survey. *World J. Gastroenterol.* **2003**, *9*, 1106–1110. [CrossRef]
23. Sibai, A.-M.; Obeid, O.; Batal, M.; Adra, N.; Khoury, D.E.; Hwalla, N. Prevalence and correlates of metabolic syndrome in an adult Lebanese population. *Glob. Heart* **2008**, *3*, 83–90. [CrossRef]
24. Dowman, J.K.; Tomlinson, J.; Newsome, P. Pathogenesis of non-alcoholic fatty liver disease. *QJM Int. J. Med.* **2009**, *103*, 71–83. [CrossRef]
25. Ricci, G.; Canducci, E.; Pasini, V.; Rossi, A.; Bersani, G.; Ricci, E.; Alvisi, V. Nutrient intake in Italian obese patients: Relationships with insulin resistance and markers of non-alcoholic fatty liver disease. *Nutrition* **2011**, *27*, 672–676. [CrossRef] [PubMed]
26. WHO | Diet, Nutrition and the Prevention of Chronic Diseases. WHO. 2020. Available online: https://www.who.int/dietphysicalactivity/publications/trs916/en/ (accessed on 10 November 2021).
27. Illouz, F.; Roulier, V.; Rod, A.; Gallois, Y.; Pellé, C.-P.; Aubé, C.; Rohmer, V.; Ritz, P.; Ducluzeau, P. Distribution of adipose tissue: Quantification and relationship with hepatic steatosis and vascular profiles of type 2 diabetic patients with metabolic syndrome. *Diabetes Metab.* **2008**, *34*, 68–74. [CrossRef] [PubMed]
28. Ghamar-Chehreh, M.E.; Khedmat, H.; Amini, M.; Taheri, S. Predictive value of having positive family history of cardio-vascular disorders, diabetes mellitus, dyslipidemia, and hypertension in non-alcoholic fatty liver disease patients. *Acta. Med. Iran* **2013**, *51*, 307–313.

29. Gerber, L.; Otgonsuren, M.; Mishra, A.; Escheik, C.; Birerdinc, A.; Stepanova, M.; Younossi, Z.M. Non-alcoholic fatty liver disease (NAFLD) is associated with low level of physical activity: A population-based study. *Aliment. Pharmacol. Ther.* **2012**, *36*, 772–781. [CrossRef] [PubMed]
30. Huang, M.A.; Greenson, J.K.; Chao, C.; Anderson, L.; Peterman, D.; Jacobson, J.; Emick, D.; Lok, A.S.; Conjeevaram, H.S. One-Year Intense Nutritional Counseling Results in Histological Improvement in Patients with Nonalcoholic Steatohepatitis: A Pilot Study. *Am. J. Gastroenterol.* **2005**, *100*, 1072–1081. [CrossRef]
31. Suzuki, A.; Lindor, K.; Saver, J.S.; Lymp, J.; Mendes, F.; Muto, A.; Okada, T.; Angulo, P. Effect of changes on body weight and lifestyle in nonalcoholic fatty liver disease. *J. Hepatol.* **2005**, *43*, 1060–1066. [CrossRef]
32. Sibai, A.M.; Obeid, O.; Batal, M.; Adra, N.; El Khoury, D.; Hwalla, N. *Non-Communicable Diseases and Behavioral Risk Factor Survey*; World Health Organization: Beirut, Lebanon, 2010.
33. De Castro, P.C.S.; Alberton, H.C.P.; Pedroso, M.L.A.; Morsoletto, D.B.G.; Junior, A.P.; Ivantes, C.A.P. Evaluation of progression of hepatic fibrosis in a group of patients with non-alcoholic fatty liver disease accompanied for 10 years. *Arq. Gastroenterol.* **2019**, *56*, 256–260. [CrossRef]
34. Åberg, F.; Danford, C.J.; Thiele, M.; Talbäck, M.; Rasmussen, D.N.; Jiang, Z.G.; Hammar, N.; Nasr, P.; Ekstedt, M.; But, A.; et al. A Dynamic Aspartate-to-Alanine Aminotransferase Ratio Provides Valid Predictions of Incident Severe Liver Disease. *Hepatol. Commun.* **2021**, *5*, 1021–1035. [CrossRef]
35. Bugianesi, E.; Zannoni, C.; Vanni, E.; Marzocchi, R.; Marchesini, G. Non-alcoholic fatty liver and insulin resistance: A cause–effect relationship? *Dig. Liver Dis.* **2004**, *36*, 165–173. [CrossRef]
36. Abdelmalek, M.F.; Suzuki, A.; Guy, C.; Unalp-Arida, A.; Colvin, R.; Johnson, R.J.; Diehl, A.M.; Nonalcoholic Steatohepatitis Clinical Research Network. Increased fructose consumption is associated with fibrosis severity in patients with nonalcoholic fatty liver disease. *Hepatology* **2010**, *51*, 1961–1971. [CrossRef]
37. Hiza, H.A.; Casavale, K.O.; Guenther, P.; Davis, C.A. Diet Quality of Americans Differs by Age, Sex, Race/Ethnicity, Income, and Education Level. *J. Acad. Nutr. Diet.* **2013**, *113*, 297–306. [CrossRef] [PubMed]
38. Rippin, H.L.; Hutchinson, J.; Greenwood, D.C.; Jewell, J.; Breda, J.J.; Martin, A.; Rippin, D.M.; Schindler, K.; Rust, P.; Fagt, S.; et al. Inequalities in education and national income are associated with poorer diet: Pooled analysis of individual participant data across 12 European countries. *PLoS ONE* **2020**, *15*, e0232447. [CrossRef]
39. Gutiérrez-Pliego, L.E.; Camarillo-Romero, E.D.S.; Montenegro-Morales, L.P.; Garduño-García, J.D.J. Dietary patterns associated with body mass index (BMI) and lifestyle in Mexican adolescents. *BMC Public Health* **2016**, *16*, 850. [CrossRef] [PubMed]
40. Sofi, F.; Casini, A. Mediterranean diet and non-alcoholic fatty liver disease: New therapeutic option around the corner? *World J. Gastroenterol.* **2014**, *20*, 7339–7346. [CrossRef] [PubMed]
41. Assy, N.; Nassar, F.; Nasser, G.; Grosovski, M. Olive oil consumption and non-alcoholic fatty liver disease. *World J. Gastroenterol.* **2009**, *15*, 1809–1815. [CrossRef]
42. Oddy, W.H.; Herbison, C.E.; Jacoby, P.; Ambrosini, G.L.; O'Sullivan, T.A.; Ayonrinde, O.T.; Olynyk, J.K.; Black, L.J.; Beilin, L.J.; Mori, T.A.; et al. The Western Dietary Pattern Is Prospectively Associated With Nonalcoholic Fatty Liver Disease in Adolescence. *Am. J. Gastroenterol.* **2013**, *108*, 778–785. [CrossRef]
43. Schwarz, J.M.; Neese, R.A.; Turner, S.; Dare, D.; Hellerstein, M.K. Short-term alterations in carbohydrate energy intake in humans. *J. Clin. Investig.* **1995**, *96*, 2735–2743. [CrossRef]
44. Solga, S.; Alkhuraishe, A.R.; Clark, J.M.; Torbenson, M.; Greenwald, A.; Diehl, A.M.; Magnuson, T. Dietary Composition and Nonalcoholic Fatty Liver Disease. *Dig. Dis. Sci.* **2004**, *49*, 1578–1583. [CrossRef]
45. Zivkovic, A.M.; German, J.B.; Sanyal, A.J. Comparative review of diets for the metabolic syndrome: Implications for nonalcoholic fatty liver disease. *Am. J. Clin. Nutr.* **2007**, *86*, 285–300. [CrossRef]
46. Yoshikawa, T.; Shimano, H.; Yahagi, N.; Ide, T.; Amemiya-Kudo, M.; Matsuzaka, T.; Nakakuki, M.; Tomita, S.; Okazaki, H.; Tamura, Y.; et al. Polyunsaturated Fatty Acids Suppress Sterol Regulatory Element-binding Protein 1c Promoter Activity by Inhibition of Liver X Receptor (LXR) Binding to LXR Response Elements. *J. Biol. Chem.* **2002**, *277*, 1705–1711. [CrossRef]
47. Nasreddine, L.; Hwalla, N.; Sibai, A.; Hamzé, M.; Parent-Massin, D. Food consumption patterns in an adult urban population in Beirut, Lebanon. *Public Health Nutr.* **2006**, *9*, 194–203. [CrossRef] [PubMed]
48. Itoh, M.; Suganami, T.; Satoh, N.; Tanimoto-Koyama, K.; Yuan, X.; Tanaka, M.; Kawano, H.; Yano, T.; Aoe, S.; Takeya, M.; et al. Increased Adiponectin Secretion by Highly Purified Eicosapentaenoic Acid in Rodent Models of Obesity and Human Obese Subjects. *Arter. Thromb. Vasc. Biol.* **2007**, *27*, 1918–1925. [CrossRef] [PubMed]
49. Sugano, M.; Hirahara, F. Polyunsaturated fatty acids in the food chain in Japan. *Am. J. Clin. Nutr.* **2000**, *71*, 189S–196S. [CrossRef]
50. Keim, N.L.; Mares-Perlman, J.A. Development of Hepatic Steatosis and Essential Fatty Acid Deficiency in Rats with Hypercaloric, Fat-Free Parenteral Nutrition. *J. Nutr.* **1984**, *114*, 1807–1815. [CrossRef]
51. Werner, A.; Havinga, R.; Bos, T.; Bloks, V.W.; Kuipers, F.; Verkade, H.J. Essential fatty acid deficiency in mice is associated with hepatic steatosis and secretion of large VLDL particles. *Am. J. Physiol. Liver Physiol.* **2005**, *288*, G1150–G1158. [CrossRef]
52. Alwayn, I.P.; Javid, P.J.; Gura, K.M.; Nosé, V.; Ollero, M.; Puder, M. Do polyunsaturated fatty acids ameliorate hepatic steatosis in obese mice by SREPB-1 suppression or by correcting essential fatty acid deficiency. *Hepatology* **2004**, *39*, 1176–1177. [CrossRef] [PubMed]
53. Cazeils, J.L.; Bouillier-Oudot, M.; Auvergne, A.; Candau, M.; Babile, R. Lipid composition of hepatocyte plasma membranes from geese overfed with corn. *Lipids* **1999**, *34*, 937–942. [CrossRef]

54. Bouziane, M.; Prost, J.; Belleville, J. Dietary protein deficiency affects n−3 and n−6 polyunsaturated fatty acids hepatic storage and very low density lipoprotein transport in rats on different diets. *Lipids* **1994**, *29*, 265–272. [CrossRef]
55. Byrne, C.D. Fatty liver: Role of inflammation and fatty acid nutrition. *Prostaglandins Leukot. Essent. Fat. Acids* **2010**, *82*, 265–271. [CrossRef]
56. Schwenzer, N.F.; Springer, F.; Schraml, C.; Stefan, N.; Machann, J.; Schick, F. Non-invasive assessment and quantification of liver steatosis by ultrasound, computed tomography and magnetic resonance. *J. Hepatol.* **2009**, *51*, 433–445. [CrossRef]
57. Abenavoli, L.; Milic, N.; Di Renzo, L.; Preveden, T.; Medić-Stojanoska, M.; De Lorenzo, A. Metabolic aspects of adult patients with nonalcoholic fatty liver disease. *World J. Gastroenterol.* **2016**, *22*, 7006–7016. [CrossRef] [PubMed]

nutrients

MDPI

Article

Dietary Patterns and Long-Term Outcomes in Patients with NAFLD: A Prospective Analysis of 128,695 UK Biobank Participants

Zhening Liu [1,†], Hangkai Huang [1,†], Jiarong Xie [1,2,3] and Chengfu Xu [1,3,*]

1 Department of Gastroenterology, The First Affiliated Hospital, Zhejiang University School of Medicine, Hangzhou 310003, China
2 Department of Gastroenterology, Ningbo First Hospital, Ningbo 315010, China
3 Zhejiang Provincial Clinical Research Center for Digestive Diseases, Hangzhou 310003, China
* Correspondence: xiaofu@zju.edu.cn
† These authors contributed equally to this work.

Abstract: Large longitudinal studies exploring the role of dietary patterns in the assessment of long-term outcomes of NAFLD are still lacking. We conducted a prospective analysis of 128,695 UK Biobank participants. Cox proportional hazards models were used to estimate the risk associated with two dietary patterns for long-term outcomes of NAFLD. During a median follow-up of 12.5 years, 1925 cases of end-stage liver disease (ESLD) and 12,466 deaths occurred in patients with NAFLD. Compared with patients in the lowest quintile, those in the highest quintile of the diet quality score was negatively associated with the risks of ESLD and all-cause mortality (HR_{Q5vsQ1}: 0.76, 95% CI: 0.66–0.87, $p < 0.001$; HR_{Q5vsQ1}: 0.84, 95% CI: 0.79–0.88, $p < 0.001$, respectively). NAFLD patients with high-quality carbohydrate patterns carried a 0.74-fold risk of ESLD and a 0.86-fold risk of all-cause mortality (HR_{Q5vsQ1}: 0.74, 95% CI: 0.65–0.86, $p < 0.001$; HR_{Q5vsQ1}: 0.86, 95% CI: 0.82–0.91, $p < 0.001$, respectively). For prudent dietary patterns rich in vegetables, fruits and fish, the adjusted HR $_{Q5vsQ1}$ (95% CI) was 0.87 (0.76–0.99) and 0.94 (0.89–0.99) for ESLD and all-cause mortality of NAFLD patients. There was a U-shaped association between the meat-rich dietary pattern and all-cause mortality in patients with NAFLD. These findings suggest that a diet characterized by a high-quality, high intake of vegetables, fruits, fish and whole grains as well as an appropriate intake of meat, was associated with a lower risk of adverse outcomes of NAFLD.

Keywords: non-alcoholic fatty liver disease; end-stage liver disease; mortality; dietary pattern

Citation: Liu, Z.; Huang, H.; Xie, J.; Xu, C. Dietary Patterns and Long-Term Outcomes in Patients with NAFLD: A Prospective Analysis of 128,695 UK Biobank Participants. *Nutrients* **2023**, *15*, 271. https://doi.org/10.3390/nu15020271

Academic Editors: Andriana Kaliora, Chara Tzavara, Charalampia Amerikanou and Luis A. Moreno

Received: 11 December 2022
Revised: 30 December 2022
Accepted: 3 January 2023
Published: 5 January 2023

1. Introduction

Nonalcoholic fatty liver disease (NAFLD) is a very prevalent but widely underappreciated liver disease that is closely related to other metabolic disorders and was therefore called metabolic dysfunction-associated fatty liver disease (MAFLD) [1]. The global prevalence of NAFLD was estimated to be 32.4% [2]. The dramatically increased disease burden of NAFLD has been propelled by the progressively severe health effects of obesity and type 2 diabetes [3]. NAFLD covers a range of liver conditions, from simple steatosis to steatohepatitis and fibrosis, the latter of which carries a higher risk of developing end-stage liver disease [4]. The incidence rate of hepatocellular carcinoma parallels the severity of NAFLD, rising from 0.15 to 14.46 and 19.13 per 1000 person-years for steatosis, fibrosis and cirrhosis, respectively [5]. Moreover, NAFLD patients have a significantly increased risk of overall mortality, and this risk changes in tandem with the histology stage of NAFLD [6]. For the majority of patients, NAFLD is a benign condition [7]. However, advanced liver disease is usually diagnosed late, and interventions at this stage are less effective than earlier treatments [8]. A critical challenge is to identity NAFLD patients at higher risk of

2.6. Statistical Analysis

Baseline sociodemographic, lifestyle and other characteristics were summarized across diet score quartiles. The categorical variables are displayed as percentages and were tested by chi-squared tests. Continuous variables are displayed as means with standard deviations (SDs) and were tested by one-way ANOVA. The associations of diet quality and derived patterns with incident ESLD, all-cause mortality and cause-specific mortality were investigated using Cox-proportional hazard models. Hazard ratios (HR) and 95% confidence intervals (CI) for each quartile of exposure were calculated. Model 1 was adjusted for age, sex, ethnicity, Townsend deprivation index (quintiles), education level (university/college degree or others), household income (less than £18,000, £18,000 to £30,999, £31,000 to £51,999, £52,000 to £100,000, greater than £100,000 or do not know/prefer not to answer), and model 2 was adjusted for model 1 plus self-reported smoking status (never, former or current smoker), sedentary behavior, body mass index, baseline diabetes, baseline hypertension, serum alanine aminotransferase, triglycerides and cholesterol. Then, we used multivariate cubic regression splines with 3 knots (10th, 50th, 90th) to visualize the potential nonlinear associations of dietary patterns with incident ESLD and all-cause mortality discovered in the Cox model above by SAS macro *%RCS_Reg*. To examine the overall statistical significance as well as the non-linearity of the exposures, we used likelihood ratio tests. We then investigated whether these associations differed by age, sex and other factors by performing a subgroup analysis and fitting an interaction term to the model. The hazard ratio of the product term was the measure of the interaction on the multiplicative scale. Sensitivity analyses were performed by excluding individuals with incident ESLD or who died within 2 years, those who had extreme BMIs (BMI < 15 or >40 kg/m^2), those who made any major changes to their diet in the last 5 years and those whose diet varied much from week to week. SAS 9.4 was used for all analyses. All statistical tests were 2-sided, and $p < 0.05$ was defined as statistically significant.

3. Results

3.1. Baseline Characteristics

The baseline characteristics of participants by diet quality score quintiles are shown in Table 1. At baseline, the participants with a higher diet quality tended to be female, older, of White ethnicity, less socially deprived and more educated. In addition, they were less often current smokers and spent less time sitting still. They also had lower levels of alanine aminotransferase, gamma glutamyl transferase, triglycerides and total cholesterol. Interestingly, those NAFLD patients with a higher diet quality were more likely to suffer from comorbid hypertension and type 2 diabetes.

3.2. Association of Diet Quality with Incident ESLD and Mortality

During a median of 12.5 years follow-up (1,569,342 person-years), 1925 ESLD and 12,466 deaths occurred. In Table 2, compared with the patients in the lowest quintile, those in the highest quintile of the diet quality had 16% lower odds of ESLD and 18% lower odds of all-cause mortality after adjustments for covariates in model 1. After further adjusting for lifestyle and biochemistry factors in model 2, the inverse association remained significant. The HRs (95% CIs) in quintiles 2–5 were 0.95 (0.83–1.08), 0.83 (0.72–0.95), 0.84 (0.73–0.96) and 0.76 (0.66–0.87) for ESLD, and 0.94 (0.89–0.99), 0.91 (0.86–0.96), 0.85 (0.80–0.90) and 0.84 (0.79–0.88) for all-cause mortality, respectively. The associations of diet quality with cause-specific mortality were concluded in Supplementary Table S4. Similarly, a higher diet quality significantly reduced the risk of liver, CVD and cancer mortality [HR_{Q5vsQ1}: 0.53 (0.33–0.85), 0.88 (0.78–0.98) and 0.85 (0.79–0.93), respectively].

Table 1. UK Biobank participants' characteristics by the diet quality score.

Variables	Q1	Q2	Q3	Q4	Q5	*p* Value
Male (%)	70.9	62.4	56.5	53.2	53.7	<0.001
Age (years)	55.7 ± 8.3	56.9 ± 8.1	57.3 ± 7.9	57.9 ± 7.6	59.1 ± 7.1	<0.001
White ethnicity (%)	93.4	91.8	92.1	93.7	94.5	<0.001
Townsend deprivation index	−0.4 ± 3.4	−0.8 ± 3.3	−1.1 ± 3.2	−1.3 ± 3.1	−1.2 ± 3.1	<0.001
College or university degree (%)	17.53	23.13	26.33	30.08	29.79	<0.001
Household income (£)						<0.001
<18,000	26.5	25.0	22.8	22.5	24.8	
18,000 to 30,999	22.2	22.1	22.5	22.5	23.6	
31,000 to 51,999	20.3	20.5	20.7	21.1	20.5	
52,000 to 100,000	12.5	13.7	15.0	15.4	13.5	
>100,000	2.2	2.6	3.1	3.6	2.8	
Sedentary behavior (%)	46.5	42.9	40.9	39.1	39.1	<0.001
Smoking status (%)						<0.001
Never	48.3	52.9	54.7	55.5	53.0	
Previous	34.0	35.7	36.6	37.1	41.0	
Current	17.7	11.4	8.7	7.3	6.0	
Alcohol						
Never or in special occasions only	29.8	29.5	28.5	26.9	28.2	
1 to 3 times/month	15.9	15.2	15.5	15.4	15.0	
1 to 4 times/week	47.7	48.6	49.4	50.6	50.3	
Daily or almost daily	6.6	6.8	6.6	7.0	6.6	
Body mass index (kg/m^2)	31.6 ± 4.7	31.8 ± 4.6	31.9 ± 4.6	31.8 ± 4.6	31.9 ± 4.6	<0.001
Waist circumference (cm)	103.5 ± 10.2	102.9 ± 10.2	102.5 ± 10.1	102.2 ± 10.1	102.3 ± 10.1	<0.001
Hypertension (%)	68.2	69.4	69.4	70.0	72.9	<0.001
Diabetes (%)	10.0	11.2	12.2	12.2	17.4	<0.001
Alanine aminotransferase (U/L)	30.3 ± 17.0	29.5 ± 16.5	29.1 ± 17.4	28.9 ± 16.1	28.8 ± 16.6	<0.001
Gamma glutamyltransferase (U/L)	52.3 ± 52.3	50.6 ± 48.8	49.3 ± 49.8	48.0 ± 45.9	47.8 ± 48.4	<0.001
Triglycerides (mmol/L)	2.5 ± 1.2	2.4 ± 1.2	2.4 ± 1.2	2.3 ± 1.1	2.3 ± 1.1	<0.001
Total cholesterol (mmol/L)	5.7 ± 1.2	5.7 ± 1.2	5.7 ± 1.2	5.7 ± 1.2	5.6 ± 1.3	<0.001

Quintiles of the diet score: Q1, ≤40.54; Q2, 40.55–48.66; Q3, 48.67–56.11; Q4, 56.12–63.61; and Q5, ≥63.62. Values are the mean (±standard deviation, SD) or percentage (%) and were examined by one-way ANOVA or chi-square test. Post hoc analysis (Bonferroni method) showed significant differences for Q2–Q5 compared with Q1, except for total cholesterol.

Table 2. HRs of ESLD and all-cause mortality for quintiles of the diet quality score.

Diet Score	ESLD			All-Cause Mortality		
	HR (95% CI)	*p* Value	p_{trend}	HR (95% CI)	*p* Value	p_{trend}
Model 1			0.001			0.004
Q1	1 (ref)			1 (ref)		
Q2	0.95 (0.83–1.09)	0.497		0.92 (0.87–0.97)	0.003	
Q3	0.86 (0.75–0.99)	0.035		0.88 (0.84–0.94)	<0.001	
Q4	0.85 (0.74–0.98)	0.027		0.84 (0.80–0.89)	<0.001	
Q5	0.84 (0.73–0.96)	0.010		0.82 (0.77–0.87)	<0.001	
Model 2			<0.001			<0.001
Q1	1 (ref)			1 (ref)		
Q2	0.95 (0.83–1.08)	0.412		0.94 (0.89–0.99)	0.021	
Q3	0.83 (0.72–0.95)	0.008		0.91 (0.86–0.96)	0.001	
Q4	0.84 (0.73–0.96)	0.013		0.85 (0.80–0.90)	<0.001	
Q5	0.76 (0.66–0.87)	<0.001		0.84 (0.79–0.88)	<0.001	

Quintiles of the diet score: Q1, ≤40.54; Q2, 40.55–48.66; Q3, 48.67–56.11; Q4, 56.12–63.61; and Q5, ≥63.62. Model 1 was adjusted for age, sex, ethnicity, Townsend deprivation index (quintiles), education level (university/college degree or others) and household income (less than £18,000, £18,000 to £30,999, £31,000 to £51,999, £52,000 to £100,000, greater than £100,000 or do not know/prefer not to answer). Model 2 was adjusted for model 1 plus self-reported smoking status (never, former or current smoker), sedentary behavior, body mass index, baseline diabetes, baseline hypertension, serum alanine aminotransferase, triglycerides and cholesterol.

3.3. Association of Dietary Patterns with Incident ESLD and Mortality

The associations between dietary patterns and ESLD risk are shown in Figure 1. We observed that a high-quality carbohydrate dietary pattern was negatively correlated with the risk of ESLD [HR_{Q5vsQ1}: 0.74 (0.65–0.86)]. In addition, a prudent dietary pattern showed a non-linear negative association with the risk of ESLD; the HRs (95% CIs) in quintiles 2–5 were 0.94 (0.82–1.08), 0.85 (0.74–0.98), 0.84 (0.73–0.97), and 0.87 (0.76–0.99). However, we did not find a significant association with the meat-rich dietary pattern.

Dietary pattern	HR (95% CI)	*P* value
Meat		
Q1	1.04 (0.90–1.20)	0.566
Q2	1.00 (0.86–1.15)	0.894
Q3	1 (ref)	
Q4	1.00 (0.85–1.15)	0.840
Q5	1.04 (0.90–1.20)	0.479
Prudent		
Q1	1 (ref)	
Q2	0.94 (0.82–1.08)	0.386
Q3	0.85 (0.74–0.98)	0.029
Q4	0.84 (0.73–0.97)	0.015
Q5	0.87 (0.76–0.99)	0.038
High quality carbohydrate		
Q1	1 (ref)	
Q2	0.92 (0.80–1.05)	0.217
Q3	0.83 (0.72–0.95)	0.008
Q4	0.81 (0.70–0.93)	0.003
Q5	0.74 (0.65–0.86)	<0.001

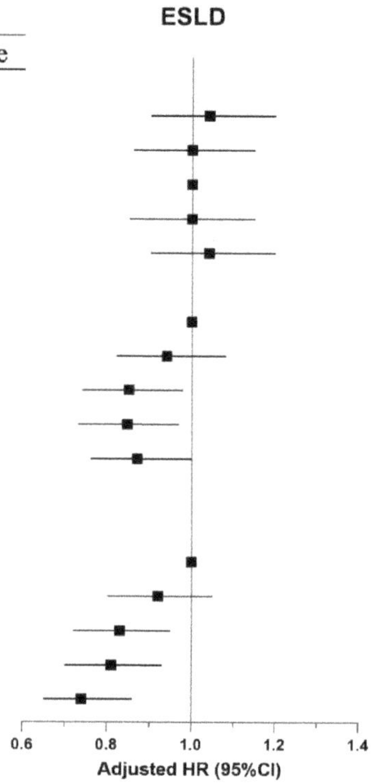

Figure 1. Association of dietary patterns with incident ESLD. For the meat diet pattern, the quintile with the lowest hazard ratio (Q3) was set as a reference. The model was adjusted for age, sex, ethnicity, Townsend deprivation index (quintiles), education level (university/college degree or others), household income (less than £18,000, £18,000 to £30,999, £31,000 to £51,999, £52,000 to £100,000, greater than £100,000 or do not know/prefer not to answer), self-reported smoking status (never, former or current smoker), sedentary behavior, body mass index, baseline diabetes, baseline hypertension, serum alanine aminotransferase, triglycerides and cholesterol.

The results of the relation between the dietary patterns and mortality are shown in Figure 2. Here, we revealed a U-shaped association of the meat-rich dietary pattern with all-cause mortality. Compared with quintile 3, the HRs (95% CIs) in quintiles 1–2 were 1.08 (1.02–1.15) and 1.06 (1.00–1.13); the HRs (95% CIs) in quintiles 4–5 were 1.09 (1.03–1.16) and 1.12 (1.05–1.18). Similar to what observed for ESLD, a prudent dietary pattern and high-quality carbohydrate dietary pattern also demonstrated negative associations with all-cause mortality. The associations of these dietary patterns with cause-specific mortality are reported in Supplementary Table S5.

Dietary pattern	HR (95% CI)	*P* value
Meat		
Q1	1.08 (1.02–1.15)	0.007
Q2	1.06 (1.00–1.13)	0.041
Q3	1 (ref)	
Q4	1.09 (1.03–1.16)	0.022
Q5	1.12 (1.05–1.18)	<0.001
Prudent		
Q1	1 (ref)	
Q2	0.94 (0.89–1.00)	0.049
Q3	0.93 (0.88–0.98)	0.009
Q4	0.91 (0.86–0.97)	0.002
Q5	0.94 (0.89–0.99)	0.032
High quality carbohydrate		
Q1	1 (ref)	
Q2	0.96 (0.91–1.02)	0.166
Q3	0.92 (0.87–0.98)	0.005
Q4	0.87 (0.83–0.92)	<0.001
Q5	0.86 (0.82–0.91)	<0.001

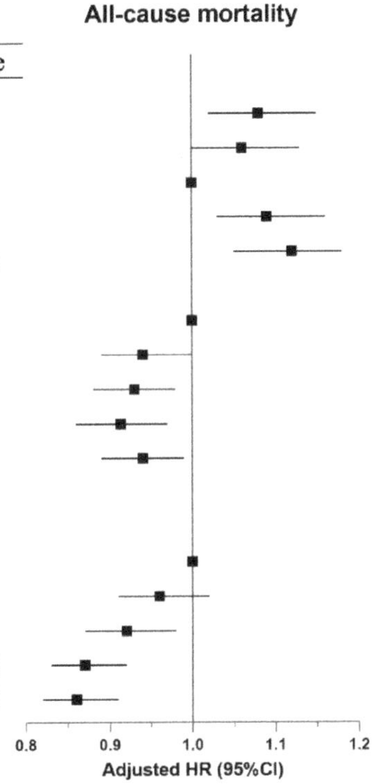

Figure 2. Association of dietary patterns with all-cause mortality. For the meat diet pattern, the quintile with the lowest hazard ratio (Q3) was set as a reference. The model was adjusted for age, sex, ethnicity, Townsend deprivation index (quintiles), education level (university/college degree or others), household income (less than £18,000, £18,000 to £30,999, £31,000 to £51,999, £52,000 to £100,000, greater than £100,000, or do not know/prefer not to answer), self-reported smoking status (never, former or current smoker), sedentary behavior, body mass index, baseline diabetes, baseline hypertension, serum alanine aminotransferase, triglycerides and cholesterol.

The analysis of cubic splines (Figure 3) also showed the U-shaped association of the meat-rich dietary pattern with all-cause mortality ($p_{non\text{-}linearity} < 0.001$) and the L-shaped association between a prudent dietary pattern and ESLD as well as all-cause mortality (All $p_{non\text{-}linearity} \leq 0.001$). For the high-quality carbohydrate dietary pattern, there was a linear association with ESLD ($p_{non\text{-}linearity} = 0.675$) and all-cause mortality ($p_{non\text{-}linearity} = 0.155$).

3.4. Subgroup Analyses and Sensitivity Analyses

The subgroup analyses for diet quality according to different risk factors are shown in Supplementary Tables S6 and S7. There were no significant differences across all investigated subgroups for ESLD. For all-cause mortality, we found that a higher diet quality was associated with a decreased risk among current/previous smokers ($p_{\text{interaction}} < 0.002$).

We performed a number of sensitivity analyses to examine the robustness of the findings. When we excluded the first 2 years of follow-up, the patients with extreme BMIs, those who made any major changes to their diet in the last 5 years, and those whose diet varied much from week to week, we found that the observed associations of diet quality with ESLD and all-cause mortality remained unchanged (Supplementary Table S8).

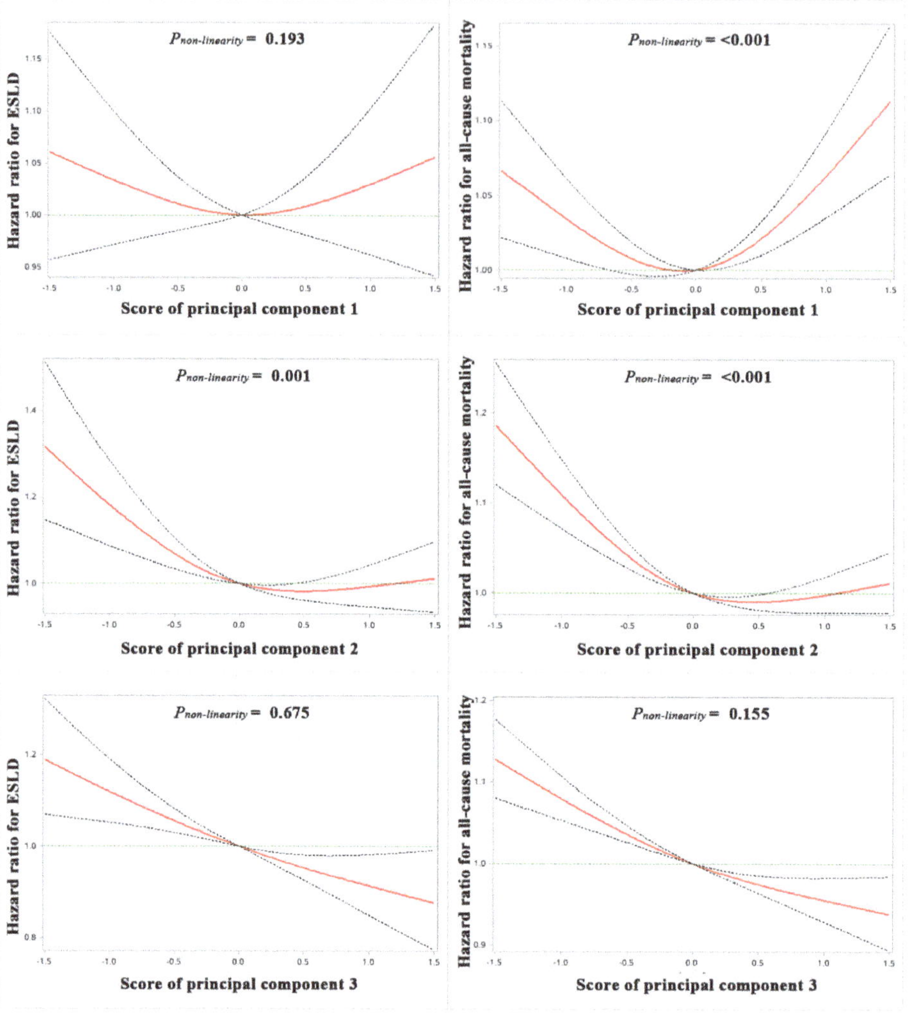

Figure 3. Restricted cubic spline analyses of the relationship between dietary patterns and the incidence of end-stage liver disease (ESLD) and all-cause mortality. Red lines represent adjusted hazard ratios [with 95% CI (dashed lines)] based on restricted cubic splines with knots at the 10th, 50th, 90th. The model was fully adjusted. Principal component 1: meat; principal component 2: prudent; principal component 3: high-quality carbohydrate.

4. Discussion

In this large longitudinal study, we observed that dietary patterns were significantly associated with the long-term outcomes of NAFLD. First, a higher a priori-derived diet quality score was inversely related to the risks of ESLD as well as all-cause mortality and liver-, CVD- and cancer-related mortality. Second, a greater adherence to a posteriori-derived dietary patterns (prudent and high-quality carbohydrate patterns) was associated with a lower risk of ESLD in NAFLD patients, whereas this association was nonsignificant for the meat-rich pattern. Third, there was a U-shaped association between the meat-rich pattern and all-cause mortality for NAFLD patients, while this association was negative for the prudent and high-quality carbohydrate patterns.

A large body of cross-sectional studies have reported an inverse relationship of the diet quality score, as assessed by AHEI [20] and MDS [21], with prevalent NAFLD. In addition, whether different dietary patterns were related to NAFLD risks was also investigated, but contradictory results were reached [17]. A cross-sectional study conducted in 229 Brazilian adults demonstrated that a prudent pattern was negatively associated with NAFLD diagnosed by ultrasonography [22]. In contrast, another cross-sectional study covering 999 Chinese patients found this association to be nonsignificant [23]. An Australian prospective cohort study including 995 adolescents observed that a Western dietary pattern high in red and processed meat, soft drinks, refined grains and sauces at age 14 was associated with a 1.59-fold risk of NAFLD three years later [24]. However, till now, these studies were mainly small cross-sectional studies and centered on the prevalent risks of NAFLD. The evidence for a relationship between dietary pattern and long-term outcomes of NAFLD remains sparce.

In this study, we used a diet quality score which was adopted in previous epidemiological studies of cardiovascular disease [25] and type 2 diabetes [26], created on the basis of recent dietary priorities of cardiometabolic health [19]. In addition, in recent years, the role of dietary patterns generated by principal component analysis has been extensively investigated in observational studies of cardiometabolic disease [27]. They are more representative of dietary patterns in a given population [28]. Therefore, we combined these two methods in this study to provide a more complete picture for the relation of diet with the long-term outcomes of NAFLD. Our analysis showed that NAFLD patients with a higher diet quality carried lowers odds of ESLD and all-cause and cause-specific mortality. In addition, a prudent dietary pattern high in vegetables, fruits and fish was negatively associated with a poor prognosis of NAFLD. This association was also observed for the high-quality carbohydrate dietary pattern which was high in whole grains and low in refined grains.

For the clinical implications of this study, it provides a more comprehensive understanding of the effects of dietary patterns on the development of severe outcomes of NAFLD. The advanced liver disease in the late course of NAFLD is associated with a severely impaired quality of life and poor prognosis [29]. Given the sheer number of NAFLD patients and the fact that advanced liver disease is usually diagnosed late, a better risk stratification of NAFLD is urgently needed [30]. The early recognition of NAFLD patients with adverse outcomes would allow policy makers and clinicians to plan and implement a more effective secondary prevention [9]. This study showed that NAFLD patients may benefit from a high diet quality and prudent and high-quality carbohydrate dietary patterns. Conversely, NAFLD patients with other diet patterns may be more likely to suffer from adverse health outcomes and warrant more close attention during regular follow-up.

There are several possible mechanisms linking the dietary patterns with the long-term outcomes of NAFLD. The prudent dietary pattern has been shown to have beneficial effects on NAFLD due to its anti-inflammatory, anti-fibrosis and antioxidant capacity [31]. Carotenoids and polyphenols are two major antioxidants that are abundant in vegetables and fruits. In experimental studies of NAFLD models, they improved insulin sensitivity, accelerated β-oxidation and repressed de novo lipogenesis [32]. Furthermore, they inhibited the activation of hepatic stellate cells and therefore ameliorated carcinogenesis [32]. Omega-3 poly-unsaturated fatty acids contained in fish oil can alleviate insulin resistance, reduce hepatic lipid accumulation and improve steatohepatitis [33,34]. The mechanism through which whole grains exert favorable impacts on NAFLD is multifaced. First, wheat bran, a more abundant compound in whole grains than in refined grains, reduced the liver triglyceride content in an in vivo model of metabolic syndrome [35]. Second, several phytochemicals that are significantly reduced after grain refining can promote the synthesis of VLDL and thus export lipids outside the liver [36]. Third, whole grains may display beneficial effects on the composition of the gut microbiota [37], which may influence the progression of NAFLD through the gut–liver axis [38].

This study has several limitations that warrant discussion. First, we selected several important diet constituents to create a dietary quality score based on the recent guidelines of cardiometabolic health. However, other components may also display a key role in the progression of NAFLD. Second, we analyzed the association between dietary patterns at baseline and the risk of adverse outcomes of NAFLD. The dietary patterns were not assessed during the follow-up. We were unable to assess the longitudinal dynamic change in dietary patterns, which may more closely reflect the habitual eating in real-world life. Third, as with all observational studies, we were unable to draw causality about the relationship between dietary patterns and long-term outcomes of NAFLD. The only way to clearly measure this relationship is through experimental designs.

5. Conclusions

In conclusion, higher diet quality and greater adherence to a prudent dietary pattern rich in vegetables, fruits and fish were associated with a lower likelihood of ESLD and mortality in NAFLD patients. High-quality carbohydrate dietary patterns showed the same association. NAFLD patients with inappropriate meat dietary patterns had a higher risk of adverse outcomes. These findings need to be confirmed with further interventional studies to assess whether the improvement of the dietary patterns is effective in the primary and secondary prevention of NAFLD.

Supplementary Materials: The following supporting information can be downloaded at: https://www.mdpi.com/article/10.3390/nu15020271/s1, Figure S1. Summary of study design and analytical strategy; Figure S2. Scree plot of the factor analysis; Table S1. Components and scaling methods of diet quality score used in the UK Biobank study; Table S2. PCA -derived dietary patterns and their factor loadings; Table S3. Criteria for the ESLD, CVD, Cancer; Table S4. HRs of cause-specific mortality for quintiles of diet quality score; Table S5. HRs of cause-specific mortality for quintiles of dietary patterns; Table S6. Subgroup analyses in diet quality and ESLD; Table S7. Subgroup analyses in diet quality and all-cause mortality; Table S8. Sensitivity analyses of the HRs for the associations of diet quality with ESLD and all-cause mortality.

Author Contributions: Study concept and design: Z.L. and C.X.; Analysis and interpretation of the data: Z.L., H.H. and J.X.; Drafting of the manuscript: Z.L., H.H., J.X. and C.X.; Critical revision of the manuscript: Z.L. and C.X. All authors have read and agreed to the published version of the manuscript.

Funding: This work was supported by the National Key Research and Development Program (2018YFA0109800), the National Natural Science Foundation of China (82070585), the Key Research and Development Program of Zhejiang Province (2020C03033).

Institutional Review Board Statement: The study was conducted according to the guidelines of the Declaration of Helsinki and approved by the North West Multicenter Research Ethics Committee (reference No. 16/NW/0274).

Informed Consent Statement: Informed consent was obtained from all subjects involved in the study.

Data Availability Statement: Data described in the manuscript, code book, and analytic code will be made available upon request pending application. This research was conducted using the UK Biobank resource under application number 79302.

Acknowledgments: We thank the participants of the UK biobank.

Conflicts of Interest: The authors declare no conflict of interest.

References

1. Lazarus, J.V.; Mark, H.E.; Villota-Rivas, M.; Palayew, A.; Carrieri, P.; Colombo, M.; Ekstedt, M.; Esmat, G.; George, J.; Marchesini, G.; et al. The global NAFLD policy review and preparedness index: Are countries ready to address this silent public health challenge? *J. Hepatol.* **2022**, *76*, 771–780. [CrossRef] [PubMed]
2. Riazi, K.; Azhari, H.; Charette, J.H.; E Underwood, F.; A King, J.; Afshar, E.E.; Swain, M.G.; E Congly, S.; Kaplan, G.G.; Shaheen, A.-A. The prevalence and incidence of NAFLD worldwide: A systematic review and meta-analysis. *Lancet Gastroenterol. Hepatol.* **2022**, *7*, 851–861. [CrossRef]

3. Stefan, N.; Cusi, K. A global view of the interplay between non-alcoholic fatty liver disease and diabetes. *Lancet Diabetes Endocrinol.* **2022**, *10*, 284–296. [CrossRef]
4. Li, J.; Ha, A.; Rui, F.; Zou, B.; Yang, H.; Xue, Q.; Hu, X.; Xu, Y.; Henry, L.; Barakat, M.; et al. Meta-analysis: Global prevalence, trend and forecasting of non-alcoholic fatty liver disease in children and adolescents, 2000–2021. *Aliment. Pharmacol. Ther.* **2022**, *56*, 396–406. [CrossRef] [PubMed]
5. Thomas, J.A.; Kendall, B.J.; Dalais, C.; Macdonald, G.A.; Thrift, A.P. Hepatocellular and extrahepatic cancers in non-alcoholic fatty liver disease: A systematic review and meta-analysis. *Eur. J. Cancer* **2022**, *173*, 250–262. [CrossRef] [PubMed]
6. Loomba, R.; Wong, R.; Fraysse, J.; Shreay, S.; Li, S.; Harrison, S.; Gordon, S.C. Nonalcoholic fatty liver disease progression rates to cirrhosis and progression of cirrhosis to decompensation and mortality: A real world analysis of Medicare data. *Aliment. Pharmacol. Ther.* **2020**, *51*, 1149–1159. [CrossRef]
7. Lazarus, J.V.; Mark, H.E.; Anstee, Q.M.; Arab, J.P.; Batterham, R.L.; Castera, L.; Cortez-Pinto, H.; Crespo, J.; Cusi, K.; Dirac, M.A.; et al. Advancing the global public health agenda for NAFLD: A consensus statement. *Nat. Rev. Gastroenterol. Hepatol.* **2022**, *19*, 60–78. [CrossRef] [PubMed]
8. European Association for the Study of the Liver (EASL); European Association for the Study of Diabetes (EASD); European Association for the Study of Obesity (EASO). EASL-EASD-EASO Clinical Practice Guidelines for the management of non-alcoholic fatty liver disease. *J. Hepatol.* **2016**, *64*, 1388–1402. [CrossRef]
9. Alexander, M.; Loomis, A.K.; van der Lei, J.; Duarte-Salles, T.; Prieto-Alhambra, D.; Ansell, D.; Pasqua, A.; Lapi, F.; Rijnbeek, P.; Mosseveld, M.; et al. Risks and clinical predictors of cirrhosis and hepatocellular carcinoma diagnoses in adults with diagnosed NAFLD: Real-world study of 18 million patients in four European cohorts. *BMC Med.* **2019**, *17*, 95. [CrossRef]
10. Younossi, Z.M.; Corey, K.E.; Lim, J.K. AGA Clinical Practice Update on Lifestyle Modification Using Diet and Exercise to Achieve Weight Loss in the Management of Nonalcoholic Fatty Liver Disease: Expert Review. *Gastroenterology* **2021**, *160*, 912–918. [CrossRef]
11. Hashemian, M.; Merat, S.; Poustchi, H.; Jafari, E.; Radmard, A.-R.; Kamangar, F.; Freedman, N.; Hekmatdoost, A.; Sheikh, M.; Boffetta, P.; et al. Red Meat Consumption and Risk of Nonalcoholic Fatty Liver Disease in a Population with Low Meat Consumption: The Golestan Cohort Study. *Am. J. Gastroenterol.* **2021**, *116*, 1667–1675. [CrossRef]
12. Zhang, S.; Fu, J.; Zhang, Q.; Liu, L.; Lu, M.; Meng, G.; Yao, Z.; Wu, H.; Xia, Y.; Bao, X.; et al. Association between habitual yogurt consumption and newly diagnosed non-alcoholic fatty liver disease. *Eur. J. Clin. Nutr.* **2019**, *74*, 491–499. [CrossRef]
13. Eslami, O.; Shidfar, F.; Maleki, Z.; Jazayeri, S.; Hosseini, A.F.; Agah, S.; Ardiyani, F. Effect of Soy Milk on Metabolic Status of Patients with Nonalcoholic Fatty Liver Disease: A Randomized Clinical Trial. *J. Am. Coll. Nutr.* **2018**, *38*, 51–58. [CrossRef] [PubMed]
14. Cespedes, E.M.; Hu, F.B. Dietary patterns: From nutritional epidemiologic analysis to national guidelines. *Am. J. Clin. Nutr.* **2015**, *101*, 899–900. [CrossRef]
15. Strate, L.L.; Keeley, B.R.; Cao, Y.; Wu, K.; Giovannucci, E.L.; Chan, A.T. Western Dietary Pattern Increases, and Prudent Dietary Pattern Decreases, Risk of Incident Diverticulitis in a Prospective Cohort Study. *Gastroenterology* **2017**, *152*, 1023–1030.e2. [CrossRef] [PubMed]
16. Schwedhelm, C.; Iqbal, K.; Knüppel, S.; Schwingshackl, L.; Boeing, H. Contribution to the understanding of how principal component analysis–derived dietary patterns emerge from habitual data on food consumption. *Am. J. Clin. Nutr.* **2018**, *107*, 227–235. [CrossRef] [PubMed]
17. Zadeh, S.H.; Mansoori, A.; Hosseinzadeh, M. Relationship between dietary patterns and non-alcoholic fatty liver disease: A systematic review and meta-analysis. *J. Gastroenterol. Hepatol.* **2021**, *36*, 1470–1478. [CrossRef]
18. Ma, J.; Hennein, R.; Liu, C.; Long, M.T.; Hoffmann, U.; Jacques, P.F.; Lichtenstein, A.H.; Hu, F.B.; Levy, D. Improved Diet Quality Associates with Reduction in Liver Fat, Particularly in Individuals with High Genetic Risk Scores for Nonalcoholic Fatty Liver Disease. *Gastroenterology* **2018**, *155*, 107–117. [CrossRef]
19. Mozaffarian, D. Dietary and Policy Priorities for Cardiovascular Disease, Diabetes, and Obesity: A Comprehensive Review. *Circulation* **2016**, *133*, 187–225. [CrossRef]
20. Vilar-Gomez, E.; Nephew, L.D.; Vuppalanchi, R.; Gawrieh, S.; Mladenovic, A.; Pike, F.; Samala, N.; Chalasani, N. High-quality diet, physical activity, and college education are associated with low risk of NAFLD among the US population. *Hepatology* **2022**, *75*, 1491–1506. [CrossRef]
21. Khalatbari-Soltani, S.; Imamura, F.; Brage, S.; De Lucia Rolfe, E.; Griffin, S.J.; Wareham, N.J.; Marques-Vidal, P.; Forouhi, N.G. The association between adherence to the Mediterranean diet and hepatic steatosis: Cross-sectional analysis of two independent studies, the UK Fenland Study and the Swiss CoLaus Study. *BMC Med.* **2019**, *17*, 19. [CrossRef]
22. Adriano, L.S.; de Carvalho Sampaio, H.A.; Arruda, S.P.M.; de Melo Portela, C.L.; de Melo, M.L.P.; Carioca, A.A.F.; Soares, N.T. Healthy dietary pattern is inversely associated with non-alcoholic fatty liver disease in elderly. *Br. J. Nutr.* **2016**, *115*, 2189–2195. [CrossRef] [PubMed]
23. Yang, C.-Q.; Shu, L.; Wang, S.; Wang, J.-J.; Zhou, Y.; Xuan, Y.-J.; Wang, S.-F. Dietary Patterns Modulate the Risk of Non-Alcoholic Fatty Liver Disease in Chinese Adults. *Nutrients* **2015**, *7*, 4778–4791. [CrossRef] [PubMed]
24. Oddy, W.H.; Herbison, C.E.; Jacoby, P.; Ambrosini, G.L.; O'Sullivan, T.A.; Ayonrinde, O.T.; Olynyk, J.K.; Black, L.J.; Beilin, L.J.; Mori, T.A.; et al. The Western Dietary Pattern Is Prospectively Associated With Nonalcoholic Fatty Liver Disease in Adolescence. *Am. J. Gastroenterol.* **2013**, *108*, 778–785. [CrossRef]

25. Said, M.A.; Verweij, N.; Van Der Harst, P. Associations of Combined Genetic and Lifestyle Risks with Incident Cardiovascular Disease and Diabetes in the UK Biobank Study. *JAMA Cardiol.* **2018**, *3*, 693–702. [CrossRef] [PubMed]
26. Zhuang, P.; Liu, X.; Li, Y.; Wan, X.; Wu, Y.; Wu, F.; Zhang, Y.; Jiao, J. Effect of Diet Quality and Genetic Predisposition on Hemoglobin A1c and Type 2 Diabetes Risk: Gene-Diet Interaction Analysis of 357,419 Individuals. *Diabetes Care* **2021**, *44*, 2470–2479. [CrossRef]
27. Zhao, J.; Li, Z.; Gao, Q.; Zhao, H.; Chen, S.; Huang, L.; Wang, W.; Wang, T. A review of statistical methods for dietary pattern analysis. *Nutr. J.* **2021**, *20*, 37. [CrossRef]
28. Murakami, K.; Shinozaki, N.; Fujiwara, A.; Yuan, X.; Hashimoto, A.; Fujihashi, H.; Wang, H.-C.; Livingstone, M.B.E.; Sasaki, S. A Systematic Review of Principal Component Analysis–Derived Dietary Patterns in Japanese Adults: Are Major Dietary Patterns Reproducible within a Country? *Adv. Nutr.* **2019**, *10*, 237–249. [CrossRef]
29. Orr, J.G.; Homer, T.; Ternent, L.; Newton, J.; McNeil, C.J.; Hudson, M.; Jones, D.E. Health related quality of life in people with advanced chronic liver disease. *J. Hepatol.* **2014**, *61*, 1158–1165. [CrossRef] [PubMed]
30. Kanwal, F.; Kramer, J.R.; Li, L.; Dai, J.; Natarajan, Y.; Yu, X.; Asch, S.M.; El-Serag, H.B. Effect of Metabolic Traits on the Risk of Cirrhosis and Hepatocellular Cancer in Nonalcoholic Fatty Liver Disease. *Hepatology* **2020**, *71*, 808–819. [CrossRef]
31. Zelber-Sagi, S.; Salomone, F.; Mlynarsky, L. The Mediterranean dietary pattern as the diet of choice for non-alcoholic fatty liver disease: Evidence and plausible mechanisms. *Liver Int. Off. J. Int. Assoc. Study Liver* **2017**, *37*, 936–949. [CrossRef] [PubMed]
32. Salomone, F.; Godos, J.; Zelber-Sagi, S. Natural antioxidants for non-alcoholic fatty liver disease: Molecular targets and clinical perspectives. *Liver Int. Off. J. Int. Assoc. Study Liver* **2016**, *36*, 5–20. [CrossRef] [PubMed]
33. Sekiya, M.; Yahagi, N.; Matsuzaka, T.; Najima, Y.; Nakakuki, M.; Nagai, R.; Ishibashi, S.; Osuga, J.; Yamada, N.; Shimano, H. Polyunsaturated fatty acids ameliorate hepatic steatosis in obese mice by SREBP-1 suppression. *Hepatology* **2003**, *38*, 1529–1539. [CrossRef] [PubMed]
34. Levy, J.R.; Clore, J.N.; Stevens, W. Dietary n-3 polyunsaturated fatty acids decrease hepatic triglycerides in Fischer 344 rats. *Hepatology* **2004**, *39*, 608–616. [CrossRef]
35. Naowaboot, J.; Piyabhan, P.; Munkong, N.; Parklak, W.; Pannangpetch, P. Ferulic acid improves lipid and glucose homeostasis in high-fat diet-induced obese mice. *Clin. Exp. Pharmacol. Physiol.* **2016**, *43*, 242–250. [CrossRef]
36. Kathirvel, E.; Morgan, K.; Nandgiri, G.; Sandoval, B.C.; Caudill, M.A.; Bottiglieri, T.; French, S.W.; Morgan, T.R. Betaine improves nonalcoholic fatty liver and associated hepatic insulin resistance: A potential mechanism for hepatoprotection by betaine. *Am. J. Physiol. Gastrointest Liver Physiol.* **2010**, *299*, G1068–G1077. [CrossRef]
37. Rose, D.J. Impact of whole grains on the gut microbiota: The next frontier for oats? *Br. J. Nutr.* **2014**, *112* (Suppl. S2), S44–S49. [CrossRef]
38. Federico, A.; Dallio, M.; Godos, J.; Loguercio, C.; Salomone, F. Targeting gut-liver axis for the treatment of nonalcoholic steatohepatitis: Translational and clinical evidence. *Transl. Res.* **2016**, *167*, 116–124. [CrossRef]

nutrients

MDPI

Article

Associations between Consumption of Dietary Fibers and the Risk of Type 2 Diabetes, Hypertension, Obesity, Cardiovascular Diseases, and Mortality in Chinese Adults: Longitudinal Analyses from the China Health and Nutrition Survey

Zhaoxia Zhang [1,2,†], Bo Chen [1,†], Jingjing Zeng [1], Menglin Fan [3], Wenlei Xu [1], Xiaying Li [4], Ying Xing [2,*] and Shaoyong Xu [1,5,*]

1 Center for Clinical Evidence-Based and Translational Medicine, Xiangyang Central Hospital, Affiliated Hospital of Hubei University of Arts and Science, Xiangyang 441021, China; zhaoxia19890306@163.com (Z.Z.); chenbofhc@163.com (B.C.); zengjj9@mail2.sysu.edu.cn (J.Z.); xu2972047610@163.com (W.X.)
2 Department of Endocrinology, Daxing Hospital, Xi'an 710000, China
3 School of Public Health, Tongji Medical College, Huazhong University of Science and Technology, Wuhan 430000, China; fanmenglin08@163.com
4 College of Medicine, Wuhan University of Science and Technology, Wuhan 430065, China; lixiaying0808@163.com
5 Department of Endocrinology, Xiangyang Central Hospital, Affiliated Hospital of Hubei University of Arts and Science, Xiangyang 441021, China
* Correspondence: yoji_xu@hotmail.com (S.X.); drxingying@hotmail.com (Y.X.); Tel.: +86-186-2968-0357 (S.X.); +86-139-919-55717 (Y.X.)
† These authors contributed equally to this work.

Citation: Zhang, Z.; Chen, B.; Zeng, J.; Fan, M.; Xu, W.; Li, X.; Xing, Y.; Xu, S. Associations between Consumption of Dietary Fibers and the Risk of Type 2 Diabetes, Hypertension, Obesity, Cardiovascular Diseases, and Mortality in Chinese Adults: Longitudinal Analyses from the China Health and Nutrition Survey. *Nutrients* **2022**, *14*, 2650. https://doi.org/10.3390/nu14132650

Academic Editors: Andriana Kaliora, Chara Tzavara and Charalampia Amerikanou

Received: 29 May 2022
Accepted: 23 June 2022
Published: 27 June 2022

Publisher's Note: MDPI stays neutral with regard to jurisdictional claims in published maps and institutional affiliations.

Abstract: Although many studies have explored the relationship between total dietary fiber intake and the risk of chronic non-communicable diseases, the results are mixed. There is also a lack of research on the association between dietary fiber intake from different food sources and disease. Using data from the China Nutrition and Health Database from 2004 to 2015, Cox proportional risk models were used to explore the associations between total dietary fiber and fiber intake from different food sources and the occurrence of type 2 diabetes, hypertension, obesity, cardiovascular disease, and all-cause mortality. After multi-factorial adjustment, the hazard ratios (95% confidence interval) of total dietary fiber intake (quartile 4 vs. quartile 1) in type 2 diabetes, hypertension, obesity, cardiovascular disease, and all-cause mortality cohorts were 1.20 (0.93, 1.55), 0.91 (0.75, 1.12), 0.93 (0.64, 1.35), 1.13 (0.60, 2.12), 1.13 (0.60, 2.12), and 1.13 (0.84, 1.52). Whole-grain fiber intake was positively associated with hypertension but not with the occurrence of other diseases. No association was observed between legume fibers, fruit fibers, and vegetable fibers in the cohorts of type 2 diabetes, hypertension, obesity, cardiovascular diseases and all-cause mortality. Our study did not find any association between total dietary fiber and dietary fiber intake from different food sources and type 2 diabetes, obesity, cardiovascular disease, and all-cause mortality in the Chinese population. The role of dietary fiber in the Chinese population may be overestimated. More extraordinary efforts are needed to further confirm the association between dietary fiber and these diseases in the Chinese population.

Keywords: dietary fiber; China nutrition and health database; chronic non-communicable diseases; population-based cohort study

1. Introduction

Chronic non-communicable diseases such as diabetes, hypertension, obesity, and cardiovascular disease (CVD) have high incidence and mortality rates worldwide, with type 2 diabetes (T2D) and cardiovascular disease accounting for 11.3% and 30% of total deaths worldwide, respectively [1,2]. The incidence of adult obesity is about 20% [3]. More than 1.39 billion people suffer from hypertension [4], and the public health risks caused by these diseases should not be disregarded.

Many experts have focused their research on preventing the onset of chronic diseases. Dietary fiber consumption has been linked to the prevention of various chronic diseases [5–9]. Dietary fiber is a type of nutrient that is difficult to digest and absorb in the small intestine, and it has several potential mechanisms for preventing chronic diseases, including (1) increasing satiety and possibly promoting weight loss; (2) encouraging intestinal microorganisms to produce short-chain fatty acids with immunomodulatory and anti-inflammatory properties; and (3) experiments in animal models have shown that dietary fiber intake is associated with lower levels of inflammatory and oxidative stress markers [10–12]. However, with the continuous refinement and classification of the concept of dietary fiber, studies have found that dietary fiber from different food sources is not all negatively associated with the development of disease [5,7,8,13].

Although more studies have explored the relationship between total dietary fiber intake and the risk of T2D, hypertension, obesity, cardiovascular disease, and all-cause mortality, the results are mixed [5,6,8,14–18]. There is a lack of research on the association between dietary fiber intake from different food sources and disease, especially hypertension, and only a few studies focus on all-cause mortality [9,13,14]. Furthermore, while T2D and cardiovascular disease research are becoming more common, the results from studies of fiber from various dietary sources are also inconsistent [5,7,13]. More importantly, no research has examined the association between dietary fiber intake from various food sources and chronic disease in the Chinese population. Therefore, considering the heterogeneity found in the total dietary fiber intake and dietary fiber intake from different food sources in different populations and its effects on the health statuses, these associations may require further analysis in Chinese populations that traditionally consume diets rich in plant foods.

This study aimed to determine the association between fiber consumption from various dietary sources and T2D, hypertension, obesity, cardiovascular disease, and all-cause mortality in the Chinese population.

2. Materials and Methods

2.1. Study Design and Participants

Data from the China Nutrition and Health Database (CHNS) were used in this study, a multi-purpose longitudinal open cohort study that began in 1989 and was followed up in 1991, 1993, 1997, 2000, 2004, 2006, 2009, 2011, and 2015. CHNS used a multi-stage stratified whole-group random sampling strategy to sample people from 15 provinces with different populations, geography, economy and public resources in the eastern, central and western regions of China. Demographic, socioeconomic, lifestyle, dietary, and health data is collected during each survey wave. Blood samples were collected and analyzed during the 2009 CHNS. The scientific rationale and design of the CHNS were previously reported [19,20]. The survey was conducted according to the guidelines of the Declaration of Helsinki and was approved by the Institutional Review Committees of the University of North Carolina at Chapel Hill (UNC-CH) and the National Institute for Nutrition and Health, Chinese Center for Disease Control and Prevention; each participant provided informed consent.

Because new food codes were implemented in 2004, CHNS data from 2004 to 2015 were used in this analysis. We first excluded participants who were under 18 years old at baseline, participants with abnormal dietary energy data (e.g., a daily energy intake <800 kcal or >4200 kcal for men and <600 kcal or >3500 kcal for women), and participants with a history of cardiovascular disease, cancer, death, or pregnancy at baseline. The initial wave of these surveys served as a baseline. Finally, the final analysis comprised a total of 9376 participants.

2.2. Dietary Assessment

Details on dietary measures are available elsewhere [21]. In brief, at the time of the baseline (2004 or 2006) survey, qualified investigators obtained individual dietary data through face-to-face interviews in each survey round. Individual diets were evaluated several times by recalling the food consumed by individuals during three 24-hour periods. Participants were asked to list all foods and beverages consumed during 24 h, with the types and amounts of items documented based on food models and images, supported by standard sizes (household containers, grams indicated on the packaging). Three consecutive days (2 weekdays and 1 weekend day) were assigned at random during the week, and each sampling unit was nearly balanced across the seven days of the week. The Chinese Food Composition Table (2004 edition) was used to compute dietary component intakes (energy, protein, fat, carbohydrate, dietary fiber), and food groups such as cereals, legumes, vegetables, and fruits were also categorized using the Chinese Food Composition Table. This study did not cover supplement consumption.

To ensure the stability of the dietary data, we used average year data for 2004 and 2006 as the baseline dietary assessment. When participants participated in only one of the survey years, the current year's dietary intake could be utilized as the baseline assessment. This research method can also be seen in other articles [13,22].

2.3. Measurement

Questionnaires were used to collect demographics and lifestyle information, such as age, gender, cigarette and alcohol use, education, physical activity, place of residence, and area. Physical activity was estimated as the duration of total physical activity and reported as the metabolic equivalent task (MET)-minutes/week. Height and weight were assessed using conventional protocols and calibrated equipment by certified medical practitioners. Body mass index (BMI) was calculated as weight (kg)/squared height (m^2). Professional researchers estimated blood pressure as the mean of three independent measurements.

2.4. Case Ascertainment

T2D. A questionnaire-based interview was used to establish diabetes status at each follow-up appointment. Blood samples were also taken and examined in 2009. T2D was characterized as meeting at least one of the following criteria in the 2009 study, according to the American Diabetes Association diagnostic criteria: (1) fasting blood glucose concentration $\geq$ 7.0 mmol/L (126 mg/dL); (2) HbA1c $\geq$ 6.5%; (3) self-reported T2D diagnosis or use of hypoglycemic medicine. T2D was defined as self-reported diabetes or use of hypoglycemic medication in 2015. Previous research demonstrated that self-reported diabetes is a relatively useful approach for determining the diabetes status of Chinese study participants [23].

Hypertension. At 10 min of rest, a standard mercury sphygmomanometer was used to monitor diastolic and systolic blood pressures on subjects' left or right arms, and the mean readings were measured three times and recorded. Hypertension was defined as (1) a systolic blood pressure $\geq$140 mm Hg; (2) a diastolic blood pressure $\geq$90 mm Hg; or (3) self-reporting of a diagnosis of hypertension or currently on oral anti-hypertensive medication during follow-up.

Obesity. Each participant's weight and height were measured by a trained health worker. BMI $\geq$ 28.0 kg/m^2 was considered obese according to the Expert Consensus on Weight Management Process for Overweight or Obese People (2021).

Cardiovascular disease. Myocardial infarction or stroke was used to characterize the cardiovascular disease. The following questions were used to determine myocardial infarction and stroke information: "Has your doctor ever told you that you have a myocardial infarction?". "Have you been diagnosed with stroke by your doctor?" cardiovascular disease was defined as answering yes to any of these questions.

All-cause mortality. The census confirmed participants' death status based on information submitted in each survey wave and the household system. The year of death was recorded if the participant died.

2.5. Statistical Analysis

Descriptive analyses were reported as the mean ± standard deviation (SD) or median (interquartile range) for continuous variables and frequency (percentage) for categorical variables. ANOVA, Kruskal–Wallis test or chi-square test showed statistical differences between quartiles.

We used Cox proportional risk models to assess the association between total dietary fiber and fiber intake from different food sources and the occurrence of type 2 diabetes, hypertension, obesity, cardiovascular disease, and all-cause mortality. The time indicator was the follow-up time from 2004 to the disease onset or cut-off date. To test for potential nonlinear associations, we tested for linear trends using the median score per quantile. All statistical tests were two-sided and performed using SAS 9.4 (SAS Institute, Cary, NC, USA). $p < 0.05$ was considered statistically significant.

3. Results

3.1. Description of the Study Population

For each particular disease, 9376 participants were eligible; the baseline data excluded cases with existing disease and cases with lost follow-up or missing outcomes, as shown in Figure 1. A total of 6886 cases in the T2D cohort, 3838 cases in the hypertension cohort, 4115 cases in the obesity cohort, 4932 cases in the cardiovascular disease cohort, and 8307 cases in the all-cause mortality cohort were finally included. The T2D cohort of participants was divided into quartiles based on dietary fiber intake, and their baseline characteristics are shown in Table 1. Notably, only 25% of participants had a daily dietary fiber intake of more than 13.5 g/day, which is well below the 25–30 g/day recommended in the Chinese Dietary Guidelines 2022 [24]. Baseline characteristics of hypertension, obesity, cardiovascular disease, and all-cause mortality cohorts are shown in Tables S1–S4 in the Supplementary Materials.

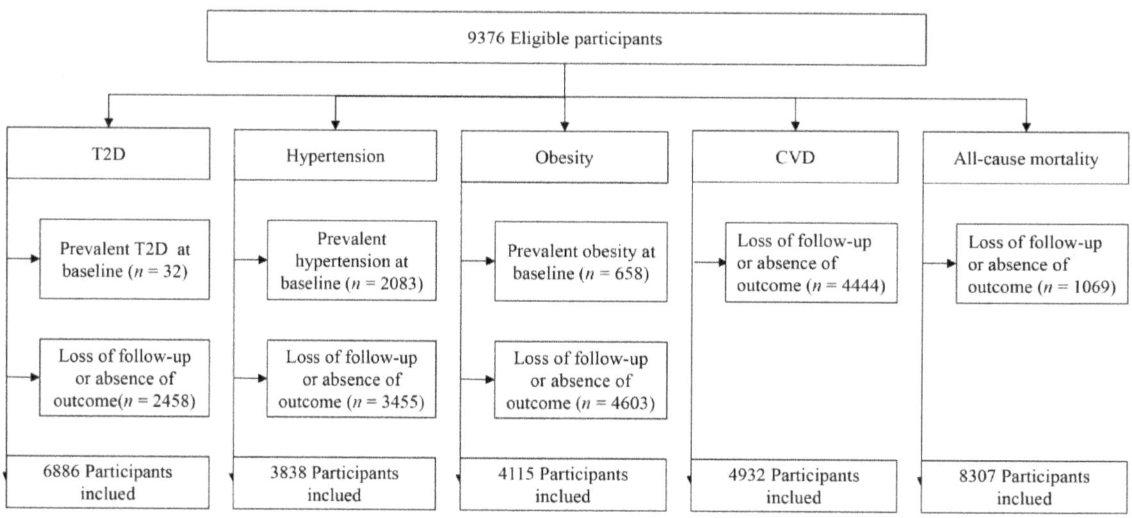

Figure 1. Participants flowchart.

Table 1. Characteristics of the study population at inclusion in the T2D cohort, according to quartiles of total dietary fiber intakes.

Characteristics	Fiber Intake (g/d)				p
	Quartile 1 (<6.21)	Quartile 2 (6.21–8.81)	Quartile 3 (8.81–13.50)	Quartile 4 (>13.50)	
Number	1611	1718	1799	1778	
Age (years)	47.62 ± 14.89	46.93 ± 13.71	47.44 ± 13.41	47.77 ± 13.34	0.307
Male, n (%)	667 (42.14)	797 (46.69)	865 (49.01)	931 (52.57)	<0.001
BMI (kg/m^2)	22.97 ± 3.33	23.04 ± 3.41	23.25 ± 3.28	23.10 ± 3.29	0.097
Waist circumference (cm)	80.04 ± 9.98	80.44 ± 9.85	81.34 ± 9.78	81.32 ± 9.58	0.001
Systolic blood pressure (mmHg)	122.46 ± 17.89	121.43 ± 17.35	121.97 ± 18.70	121.04 ± 17.29	0.136
Diastolic blood pressure (mmHg)	79.24 ± 10.79	78.58 ± 10.47	78.57 ± 11.94	78.11 ± 10.92	0.048
Physical activity (MET-MIN/day)	274.28 (0.00–1531.29)	978.96 (0.00–1992.86)	993.01 (0.00–2090.14)	1079.04 (0.00–2220.01)	<0.001
Smoking, n (%)					
Yes	461 (29.21)	555 (32.44)	613 (34.61)	628 (35.44)	0.006
No	1117 (70.79)	1156 (67.56)	1158 (65.39)	1144 (64.56)	
Alcohol drinking, n (%)					
Yes	444 (28.14)	555 (32.44)	638 (36.09)	649 (36.60)	<0.001
No	1134 (71.86	1156 (67.56)	1130 (63.91)	1124 (63.40)	
Educational level, n (%)					0.004
Primary school or below	699 (44.16)	768 (44.99)	835 (47.34)	916 (51.72)	
Junior high school	497 (31.40)	558 (32.69)	571 (32.37)	515 (29.08)	
Senior high school	237 (14.97)	229 (13.42)	214 (12.13)	211 (11.91)	
College and above	150 (9.48)	152 (8.90)	144 (8.16)	129 (7.28)	
Regions, n (%)					
Urban	600 (37.24)	535 (31.14)	531 (29.85)	444 (24.97)	<0.001
Rural	1011 (62.76)	1183 (68.86)	1248 (70.15)	1334 (75.03)	
Total energy intake (kcal/day)	1869.11 ± 547.14	2115.88 ± 550.74	2280.64 ± 576.40	2479.94 ± 598.07	<0.001
Total carbohydrate intake (g/d)	263.19 ± 79.91	311.06 ± 85.94	338.77 ± 100.03	378.01 ± 108.77	<0.001
Total protein intake (g/d)	55.26 ± 18.97	62.11 ± 19.08	69.03 ± 20.59	76.46 ± 25.01	<0.001
Total fatty intake (g/d)	63.99 ± 35.09	66.95 ± 35.87	68.67 ± 35.74	70.90 ± 37.45	<0.001
Na intake (mg/d)	60.91 (14.41–485.21)	52.81 (16.20–829.61)	60.60 (16.63–1130.85)	49.61 (16.74–850.52)	<0.001
Whole-grain fiber intake (g/d)	0.37 ± 0.79	0.72 ± 1.26	1.18 ± 1.99	2.58 ± 5.49	<0.001
Legume fiber intake (g/d)	0.11 ± 0.32	0.33 ± 0.74	0.76 ± 1.53	2.56 ± 5.52	<0.001
Vegetable fiber intake (g/d)	2.01 ± 0.96	3.37 ± 1.32	4.82 ± 2.36	10.86 ± 9.53	<0.001
Fruit fiber intake (g/d)	0.10 ± 0.33	0.21 ± 0.59	0.37 ± 0.93	0.48 ± 1.37	<0.001

Descriptive analyses of continuous variables were conducted by means ± standard deviations (SD) or medians (interquartile range), and categorical variables were described by number (percentage). Analysis of variance or Kruskal–Wallis test was used for continuous variables, and chi-square test was used for categorical variables. BMI: body mass index.

3.2. *Associations between Total Dietary Fiber Intake and T2D, Hypertension, Obesity, CVD, and All-Cause Mortality*

Among the 6886 participants with T2D followed up, 650 cases developed T2D, and the incidence density (1000 person-years) of T2D by quartiles of total dietary fiber intake was 7.81, 9.28, 9.38, and 9.61 from quartile 1 (Q1) to quartile 4 (Q4), respectively, with a hazard ratio $(HR)_{Q4\ vs.\ Q1}$ and 95% confidence interval (95% CI) of 1.22 (0.97, 1.53), $p > 0.05$. Among the 3838 participants with hypertension at follow-up, a total of 1178 developed hypertension, and the incidence density (1000 person-years) of hypertension by quartiles of total dietary fiber intake was 29.21, 29.35, 28.78, and 28.81 for Q1–Q4. $HR_{Q4\ vs.\ Q1}$ was 0.99 (0.84, 1.17), $p > 0.05$. In the follow-up cohort of 4115 cases with obesity, 379 cases developed obesity, and the incidence density (1000 person-years) of obesity in Q1–Q4 was 7.43, 8.67, 9.01, and 9.23, and $HR_{Q4\ vs.\ Q1}$ was 1.11 (95% CI: 0.82, 1.50), $p > 0.05$. In the follow-up cohort of 4932 cases with cardiovascular disease cohort, a total of 127 cases developed cardiovascular disease, and the incidence density (1000 person-years) of cardiovascular disease in Q1–Q4 was 1.79, 2.47, 2.48, and 2.23, respectively; $HR_{Q4\ vs.\ Q1}$ was 1.25 (95% CI: 0.73, 2.15), $p > 0.05$. In the followed-up cohort of 8307 all-cause deaths, a total of 468 cases of mortality occurred, and the incidence density (1000 person-years) of mortality from Q1–Q4, it was 6.16, 4.68, 4.45, and 5.72, respectively, with an $HR_{Q4\ vs.\ Q1}$ of 0.93 (95% CI: 0.73, 1.18), $p > 0.05$.

After multi-factorial adjustment, $HR_{Q4\ vs.\ Q1}$ was 1.20 (95% CI: 0.93,1.55), 0.91 (95% CI: 0.75, 1.12), 0.93 (95% CI: 0.64, 1.35), 1.13 (95% CI: 0.60, 2.12), and 1.13 (95%CI: 0.84, 1.52) in the cohorts with diabetes, hypertension, obesity, cardiovascular disease, and all-cause mortality, respectively; *p*-trend > 0.05 (see Table 2, Figure 2).

Table 2. Multi-variable adjusted HRs (95% CI) of T2D, hypertension, obesity, CVD, and all-cause mortality according to quartiles of total dietary fiber.

	Total Fiber Intake				
	Quartile 1	**Quartile 2**	**Quartile 3**	**Quartile 4**	***p*-Trend**
		T2D			
Number of cases	132	166	174	178	
Person-years	16,902	17,888	18,545	18,515	
Incidence density (1000 person-years)	7.81	9.28	9.38	9.61	
T2D [a]	1.0	1.18 (0.94, 1.48)	1.19 (0.95, 1.50)	1.22 (0.97, 1.53)	0.12
T2D [b]	1.0	1.12 (0.88, 1.43)	1.13 (0.88, 1.44)	1.20 (0.93, 1.55)	0.08
		Hypertension			
Number of cases	270	297	307	304	
Person-years	9241	10,116	10,667	10,552	
Incidence density (1000 person-years)	29.21	29.35	28.78	28.81	
Hypertension [a]	1.0	1.01 (0.86, 1.19)	0.99 (0.84, 1.17)	0.99 (0.84, 1.17)	0.63
Hypertension [c]	1.0	0.98 (0.82, 1.18)	0.86 (0.71, 1.04)	0.91 (0.75, 1.12)	0.25
		Obesity			
Number of cases	75	98	109	97	
Person-years	10,087	11,308	12,089	11,781	
Incidence density (1000 person-years)	7.43	8.67	9.01	8.23	
Obesity [a]	1.0	1.17 (0.86, 1.57)	1.21 (0.90, 1.63)	1.11 (0.82, 1.50)	0.14
Obesity [c]	1.0	1.09 (0.78, 1.52)	1.02 (0.73, 1.44)	0.93 (0.64, 1.35)	0.40

Table 2. *Cont.*

	Total Fiber Intake				
	Quartile 1	**Quartile 2**	**Quartile 3**	**Quartile 4**	***p*-Trend**
		CVD			
Number of cases	22	38	35	32	
Person-years	12,276	15,363	14,124	14,289	
Incidence density (1000 person-years)	1.79	2.47	2.48	2.23	
CVD [a]	1.0	1.56 (0.93, 2.64)	1.38 (0.81, 2.36)	1.25 (0.73, 2.15)	0.34
CVD [c]	1.0	1.56 (0.90, 2.70)	1.24 (0.70, 2.23)	1.13 (0.60, 2.12)	0.07
		All-cause mortality			
Number of cases	133	105	101	129	
Person-years	21,571	22,398	22,663	22,543	
Incidence density (1000 person-year)	6.16	4.68	4.45	5.72	
All-cause mortality [a]	1.0	0.76 (0.59, 0.98)	0.72 (0.56, 0.94)	0.93 (0.73, 1.18)	0.11
All-cause mortality [c]	1.0	1.05 (0.79, 1.38)	0.95 (0.70, 1.27)	1.13 (0.84, 1.52)	0.14

HRs were examined using Cox proportional hazard models. [a]: Confounding factors were not adjusted. [b]: Adjusted for age, sex, BMI, education, regions, physical activity, smoking status, alcohol drinking, total energy intake, total carbohydrate intake, protein intake, and fatty intake. [c]: Adjusted for age, sex, BMI, education, regions, physical activity, smoking status, alcohol drinking, total energy intake, total carbohydrate intake, protein intake, fatty intake, systolic blood pressure, diastolic blood pressure, and Na intake.

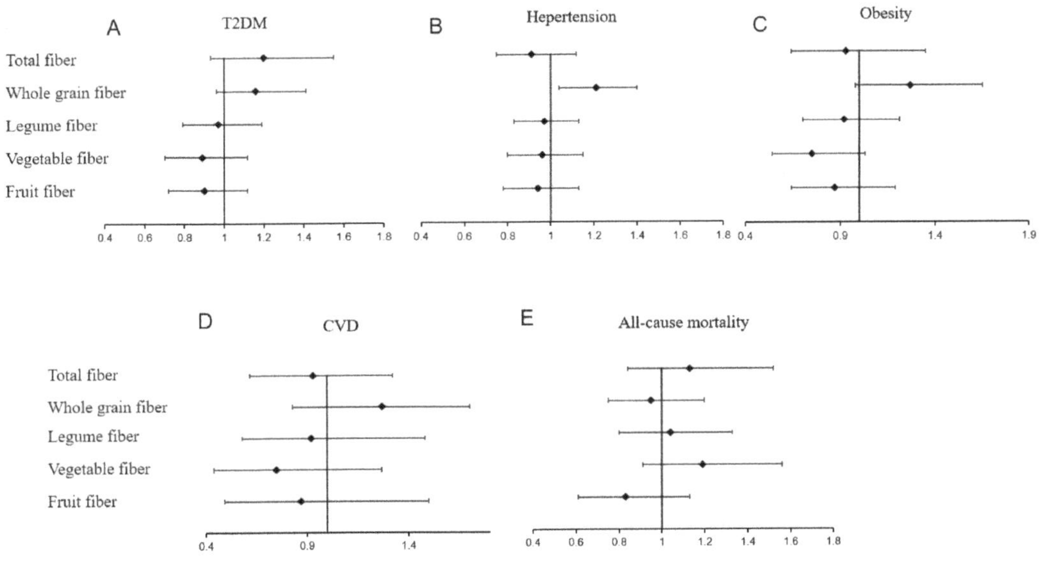

Figure 2. Associations between the consumption of dietary fibers from different sources and (**A**) T2DM, (**B**) hypertension, (**C**) obesity, (**D**) CVD, (**E**) mortality from all causes.

3.3. Associations between Whole-Grain Fiber Intake and T2D, Hypertension, Obesity, CVD, and All-Cause Mortality

The whole-grain fiber intake was divided into terciles (T1–T3). In the cohort of T2D, hypertension, obesity, CVD and all-cause mortality, the $HR_{T3\ vs.\ T1}$ were 1.31 (95% CI: 1.09, 1.57), 1.29 (95% CI: 1.13, 1.47), 1.62 (95% CI: 1.28, 2.05), 1.43 (95% CI: 0.97, 2.12), and 0.87 (95% CI: 0.69, 1.08), $p > 0.05$. With the addition of fruit fiber, vegetable fiber, and legume fiber on the basis of the adjustment for multiple factors of total dietary fiber, $HR_{T3\ vs.\ T1}$ in the cohorts of T2D, hypertension, obesity, CVD, and all-cause mortality were 1.16 (95% CI: 0.96, 1.41), 1.21 (95% CI: 1.04, 1.40), 1.27 (95% CI: 0.98, 1.65), 1.29 (95% CI: 0.85, 1.96), and 0.95 (95% CI: 0.75, 1.20), $p > 0.05$ (Table 3, Figure 2). The results show that whole-grain fiber is positively associated with the incidence of T2D, hypertension, and obesity in the unadjusted model and not with the incidence of cardiovascular disease and all-cause mortality. After multi-factorial adjustment, whole-grain fiber intake was positively associated with hypertension but not with the onset of other diseases.

Table 3. Multi-variable adjusted HRs (95%CI) of T2D, hypertension, obesity, CVD, and all-cause mortality according to terciles of whole-grain fiber.

	Whole-Grain Fiber			
	Tercile 1	**Tercile 2**	**Tercile 3**	***p*-Trend**
T2D [a]	1.00	1.33 (1.11, 1.62)	1.31 (1.09, 1.57)	0.06
T2D [b1]	1.00	1.16 (0.94, 1.43)	1.16 (0.96, 1.41)	0.12
Hypertension [a]	1.00	1.13 (0.96, 1.31)	1.29 (1.13, 1.47)	0.04
Hypertension [c1]	1.00	1.05 (0.88, 1.25)	1.21 (1.04, 1.40)	0.03
Obesity [a]	1.00	1.67 (1.29, 2.15)	1.62 (1.28, 2.05)	0.73
Obesity [c1]	1.00	1.23 (0.91, 1.65)	1.27 (0.98, 1.65)	0.55
CVD [a]	1.00	1.04 (0.64, 1.70)	1.43 (0.97, 2.12)	0.16
CVD [c1]	1.00	0.94 (0.56, 1.59)	1.29 (0.85, 1.96)	0.52
All-cause mortality [a]	1.00	0.83 (0.65, 1.06)	0.87 (0.69, 1.08)	0.14
All-cause mortality [c1]	1.00	0.98 (0.75, 1.28)	0.95 (0.75, 1.20)	0.09

[a]: Confounding factors were not adjusted. [b1]: HRs were examined using Cox proportional hazard models. adjusted for age, sex, BMI, education, regions physical activity, smoking status, alcohol drinking, total energy intake, total carbohydrate intake, protein intake, fatty intake, legume fiber, fruit fiber, and vegetable fiber. [c1]: HRs were examined using Cox proportional hazard models adjusted for age, sex, BMI, education, regions, physical activity, smoking status, alcohol drinking, total energy intake, total carbohydrate intake, protein intake, fatty intake, systolic blood pressure, diastolic blood pressure, Na intake, legume fiber, fruit fiber, and vegetable fiber.

3.4. Associations between Legume Fiber Intake and T2D, Hypertension, Obesity, CVD, and All-Cause Mortality

In the terciles of legume fiber intake, The $HR_{T3\ vs.\ T1}$ in the cohorts of T2D, hypertension, obesity, CVD, and all-cause mortality were 1.12 (95% CI: 0.93, 1.33), 1.07 (95% CI: 0.94, 1.22), 1.06 (95%CI: 0.84, 1.33), 1.06 (95%CI: 0.70, 1.60), and 0.86 (95%CI: 0.69, 1.07), $p > 0.05$. After adding fruit fiber, vegetable fiber, and whole-grain fiber to adjust for the multiple factors of total dietary fiber, $HR_{T3\ vs.\ T1}$ in T2D, hypertension, obesity, CVD and all-cause mortality cohorts were 0.97 (95% CI: 0.79, 1.19), 0.97 (95% CI: 0.83, 1.13), 0.92 (95% CI: 0.70, 1.21), 0.90 (95% CI: 0.56, 1.46), and 1.04 (95% CI: 0.80, 1.33), $p > 0.05$ (Table 4, Figure 2). The results show that legume fiber was not associated with the development of T2D, hypertension, obesity, CVD and all-cause mortality.

Table 4. Multi-variable adjusted HR (95% CI) of T2D, hypertension, obesity, CVD, and all-cause mortality according to terciles of legume fiber.

	Legume Fiber			
	Tercile 1	**Tercile 2**	**Tercile 3**	***p*-Trend**
T2D [a]	1.00	1.15 (0.92, 1.43)	1.12 (0.93, 1.33)	0.43
T2D [b2]	1.00	1.09 (0.87, 1.38)	0.97 (0.79, 1.19)	0.24
Hypertension [a]	1.00	0.95 (0.79, 1.15)	1.07 (0.94, 1.22)	0.37
Hypertension [c2]	1.00	1.01 (0.82, 1.24)	0.97 (0.83, 1.13)	0.16
Obesity [a]	1.00	0.84 (0.59, 1.20)	1.06 (0.84, 1.33)	0.20
Obesity [c2]	1.00	0.84 (0.57, 1.25)	0.92 (0.70, 1.21)	0.08
CVD [a]	1.00	1.06 (0.63, 1.76)	1.06 (0.70, 1.60)	0.12
CVD [c2]	1.00	1.09 (0.64, 1.85)	0.90 (0.56, 1.46)	0.67

Table 4. *Cont.*

	Legume Fiber			
	Tercile 1	Tercile 2	Tercile 3	*p*-Trend
All-cause mortality [a]	1.00	0.91 (0.70, 1.20)	0.86 (0.69, 1.07)	0.19
All-cause mortality [c2]	1.00	1.07 (0.81, 1.42)	1.04 (0.80, 1.33)	0.46

[a]: Confounding factors were not adjusted. [b2]: HRs were examined using Cox proportional hazard models. Adjusted for age, sex, BMI, education, regions physical activity, smoking status, alcohol drinking, total energy intake, total carbohydrate intake, protein intake, fatty intake, whole-grain fiber, fruit fiber, and vegetable fiber. [c2]: HRs were examined using Cox proportional hazard models. Adjusted for age, sex, BMI, education, regions, physical activity, smoking status, alcohol drinking, total energy intake, total carbohydrate intake, protein intake, fatty intake, systolic blood pressure, diastolic blood pressure, Na intake, whole-grain fiber, fruit fiber, and vegetable fiber.

3.5. Associations between Vegetable Fiber Intake and T2D, Hypertension, Obesity, CVD, and All-Cause Mortality

After vegetable fiber intake by quartiles classification, $HR_{Q4\ vs.\ Q1}$ in the cohorts of T2D, hypertension, obesity, CVD, and all-cause mortality were 0.85 (95% CI: 0.68, 1.06), 0.90 (95% CI: 0.77, 1.06), 0.72 (95% CI: 0.54, 0.96), 0.85 (95% CI: 0.53, 1.36), and 1.14 (95% CI: 0.89, 1.46), respectively, $p > 0.05$. After adding fruit fiber, legume fiber, and whole-grain fiber to adjust for multiple factors of total dietary fiber, $HR_{Q3\ vs.\ Q1}$ in the cohorts of T2D, hypertension, obesity, CVD and all-cause mortality were 0.89 (95% CI: 0.70, 1.12), 0.96 (95% CI: 0.80, 1.15), 0.75 (95% CI: 0.54, 1.03), 0.77 (95% CI: 0.46, 1.29), 1.19 (95% CI: 0.91, 1.56), respectively, $p > 0.05$ (Table 5, Figure 2). The results show that vegetable fiber was not associated with the onset of T2D, hypertension, obesity, CVD and all-cause mortality.

Table 5. Multivariable adjusted HRs (95% CI) of T2D, hypertension, obesity, CVD, and all-cause mortality according to quartiles of vegetable fiber.

	Vegetable Fiber				
	Quartile 1	Quartile 2	Quartile 3	Quartile 4	*p*-Trend
T2D [a]	1.00	0.85 (0.68, 1.06)	0.97 (0.79, 1.20)	0.85 (0.68, 1.06)	0.24
T2D [b3]	1.00	0.83 (0.66, 1.05)	0.97 (0.77, 1.21)	0.89 (0.70, 1.12)	0.34
Hypertension [a]	1.00	0.93 (0.79, 1.09)	0.92 (0.78, 1.08)	0.90 (0.77, 1.06)	0.18
Hypertension [c3]	1.00	1.01 (0.85, 1.20)	0.96 (0.81, 1.15)	0.96 (0.80, 1.15)	0.12
Obesity [a]	1.00	0.72 (0.54, 0.96)	0.84 (0.64, 1.10)	0.72 (0.54, 0.96)	0.46
Obesity [c3]	1.00	0.78 (0.57, 1.08)	0.87 (0.64, 1.19)	0.75 (0.54, 1.03)	0.71
CVD [a]	1.00	0.72 (0.43, 1.18)	0.83 (0.51, 1.33)	0.85 (0.53, 1.36)	0.19
CVD [c3]	1.00	0.70 (0.41, 1.18)	0.75 (0.45, 1.26)	0.77 (0.46, 1.29)	0.83
All-cause mortality [a]	1.00	0.97 (0.75, 1.25)	0.88 (0.67, 1.14)	1.14 (0.89, 1.46)	0.16
All-cause mortality [c3]	1.00	1.01 (0.77, 1.34)	1.14 (0.85, 1.51)	1.19 (0.91, 1.56)	0.47

[a]: Confounding factors were not adjusted. [b3]: HRs were examined using Cox proportional hazard models. adjusted for age, sex, BMI, education, regions, physical activity, smoking status, alcohol drinking, total energy intake, total carbohydrate intake, protein intake, fatty intake, whole-grain fiber, fruit fiber, and legume fiber. [c3]: HRs were examined using Cox proportional hazard models. Adjusted for age, sex, BMI, education, regions, physical activity, smoking status, alcohol drinking, total energy intake, total carbohydrate intake, protein intake, fatty intake, systolic blood pressure, diastolic blood pressure, Na intake, whole-grain fiber, fruit fiber, and legume fiber.

3.6. Associations between Fruit Fiber Intake and T2D, Hypertension, Obesity, CVD, and All-Cause Mortality

Because the number of people who did not eat fruit fiber was more than half, the fruit fiber intake was divided into the fruit fiber intake group and the non-fruit fiber intake group. Compared to the non-fruit fiber intake group, the HR in the intake group in the cohorts of T2D, hypertension, obesity, CVD, and all-cause mortality were 0.98 (95% CI: 0.80, 1.20), 0.86 (95% CI: 0.73, 1.02), 1.01 (95% CI: 0.77, 1.33), 0.89 (95%CI: 0.55, 1.46), and 0.58 (95%CI: 0.44, 0.77). After adding the factors of legume fiber, vegetable fiber, and whole-grain fiber to adjust for the multiple factors of total dietary fiber, the HR of the intake group were 0.90 (95% CI: 0.72, 1.12), 0.94 (95% CI: 0.78, 1.13), 0.87 (95% CI: 0.64, 1.19), 0.91 (95% CI: 0.53, 1.54), 0.83 (95% CI: 0.61, 1.13) (Table 6, Figure 2). The results show that fruit fiber was negatively associated with the onset of all-cause mortality in the unadjusted model but not with other diseases. However, after adding vegetable fiber, legume fiber, and whole-grain fiber to adjust for multiple factors of total dietary fiber, fruit fiber was not associated with the onset of T2D, hypertension, obesity, CVD, and all-cause mortality.

Table 6. Multivariable adjusted HRs (95%CI) of T2D, hypertension, obesity, CVD, and all-cause mortality according to the classification of intake of fruit fiber or not.

	Fruit Fiber	
	No Intake Group	**Intake Group**
T2D [a]	1.00	0.98 (0.80, 1.20)
T2D [b4]	1.00	0.90 (0.72, 1.12)
Hypertension [a]	1.00	0.86 (0.73, 1.02)
Hypertension [c4]	1.00	0.94 (0.78, 1.13)
Obesity [a]	1.00	1.01 (0.77, 1.33)
Obesity [c4]	1.00	0.87 (0.64, 1.19)
CVD [a]	1.00	0.89 (0.55, 1.46)
CVD [c4]	1.00	0.91 (0.53, 1.54)
All-cause mortality [a]	1.00	0.58 (0.44, 0.77)
All-cause mortality [c4]	1.00	0.83 (0.61, 1.13)

[a]: Confounding factors were not adjusted. [b4]: HRs were examined using Cox proportional hazard models. adjusted for age, sex, BMI, education, regions, physical activity, smoking status, alcohol drinking, total energy intake, total carbohydrate intake, protein intake, fatty intake, whole-grain fiber, vegetable fiber, and legume fiber. [c4]: HRs were examined using Cox proportional hazard models. adjusted for age, sex, BMI, education, regions, physical activity, smoking status, alcohol drinking, total energy intake, total carbohydrate intake, protein intake, fatty intake, systolic blood pressure, diastolic blood pressure, Na intake, whole-grain fiber, vegetable fiber, and legume fiber.

4. Discussion

In a large prospective cohort of Chinese adults, we concluded that total dietary fiber and dietary fiber intake from different food sources were not significantly associated with chronic diseases, such as T2D, hypertension, obesity, cardiovascular disease, and all-cause mortality.

There are inconsistent findings from previous studies on dietary fiber intake and the incidence of type 2 diabetes, hypertension, obesity, cardiovascular disease, and all-cause mortality. On the relationship between dietary fiber intake and the risk of developing type 2 diabetes, a meta-analysis conducted in 2015 [5], which included 18 cohort studies, suggested that dietary fiber intake was associated with a lower risk of diabetes (HR: 0.82; 95% CI: 0.69, 0.97) but was no longer statistically significant after adjusting for BMI. However, a study in Japan in (2021) [25] and a study in France in (2020) [13] suggested a negative association between fiber intake and T2D. On the relationship between dietary fiber intake and the risk of developing hypertension, two earlier studies showed that fiber intake was not associated with developing hypertension [26,27]. However, a 2021 US study [28] suggested dietary fiber was independently associated with a reduced risk of diastolic hypertension (OR = 0.848, 95% CI 0.770, 0.934) and systolic hypertension (OR = 0.906, 95% CI 0.826, 0.993) after adjustments were made for confounding factors. On the relationship between dietary fiber intake and the risk of cardiovascular disease, a 2016 Iranian study [7] and a 2022 study from NHNES [29] suggested a negative association between fiber intake and the development of cardiovascular disease, but a 2020 French study [13] concluded that dietary fiber intake was not associated with the development of cardiovascular disease (OR = 0.86, 95% CI: 0.70, 1.06). On the relationship between dietary fiber intake and the risk of obesity development, a 2010 European study [8] suggested that total fiber intake was negatively associated with increased body weight and waist circumference. For total fiber intake above 10 g/day, the combined estimated change in body weight was −39 g/year (95% CI: −71, −7), and the change in waist circumference was −0.08 cm/year (95% CI:

−0.11, −0.05). Studies on fiber intake and all-cause mortality yielded inconsistent findings, such as a meta-analysis in 2015 [30], and a 2020 Japanese study [17], which suggested a negative association between dietary fiber intake and all-cause mortality. However, a 2020 French cohort study [13] suggested that dietary fiber was not associated with all-cause mortality (OR = 0.98, 95% CI: 0.72, 1.33).

Our results only partially reproduce some of the findings of previous studies, which may be related to demographic differences (e.g., country, race, age, sex), and differences in dietary patterns. In addition, the dietary fiber intake in our population is relatively low and may be lower than an intake that would provide significant health benefits. For example, a meta-analysis that included 13 prospective studies [31] and a French cohort study [6] showed that when the total dietary fiber intake was higher than 25 g/day, it was negatively associated with developing T2D and hypertension. In contrast, the highest quartile of dietary fiber intake in our study population was >13.5 g/day, which is much lower than the dietary fiber intake in previous study populations, which may be one of the reasons why our study yielded no association.

Ample evidence on dietary fiber suggests that whole-grain fiber may be more likely than other fibers to reduce the risk of developing diabetes and obesity [8,32–36]. Many studies suggested that the beneficial effects of whole-grain fiber may be due to the fiber co-intake of other nutrients (e.g., magnesium and vitamins B1, C, and E), while the lower glycemic index of higher whole-grain fiber diets may have reduced the risk of diabetes and obesity development [35]. However, whole-grain fiber was not associated with either diabetes or obesity in our study. This result may be due to our study population's relatively low intake of whole-grain fiber, with the Chinese population generally consuming refined grains with shallow fiber content. Some studies reported that the effect of cereal fiber on reducing the risk of developing diabetes and obesity is mainly related to the intake of whole-grains [11], while other characteristics of refined grains, such as the nature of their high glycemic index, may influence the observed results.

Some studies reported that high dietary fiber consumption decreased the risk of hypertension or BP [37–39], although other research studies reported that fiber intake was not significantly associated with hypertension [40,41]. Compared with the lowest tercile, the HR (95% CI) of hypertension for the highest tercile intakes of cereal fiber was 0.80 (0.67, 0.96) in US research [37]. The data from SWAN Study suggest that dietary fiber intake, especially from grains, contributes to a lower risk of systolic and diastolic BP in middle-aged women [28]. However, our study shows that whole-grain fiber is positively associated with the incidence of hypertension, which is inconsistent with previous findings. Differences in results may be related to sample size, ethnicity, dietary patterns, and the environment of the study population. More research is needed to confirm this result.

Some prospective cohort studies suggest that vegetable and fruit fibers may reduce the risk of cardiovascular disease more than other fibers [7,14]. It has been proposed that vegetable and fruit fibers, due to their higher content of soluble and insoluble fibers, can reduce the activity of fibrinogen activator inhibitor type 1 and coagulation factor VII [42–45] and affect gut microbiota, modifying the inflammatory response of the body [46]. These may be some of the mechanisms through which they reduce the risk of cardiovascular disease. However, our study did not yield relevant conclusions, which may be influenced by the method of investigation, and differences in the amount and type of fruits and vegetables consumed in different seasons may impact our findings.

Prospective studies in the United States and Europe have shown that cereal fiber intake is significantly associated with a lower total mortality [47,48]. In contrast, studies from Japan concluded that cereal fiber intake was not associated with mortality, but fiber intake in legumes, vegetables, and fruits was significantly and negatively associated with total mortality [17]. However, our study concluded that neither total dietary fiber nor fiber intake from various food sources was significantly associated with mortality. Several studies have reached different conclusions, with genetic differences among countries and races and differences in dietary habits contributing to the final results, in addition to the low dietary fiber intake of our study population.

Studies on dietary fiber and chronic diseases such as hypertension, obesity, cardiovascular disease, and all-cause mortality in the Chinese population are lacking. A 2021 study [49] involving 3250 middle-aged and elderly participants in Hangzhou reported that dietary fiber intake was associated with a reduced risk of newly diagnosed T2D (odds ratio (OR) = 0.70, 95% CI: 0.49, 1.0), and another study [50] reported that dietary fiber intake was associated with a reduced risk of prediabetes in a population in Tianjin (OR = 0.85, 95% CI: 0.75, 0.98). We found that the highest quantile of dietary fiber intake in both study populations (>15.1 g and >21.4 g, respectively) was higher than in our study population (>13.5 g). This further confirms that the dietary fiber intake of our study population is too small to reduce the risk of disease. Of course, the region, age, gender, and dietary habits of the study population also contribute to a different fiber intake.

The strength of our study is it's prospective study, the large sample size, and the systematic exploration of the relationship between dietary fiber intake and various chronic diseases and all-cause mortality in the Chinese population. Another major strength of our study is the detailed collection of dietary intakes, collected through repeatedly validated 24-hour dietary records based on an extensive database of more than 6900 food items. This allowed us to examine the associations between different food groups and various chronic diseases [51]. However, some limitations should also be acknowledged. Firstly, limited by the database, although our model adjusted for various potential confounders, residual confounding by the family history of the disease, certain medical conditions and drugs variables, or metabolic factors may persist. Secondly, due to the limitations of the cohort study itself, dietary intake may be influenced by economic and social development, which may affect the associations with disease. However, to ensure the stability of the dietary data, we used average year data for 2004 and 2006 as the baseline dietary assessment. Lastly, in our study, disease detection was primarily based on self-report, and despite this relatively valid method, misclassification bias could not be ruled out.

5. Conclusions

Our study did not find an association between total dietary fiber and dietary fiber intake from various food sources with type 2 diabetes, hypertension, obesity, cardiovascular disease, and all-cause mortality in the Chinese population. However, our study shows that whole-grain fiber is positively associated with the incidence of hypertension. The role of dietary fiber in the Chinese population may be overestimated. To further confirm the association between dietary fiber and these diseases in the Chinese population, more extraordinary efforts are needed in the future to increase the intake of dietary fiber in the Chinese population and to try to diversify the food groups in the dietary pattern (whole-grains, legumes, vegetables, fruits, and meat).

Supplementary Materials: The following supporting information can be downloaded at: https://www.mdpi.com/article/10.3390/nu14132650/s1, Table S1: Characteristics of the study population at inclusion in the hypertension cohort, according to quartiles of total dietary fiber intakes. Table S2: Characteristics of the study population at inclusion in the obesity cohort, according to quartiles of total dietary fiber intakes. Table S3: Characteristics of the study population at inclusion in the CVD cohort, according to quartiles of total dietary fiber intakes. Table S4: Characteristics of the study population at inclusion in the all-cause mortality cohort, according to quartiles of total dietary fiber intakes.

Author Contributions: Conceptualization, S.X. and Y.X.; methodology, S.X.; validation, M.F. and J.Z.; formal analysis, B.C.; writing—original draft preparation, Z.Z. and B.C.; writing—review and editing, Z.Z. and B.C.; visualization, B.C.; supervision, J.Z., X.L., and W.X.; project administration, Y.X.; funding acquisition, S.X. All authors have read and agreed to the published version of the manuscript.

Funding: The study was partly supported by the Young Talents Project of Hubei Provincial Health Commission, China (Grand number WJ2021Q012); Science and Technology Research Key Project of Education Department of Hubei Province, China (Grand number D20212602); Sanuo Diabetes Charity Foundation, China; and Xiangyang Science and Technology Plan Project, China (Grand number 2019ZD12).

Institutional Review Board Statement: On the basis of the guidelines of the Declaration of Helsinki, this study was approved by the Institutional Review Committees of UNC-CH and the NINH, China CDC.

Informed Consent Statement: All subjects involved in the study signed an informed consent statement.

Data Availability Statement: Data from CHNS described in the manuscript will be made publicly and freely available without restriction at China Health and Nutrition Survey. Available online: https://www.cpc.unc.edu/projects/china/data/datasets/index.html (accessed on 12 May 2022).

Acknowledgments: Data in this research were from China Health and Nutrition Survey (CHNS). The authors are grateful to all subjects who participated in the nationwide population-based study. We also thank the National Institute for Nutrition and Health, China Center for Disease Control and Prevention. All authors have consented to the acknowledgement.

Conflicts of Interest: The authors declare no conflict of interest.

References

1. Saeedi, P.; Salpea, P.; Karuranga, S.; Petersohn, I.; Malanda, B.; Gregg, E.W.; Unwin, N.; Wild, S.H.; Williams, R. Mortality attributable to diabetes in 20–79 years old adults, 2019 estimates: Results from the international diabetes federation diabetes atlas, 9th edition. *Diabetes Res. Clin. Pract.* **2020**, *162*, 108086. [CrossRef]
2. Watkins, D.A.; Johnson, C.O.; Colquhoun, S.M.; Karthikeyan, G.; Beaton, A.; Bukhman, G.; Forouzanfar, M.H.; Longenecker, C.T.; Mayosi, B.M.; Mensah, G.A.; et al. Global, regional, and national burden of rheumatic heart disease, 1990–2015. *N. Engl. J. Med.* **2017**, *377*, 713–722. [CrossRef]
3. Blüher, M. Obesity: Global epidemiology and pathogenesis. *Nat. Rev. Endocrinol.* **2019**, *15*, 288–298. [CrossRef]
4. Mills, K.T.; Bundy, J.D.; Kelly, T.N.; Reed, J.E.; Kearney, P.M.; Reynolds, K.; Chen, J.; He, J. Global disparities of hypertension prevalence and control: A systematic analysis of population-based studies from 90 countries. *Circulation* **2016**, *134*, 441–450. [CrossRef]
5. InterAct Consortium. Dietary fibre and incidence of type 2 diabetes in eight European countries: The EPIC-InterAct Study and a meta-analysis of prospective studies. *Diabetologia* **2015**, *58*, 1394–1408. [CrossRef]
6. Lairon, D.; Arnault, N.; Bertrais, S.; Planells, R.; Clero, E.; Hercberg, S.; Boutron-Ruault, M.C. Dietary fiber intake and risk factors for cardiovascular disease in French adults. *Am. J. Clin. Nutr.* **2005**, *82*, 1185–1194. [CrossRef]
7. Mirmiran, P.; Bahadoran, Z.; Khalili Moghadam, S.; Zadeh Vakili, A.; Azizi, F. A Prospective study of different types of dietary fiber and risk of cardiovascular disease: Tehran lipid and glucose study. *Nutrients* **2016**, *8*, 686. [CrossRef]
8. Du, H.; van der A, D.L.; Boshuizen, H.C.; Forouhi, N.G.; Wareham, N.J.; Halkjaer, J.; Tjønneland, A.; Overvad, K.; Jakobsen, M.U.; Boeing, H.; et al. Dietary fiber and subsequent changes in body weight and waist circumference in European men and women. *Am. J. Clin. Nutr.* **2010**, *91*, 329–336. [CrossRef]
9. Buil-Cosiales, P.; Zazpe, I.; Toledo, E.; Corella, D.; Salas-Salvadó, J.; Diez-Espino, J.; Ros, E.; Navajas, J.F.-N.; Santos-Lozano, J.M.; Arós, F.; et al. Fiber intake and all-cause mortality in the Prevención con Dieta Mediterránea (PREDIMED) study. *Am. J. Clin. Nutr.* **2014**, *100*, 1498–1507. [CrossRef]
10. Xu, H.; Huang, X.; Risérus, U.; Krishnamurthy, V.M.; Cederholm, T.; Arnlöv, J.; Lindholm, B.; Sjögren, P.; Carrero, J.J. Dietary fiber, kidney function, inflammation, and mortality risk. *Clin. J. Am. Soc. Nephrol.* **2014**, *9*, 2104–2110. [CrossRef]
11. Miller, S.J.; Batra, A.K.; Shearrer, G.E.; House, B.T.; Cook, L.T.; Pont, S.J.; Goran, M.I.; Davis, J.N. Dietary fibre linked to decreased inflammation in overweight minority youth. *Pediatr. Obes.* **2016**, *11*, 33–39. [CrossRef]
12. Erkkilä, A.T.; Lichtenstein, A.H. Fiber and cardiovascular disease risk: How strong is the evidence? *J. Cardiovasc. Nurs.* **2006**, *21*, 3–8. [CrossRef]
13. Partula, V.; Deschasaux, M.; Druesne-Pecollo, N.; Latino-Martel, P.; Desmetz, E.; Chazelas, E.; Kesse-Guyot, E.; Julia, C.; Fezeu, L.K.; Galan, P.; et al. Associations between consumption of dietary fibers and the risk of cardiovascular diseases, cancers, type 2 diabetes, and mortality in the prospective NutriNet-Santé cohort. *Am. J. Clin. Nutr.* **2020**, *112*, 195–207. [CrossRef]
14. Threapleton, D.E.; Greenwood, D.C.; Evans, C.E.; Cleghorn, C.L.; Nykjaer, C.; Woodhead, C.; Cade, J.E.; Gale, C.P.; Burley, V.J. Dietary fibre intake and risk of cardiovascular disease: Systematic review and meta-analysis. *BMJ* **2013**, *347*, f6879. [CrossRef]
15. Yang, Y.; Zhao, L.G.; Wu, Q.J.; Ma, X.; Xiang, Y.B. Association between dietary fiber and lower risk of all-cause mortality: A meta-analysis of cohort studies. *Am. J. Epidemiol.* **2015**, *181*, 83–91. [CrossRef]
16. Dominguez, L.J.; Bes-Rastrollo, M.; Toledo, E.; Gea, A.; Fresán, U.; Barbagallo, M.; Martínez-González, M.A. Dietary fiber intake and mortality in a Mediterranean population: The "Seguimiento Universidad de Navarra" (SUN) project. *Eur. J. Nutr.* **2019**, *58*, 3009–3022. [CrossRef]

17. Katagiri, R.; Goto, A.; Sawada, N.; Yamaji, T.; Iwasaki, M.; Noda, M.; Iso, H.; Tsugane, S. Dietary fiber intake and total and cause-specific mortality: The Japan Public Health Center-based prospective study. *Am. J. Clin. Nutr.* **2020**, *111*, 1027–1035. [CrossRef]
18. Steffen, L.M.; Jacobs, D.R., Jr.; Stevens, J.; Shahar, E.; Carithers, T.; Folsom, A.R. Associations of whole-grain, refined-grain, and fruit and vegetable consumption with risks of all-cause mortality and incident coronary artery disease and ischemic stroke: The Atherosclerosis Risk in Communities (ARIC) Study. *Am. J. Clin. Nutr.* **2003**, *78*, 383–390. [CrossRef]
19. Popkin, B.M.; Du, S.; Zhai, F.; Zhang, B. Cohort profile: The China Health and Nutrition Survey–monitoring and understanding socio-economic and health change in China, 1989–2011. *Int. J. Epidemiol.* **2010**, *39*, 1435–1440. [CrossRef]
20. Zhang, B.; Zhai, F.Y.; Du, S.F.; Popkin, B.M. The China Health and Nutrition Survey, 1989–2011. *Obes. Rev.* **2014**, *15*, 2–7. [CrossRef]
21. Zhai, F.Y.; Du, S.F.; Wang, Z.H.; Zhang, J.G.; Du, W.W.; Popkin, B.M. Dynamics of the Chinese diet and the role of urbanicity, 1991–2011. *Obes. Rev.* **2014**, *15*, 16–26. [CrossRef]
22. Kim, H.; Lee, K.; Rebholz, C.M.; Kim, J. Plant-based diets and incident metabolic syndrome: Results from a South Korean prospective cohort study. *PLoS Med.* **2020**, *17*, e1003371. [CrossRef]
23. Yuan, X.; Liu, T.; Wu, L.; Zou, Z.Y.; Li, C. Validity of self-reported diabetes among middle-aged and older Chinese adults: The China health and retirement longitudinal study. *BMJ Open* **2015**, *5*, e006633. [CrossRef]
24. China Nutrition Association. The Chinese Dietary Guidline 2022. Available online: http://dg.cnsoc.org/ (accessed on 23 April 2022).
25. Kimura, Y.; Yoshida, D.; Hirakawa, Y.; Hata, J.; Honda, T.; Shibata, M.; Sakata, S.; Uchida, K.; Kitazono, T.; Ninomiya, T. Dietary fiber intake and risk of type 2 diabetes in a general Japanese population: The Hisayama Study. *J. Diabetes Investig.* **2021**, *12*, 527–536. [CrossRef]
26. Witteman, J.C.; Willett, W.C.; Stampfer, M.J.; Colditz, G.A.; Sacks, F.M.; Speizer, F.E.; Rosner, B.; Hennekens, C.H. A prospective study of nutritional factors and hypertension among US women. *Circulation* **1989**, *80*, 1320–1327. [CrossRef]
27. Ascherio, A.; Rimm, E.B.; Giovannucci, E.L.; Colditz, G.A.; Rosner, B.; Willett, W.C.; Sacks, F.; Stampfer, M.J. A prospective study of nutritional factors and hypertension among US men. *Circulation* **1992**, *86*, 1475–1484. [CrossRef]
28. Du, P.; Luo, K.; Wang, Y.; Xiao, Q.; Xiao, J.; Li, Y.; Zhang, X. Intake of dietary fiber from grains and the risk of hypertension in late midlife women: Results from the SWAN study. *Front. Nutr.* **2021**, *8*, 730205. [CrossRef]
29. Zhang, S.; Tian, J.; Lei, M.; Zhong, C.; Zhang, Y. Association between dietary fiber intake and atherosclerotic cardiovascular disease risk in adults: A cross-sectional study of 14,947 population based on the National Health and Nutrition Examination Surveys. *BMC Public Health* **2022**, *22*, 1076. [CrossRef]
30. Liu, L.; Wang, S.; Liu, J. Fiber consumption and all-cause, cardiovascular, and cancer mortalities: A systematic review and meta-analysis of cohort studies. *Mol. Nutr. Food Res.* **2015**, *59*, 139–146. [CrossRef]
31. Khan, K.; Jovanovski, E.; Ho, H.; Marques, A.; Zurbau, A.; Mejia, S.B.; Sievenpiper, J.L.; Vuksan, V. The effect of viscous soluble fiber on blood pressure: A systematic review and meta-analysis of randomized controlled trials. *Nutr. Metab. Cardiovasc. Dis. NMCD.* **2018**, *28*, 3–13. [CrossRef]
32. Yao, B.; Fang, H.; Xu, W.; Yan, Y.; Xu, H.; Liu, Y.; Mo, M.; Zhang, H.; Zhao, Y. Dietary fiber intake and risk of type 2 diabetes: A dose-response analysis of prospective studies. *Eur. J. Epidemiol.* **2014**, *29*, 79–88. [CrossRef] [PubMed]
33. Salmerón, J.; Ascherio, A.; Rimm, E.B.; Colditz, G.A.; Spiegelman, D.; Jenkins, D.J.; Stampfer, M.J.; Wing, A.L.; Willett, W.C. Dietary fiber, glycemic load, and risk of NIDDM in men. *Diabetes Care* **1997**, *20*, 545–550. [CrossRef] [PubMed]
34. Schulze, M.B.; Liu, S.; Rimm, E.B.; Manson, J.E.; Willett, W.C.; Hu, F.B. Glycemic index, glycemic load, and dietary fiber intake and incidence of type 2 diabetes in younger and middle-aged women. *Am. J. Clin. Nutr.* **2004**, *80*, 348–356. [CrossRef] [PubMed]
35. Koh-Banerjee, P.; Franz, M.; Sampson, L.; Liu, S.; Jacobs, D.R., Jr.; Spiegelman, D.; Willett, W.; Rimm, E. Changes in whole-grain, bran, and cereal fiber consumption in relation to 8-y weight gain among men. *Am. J. Clin. Nutr.* **2004**, *80*, 1237–1245. [CrossRef]
36. Bakker, S.J.; Hoogeveen, E.K.; Nijpels, G.; Kostense, P.J.; Dekker, J.M.; Gans, R.O.; Heine, R.J. The association of dietary fibres with glucose tolerance is partly explained by concomitant intake of thiamine: The hoorn study. *Diabetologia* **1998**, *41*, 1168–1175. [CrossRef]
37. Sun, B.; Shi, X.; Wang, T.; Zhang, D. Exploration of the association between dietary fiber intake and hypertension among U.S. adults using 2017 american college of cardiology/american heart association blood pressure guidelines: NHANES 2007–2014. *Nutrients* **2018**, *10*, 1091. [CrossRef]
38. Kochar, J.; Gaziano, J.M.; Djoussé, L. Breakfast cereals and risk of hypertension in the Physicians' Health Study I. *Clin. Nutr.* **2012**, *31*, 89–92. [CrossRef]
39. Borgi, L.; Muraki, I.; Satija, A.; Willett, W.C.; Rimm, E.B.; Forman, J.P. Fruit and vegetable consumption and the incidence of hypertension in three prospective cohort studies. *Hypertension* **2016**, *67*, 288–293. [CrossRef]
40. Masala, G.; Bendinelli, B.; Versari, D.; Saieva, C.; Ceroti, M.; Santagiuliana, F.; Caini, S.; Salvini, S.; Sera, F.; Taddei, S.; et al. Anthropometric and dietary determinants of blood pressure in over 7000 Mediterranean women: The European prospective investigation into cancer and nutrition-florence cohort. *J. Hypertens.* **2008**, *26*, 2112–2120. [CrossRef]
41. Davy, B.M.; Melby, C.L.; Beske, S.D.; Ho, R.C.; Davrath, L.R.; Davy, K.P. Oat consumption does not affect resting casual and ambulatory 24-h arterial blood pressure in men with high-normal blood pressure to stage I hypertension. *J. Nutr.* **2002**, *132*, 394–398. [CrossRef]

42. Anderson, J.W.; Tietyen-Clark, J. Dietary fiber: Hyperlipidemia, hypertension, and coronary heart disease. *Am. J. Gastroenterol.* **1986**, *81*, 907–919. [PubMed]
43. Anderson, J.W.; Chen, W.J. Plant fiber. Carbohydrate and lipid metabolism. *Am. J. Clin. Nutr.* **1979**, *32*, 346–363. [CrossRef] [PubMed]
44. Hamaker, B.R.; Tuncil, Y.E. A perspective on the complexity of dietary fiber structures and their potential effect on the gut microbiota. *J. Mol. Biol.* **2014**, *426*, 3838–3850. [CrossRef]
45. Parnell, J.A.; Reimer, R.A. Prebiotic fiber modulation of the gut microbiota improves risk factors for obesity and the metabolic syndrome. *Gut Microbes.* **2012**, *3*, 29–34. [CrossRef] [PubMed]
46. Kuo, S.M. The interplay between fiber and the intestinal microbiome in the inflammatory response. *Adv. Nutr.* **2013**, *4*, 16–28. [CrossRef] [PubMed]
47. Chuang, S.C.; Norat, T.; Murphy, N.; Olsen, A.; Tjønneland, A.; Overvad, K.; Boutron-Ruault, M.C.; Perquier, F.; Dartois, L.; Kaaks, R.; et al. Fiber intake and total and cause-specific mortality in the European Prospective Investigation into Cancer and Nutrition cohort. *Am. J. Clin. Nutr.* **2012**, *96*, 164–174. [CrossRef] [PubMed]
48. Ark, Y.; Subar, A.F.; Hollenbeck, A.; Schatzkin, A. Dietary fiber intake and mortality in the NIH-AARP diet and health study. *Arch. Intern. Med.* **2011**, *171*, 1061–1068.
49. Jin, F.; Zhang, J.; Shu, L.; Han, W. Association of dietary fiber intake with newly-diagnosed type 2 diabetes mellitus in middle-aged Chinese population. *Nutr. J.* **2021**, *20*, 81. [CrossRef]
50. Zhang, S.; Meng, G.; Zhang, Q.; Liu, L.; Yao, Z.; Wu, H.; Gu, Y.; Wang, Y.; Zhang, T.; Wang, X.; et al. Dietary fiber intake and risk of prediabetes in China: Results from the TCLSIH Cohort Study. *Br. J. Nutr.* **2021**, 1–20. [CrossRef]
51. Kim, H.; Rebholz, C.M.; Garcia-Larsen, V.; Steffen, L.M.; Coresh, J.; Caulfield, L.E. Operational differences in plant-based diet indices affect the ability to detect associations with incident hypertension in middle-aged US adults. *J. Nutr.* **2020**, *150*, 842–850. [CrossRef]

nutrients

MDPI

Article

The Relationship among Bowel ^{18}F-FDG PET Uptake, Pathological Complete Response, and Eating Habits in Breast Cancer Patients Undergoing Neoadjuvant Chemotherapy

Paola Tiberio [1], Lidija Antunovic [2], Mariangela Gaudio [1,3], Alessandro Viganò [4], Manuela Pastore [1], Chiara Miggiano [1,3], Flavia Jacobs [1,3], Chiara Benvenuti [1,3], Elisabetta Farina [4], Arturo Chiti [2,3], Armando Santoro [1,3] and Rita De Sanctis [1,3,*]

1 Medical Oncology and Hematology Unit, IRCCS Humanitas Research Hospital, Rozzano, 20089 Milan, Italy
2 Nuclear Medicine Unit, IRCCS Humanitas Research Hospital, Rozzano, 20089 Milan, Italy
3 Department of Biomedical Sciences, Humanitas University, Pieve Emanuele, 20072 Milan, Italy
4 IRCCS Fondazione Don Carlo Gnocchi, 20148 Milan, Italy
* Correspondence: rita.de_sanctis@hunimed.eu; Tel.: +39-028-224-7230

Abstract: Recently, the impact of patients' eating habits on both breast cancer (BC) management and inflammation have been proven. Here, we investigated whether inflammatory habits could correlate with baseline bowel ^{18}F-fluorodeoxyglucose (FDG) uptake and the latter, in turn, with pathological Complete Response (pCR) to neoadjuvant chemotherapy (NAC). We included stage I–III BC undergoing standard NAC at IRCCS Humanitas Research Hospital, Italy. Patients fulfilled a survey concerning eating/lifestyle behaviors and performed a staging ^{18}F-FDG positrone emission tomography/computed tomography (PET/CT). In the absence of data on the effects of individual foods, we aggregated drink and food intake for their known inflammatory properties. Data were recorded for 82 women (median age, 48). We found positive correlations between colon mean standardized uptake value (SUV_{mean}) and pro-inflammatory drinks (alcohol and spirits; $r = +0.33$, $p < 0.01$) and foods (red and cured meats; $r = +0.25$, $p = 0.04$), and a significant negative correlation between rectum SUV_{mean} and anti-inflammatory foods (fruits and vegetables; $r = -0.23$, $p = 0.04$). Furthermore, colon SUV_{mean} was significantly lower in patients with pCR compared to non pCR ($p = 0.02$). Our study showed, for the first time, that patients' eating habits affected bowel ^{18}F-FDG uptake and that colon SUV_{mean} correlated with pCR, suggesting that PET scan could be an instrument for identifying patients presenting unhealthy behaviors.

Keywords: breast cancer; neoadjuvant chemotherapy; bowel ^{18}F-FDG PET uptake; nutrition; bowel inflammation; pathologic complete response

Citation: Tiberio, P.; Antunovic, L.; Gaudio, M.; Viganò, A.; Pastore, M.; Miggiano, C.; Jacobs, F.; Benvenuti, C.; Farina, E.; Chiti, A.; et al. The Relationship among Bowel ^{18}F-FDG PET Uptake, Pathological Complete Response, and Eating Habits in Breast Cancer Patients Undergoing Neoadjuvant Chemotherapy. *Nutrients* **2023**, *15*, 211. https://doi.org/10.3390/nu15010211

Academic Editors: Andriana Kaliora, Chara Tzavara and Charalampia Amerikanou

Received: 16 December 2022
Revised: 28 December 2022
Accepted: 28 December 2022
Published: 1 January 2023

1. Introduction

Breast cancer (BC) is the most common neoplasm and the primary cause of cancer death in women worldwide. Despite its high incidence, there is a progressive decrease in cancer mortality and a consequently ever increasing number of cancer survivors [1]. However, many parameters could influence tumor development and survivors' quality of life. Among modifiable risk factors, eating habits, body weight, and lifestyle behaviors have been deeply investigated [2–8]. Increasing evidence suggested that diet plays an important role in cancer development, progression, and prevention, including BC [9,10]. Several studies showed that healthy diet and exercise might improve overall survival and quality of life after BC diagnosis by reducing chemotherapy side effects, limiting comorbidities, and enhancing therapeutic efficacy [11–15].

[18]F-fluorodeoxyglucose (FDG) positrone emission tomography/computed tomography (PET/CT) is a functional imaging technique with extensive use in oncology for staging, as well as for assessment of cancer relapse and response to therapy [16–18], but also in

different non-oncological diseases related to infection and inflammation. Interestingly, recent studies have also shown that ^{18}F-FDG PET/CT may be useful for detecting benign bowel inflammatory activity [19–21].

Despite the large number of studies investigating the association between diet and cancer risk, as well as exploring changes in eating habits in cancer survivors, to date there are no investigation regarding the possible influence of baseline dietary patterns on response to therapy in BC. In this study, we aimed to investigate whether, in BC patients, diet could correlate with bowel ^{18}F-FDG uptake and the latter, in turn, with pathological Complete Response (pCR) to standard neoadjuvant chemotherapy (NAC).

2. Materials and Methods

2.1. Study Design and Participants

We performed a prospective mono-centric longitudinal observational proof-of-principle study, enrolling women who underwent standard NAC for BC at IRCCS Humanitas Research Hospital in Rozzano, Italy.

The inclusion criteria were:

(a) willingness to participate to the study;
(b) age $\geq$ 18 years old;
(c) female gender;
(d) histopathologically confirmed diagnosis of BC;
(e) clinical stage T1c-T4, N0-N3, M0 at presentation;
(f) Eastern Cooperative Oncology Group (ECOG) Performance Status 0–1;
(g) baseline left ventricular ejection fraction $\geq$ 55%;
(h) adequate hematologic, liver and hepatic function;
(i) ability to give informed consent according to International Conference on Harmonization /European Union Good Clinical Practice, and national/local regulation.

The exclusion criteria were:

a. inability to respond to survey;
b. prior history of invasive BC;
c. stage IV BC;
d. prior systemic therapy for BC;
e. previous therapy with anthracyclines/taxanes for any malignancy;
f. use of immunomodulatory agents at the time of enrolment/during the previous 2 months;
g. use of antibiotics at the time of enrolment/during the previous month;
h. history of other malignancy within 5 years prior to the enrolment;
i. pregnancy/breastfeeding/intention of becoming pregnant during the study.

At baseline, participants performed a whole-body staging ^{18}F-FDG PET/CT scan according to the recommendations of the European Association of Nuclear Medicine (EANM) guidelines [22]. NAC was administered according to international and national clinical guidelines after a multidisciplinary discussion on each single case. NAC regimens included anthracycline-based chemotherapy followed by (i) weekly carboplatin and paclitaxel in case of triple-negative BC, (ii) docetaxel and trastuzumab in case of human epidermal growth factor receptor 2 (HER2) positive BC, or (iii) docetaxel alone in case of luminal-like disease. Data on pathological response were collected and regular follow-up of the patients was performed. A flowchart of the recruitment and follow-up process is reported in Supplementary Materials Figure S1. The study was approved by the IRCCS Humanitas Research Hospital Ethics Committee (Protocol identifying number ONC/OSS-02/2019). All patients signed the informed consent form in accordance with the Declaration of Helsinki.

2.2. Survey Design

At baseline, before NAC therapy, all enrolled women fulfilled a survey developed by our dietitian nutritionist following national and international guidelines, which concerned their eating habits and the frequency of their physical activity. Specifically, patients have been asked to describe their diet as omnivorous/varied, vegetarian, or vegan and to report weekly frequencies of the consumption of 16 food items: milk, dairy products, alcoholic drinks, spirits, white meat, read meat, eggs, fruits, vegetables, fish, pulses, cereals, cured meat, salty snacks, sweet snacks/drinks, and nuts. The survey also investigated whether BC patients performed exercise regularly (stated as weekly frequency and type of exercise; a copy of the survey is reported in Supplementary Materials Figure S2). The questionnaire was self-administered, easy to understand, and provided semi-quantitative data. Food consumption frequencies of BC patients were analyzed in comparison with an ideal healthy diet (2000 kcal/die) [23] and results were reported as "more", "correct", or "less". When no responses were provided, we indicated it as "missing data".

2.3. ^{18}F-FDG PET/CT Acquisition Protocol

Fasting for at least 6 h prior radiopharmaceutical injection and rest (restrained from excess physical activity and talking) were required as preparation for ^{18}F-FDG PET/CT imaging. Prior to radiotracer injection, blood glucose level measurements were obtained if serum glucose concentration was lower than 200 mg/dL, an intravenous injection of ~6 Megabecquerel (Mbq)/kg of ^{18}F-FDG was performed. Post-injection, a one-hour interlude was mandatory for all participants.

Subsequently, each patient was scanned using one of two integrated PET-CT scanners: a Siemens Biograph LS 6 scanner (Siemens, Munich, Germany), or a GE Discovery PET-CT 690 (General Electric Healthcare, Waukesha, WI, USA). After attenuation correction, images were reconstructed obtaining axial, sagittal, coronal CT, PET, and PET/CT fused images.

On axial images, an experienced nuclear medicine physician designed regions (ROI) to extract values of semi-quantitative parameters of radiotracer mean standardized uptake value (SUV_{mean}). Two ROIs were positioned on the area of highest uptake respectively in the rectum-sigmoid district and in the remaining part of the colon.

2.4. Statistical Analysis

Results were presented as means ± standard deviations, medians and ranges, or percentages of the total. Parametric or non-parametric tests were used according to the data mean distribution.

Correlation analysis was performed by using Pearson's coefficient between SUV_{mean} of colon and rectum on one side, and dietary habits, frequency of physical activity, body mass index (BMI), and smoking habit on the other. Concerning eating habits, in the absence of data on the effects of individual foods, weekly intake of well-known pro- and anti-inflammatory drinks or foods were aggregated. Specifically, weekly frequency of consumption of alcoholic drinks and spirits were added together and referred to as "pro-inflammatory drinks". In the same way, we putted together intakes of red and cured meats ("pro-inflammatory foods") and of fruits and vegetable ("anti-inflammatory foods"). We performed a multiple comparison correction by False Discovery Rate (FDR).

Differences between values of bowel SUV_{mean} in patients obtaining pCR and non pCR were tested using a two-sided *t*-test. Analyses were run separately for the SUV_{mean} values of the rectum and of the colon.

Moreover, multivariate approach was carried out using Discriminant Function Analysis (DFA), which is a powerful tool to build associative models with categorical outcome of interest, as in this case (pCR versus non pCR). A deeper description of DFA is provided elsewhere [24–27]. In brief, DFA estimates the linear combination of selected covariates able to split single cases into groups according to an outcome of interest. DFA provides a model in which the variables associated to the outcome are listed according to their weight in decreasing order. Significance level was set at <0.05 after proper correction. Statistics were performed with STATISTICA, version 7, StatSoft, OK, USA.

3. Results

3.1. Patients' and Tumors' Characteristics

Demographic details of the 82 patients enrolled in the study are depicted in Tables 1 and 2. Both average and median age were 48 years (range, 25–72 years). Focusing on BC risk factors, 18.3% of the population were smokers, whereas 64.6% never smoked and the remaining 17.1% stopped smoking before BC diagnosis (median time from quitting smoking, 10 years). Most women were premenopausal (57.3%), and the average BMI was 23.68 (range, 16.18–35.82). An amount of 61% of women had a normal BMI, 4.9% were underweight, and 30.5% were overweight, whereas only 3.7% were obese. Eleven patients had a gastrointestinal comorbidity and only one had insulin-dependent diabetes mellitus. We conducted an analysis of variance ANOVA to test whether comorbidities could influence levels of bowel ^{18}F-FDG uptake with no significant differences ($p = 0.50$). All patients were diagnosed with stage IA–IIIB BC and most tumors were high grade (43 patients had G3 tumor). Furthermore, 45 women were diagnosed with a HER2 + BC, 29 had a triple-negative breast cancer (TNBC), and the remaining presented luminal-like tumors (one Luminal A and seven Luminal B). At the end of NAC, 37 patients reached pCR, whereas 45 did not.

3.2. Eating Habits

Almost all enrolled women were omnivorous, two patients followed a vegetarian diet, whereas only one was vegan. In Figure 1 and Table 3, eating habits of participants alone and compared to an ideal healthy diet [23] are illustrated, respectively. Most women showed a correct intake of alcohol and spirits (weekly frequency equal to 0), as well as fruits, vegetables, and cereals. However, weekly consumption of milk, diary supplements, eggs, fish, pulses, and nuts was under the healthy recommended level [23]. On the contrary, there was an overconsumption of cured meats and snacks. Focusing on the strongest dietary recommendation for cancer patients (i.e., a correct intake of white meat, fruits and vegetables, fish, and pulses or cereals) [28], we noticed that only 3.7% of patients followed a healthy diet before BC diagnosis. Interestingly, 52.4% of patients declared that they usually exercise regularly (Table 4).

Table 1. Demographic and clinical BC patients' characteristics.

		Patients (n = 82)	
		n	%
Age			
	<50	46	56.1
	50–64	29	35.4
	≥65	7	8.5
Smoke			
	no	53	64.6
	yes	15	18.3
	former	14	17.1
Menopause			
	no	47	57.3
	yes	28	34.1
	peri	7	8.5
BMI			
	<18.5	4	4.9
	18.5–24.9	50	61.0
	25–29.9	25	30.5
	≥30	3	3.7
Comorbidies			
	None	54	65.9
	Intestinal	8	9.8
	Others	17	20.7
	Intestinal + others	3	3.7

BC, breast cancer; BMI, body mass index.

Table 2. Histopathological characteristics of BC (n = 82).

	no pCR (n = 45)		pCR (n = 37)	
	n	%	n	%
Stage				
IA	2	4.4	6	16.2
IIA	17	37.8	19	51.4
IIB	18	40.0	9	24.3
IIIA	6	13.3	3	8.1
IIIB	1	2.2	0	0
IIIC	1	2.2	0	0
Grade				
G1	0	0	0	0
G2	16	35.6	9	24.3
G3	29	64.4	28	75.7
Subtype				
Luminal A	1	2.2	0	0
Luminal B	3	6.7	4	10.8
HER2+	27	60.0	18	48.6
TNBC	14	31.1	15	40.5

pCR, pathological Complete Response; HER2+, human epidermal growth factor receptor 2 positive; TNBC, triple-negative breast cancer.

Table 3. Median of weekly frequency of patients' food consumption. Range in the brackets.

	Weekly Frequency Consumption Median
Milk	1 (0–7)
Dairy products	2 (0–7)
Alcoholic drinks	0 (0–7)
Spirits	0 (0–2)
White meat	2 (0–6)
Red meat	1 (0–3)
Eggs	1 (0–5)
Fruit	7 (0–7)
Vegetables	7 (1–7)
Fish	1.5 (0–6)
Pulses	1 (0–7)
Cereals	7 (0–7)
Cured meats	2 (0–5)
Salty snacks	1 (0–7)
Sweet snacks/drinks	2.5 (0–7)
Nuts	1 (0–7)

Table 4. Weekly frequencies of patients' physical activity.

	Patients (n = 82)	
	n	%
Exercise weekly frequency		
0	37	45.1
1–3	26	31.7
>3	17	20.7
missing	2	2.4

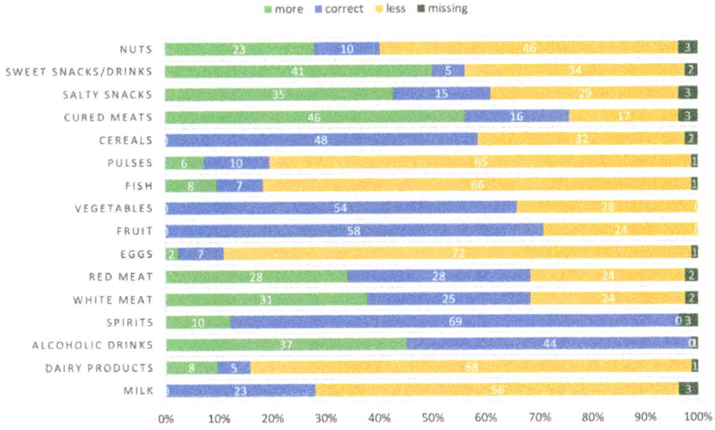

Figure 1. Distribution of patients' food consumption frequencies compared to an ideal healthy diet. "Missing" indicate that no responses were provided. White numbers in the bar represent number of patients.

3.3. *Correlation between Eating and Exercise Habits and Bowel ^{18}F-FDG Uptake*

We then investigated whether increased levels of bowel ^{18}F-FDG uptake observed at the staging PET scan could be influenced by patients' lifestyle, considering both eating and exercise habits. After FDR correction, we found a positive correlation between baseline colon SUV_{mean} and pro-inflammatory drinks (r = +0.33, $p < 0.01$) and foods (r = +0.25, $p = 0.04$). A significant negative correlation was also observed between baseline rectum SUV_{mean} and anti-inflammatory foods (r = −0.23, $p = 0.04$) (Table 5). No statistically significant associations were seen with BMI, smoking habits, or physical activity.

Table 5. Pearson's correlation (r) e relative *p*-values (*p*) between bowel ^{18}F-FDG uptake and patients' habits.

	Colon SUV_{mean}		Rectum SUV_{mean}	
	r	*p*	r	*p*
Pro-inflammatory drinks	+0.33	<0.01	+0.14	0.51
Pro-inflammatory foods	+0.25	0.04	+0.02	0.86
Anti-inflammatory foods	−0.21	0.05	−0.23	0.04
Physical activity	−0.20	0.46	−0.11	0.64
Smoke	−0.30	0.81	−0.07	0.96
BMI	+0.17	0.79	+0.15	0.83

SUV_{mean}, mean standardized uptake value; BMI, body mass index.

3.4. *Association between Bowel ^{18}F-FDG Uptake and Response to Therapy*

At baseline, rectum SUV_{mean} did not differ between patients who experienced a pCR (1.99 ± 0.59) and patients who did not (2.13 ± 1.11) ($p = 0.48$). On the other hand, colon SUV_{mean} was significantly lower in patients who experienced pCR after NAC (1.58 ± 0.56; Figure 2) compared with non pCR patients (2.05 ± 1.17; Figure 3) ($p = 0.02$; Figure 4).

The multivariate approach confirmed results from the univariate analysis ($F(6.67) = 2.49$; $p < 0.03$). DFA model pinpointed the factors that were significantly associated to pCR: colon

SUV_{mean}, cured meats, rectum SUV_{mean}, fruits, alcoholic drinks, and red meat. Together, these factors could explain up to 98.1% of the variance, and the first two factors (i.e., colon SUV_{mean} and cured meat) explained by themselves more than 90% of the variance (see Supplementary Materials Table S1).

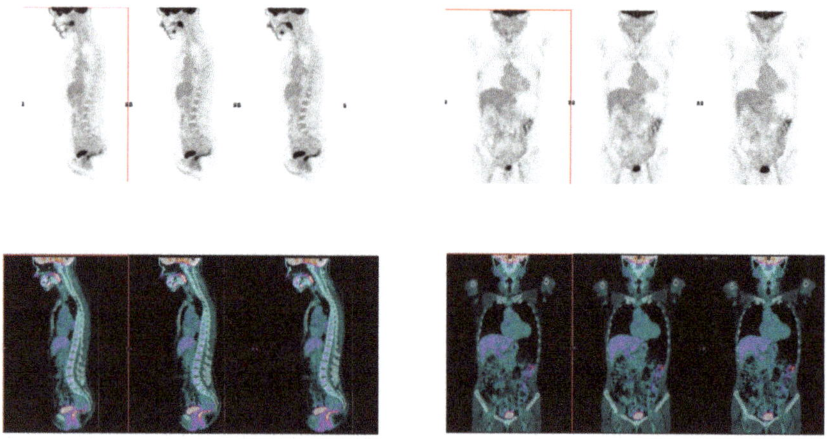

Figure 2. Sagittal (left panel) and coronal (right panel) images of whole-body ^{18}F-FDG PET/CT scan. PET (upper panels) and fused PET/CT (lower panels) images, showing faint diffuse ^{18}F-FDG colon and rectum uptake in a BC patient who reached pCR. FDG, fluorodeoxyglucose; PET/CT, positron emission tomography/computed tomography; BC, breast cancer; pCR, pathological Complete Response; PA: posterior-anterior, ; LR: left-right.

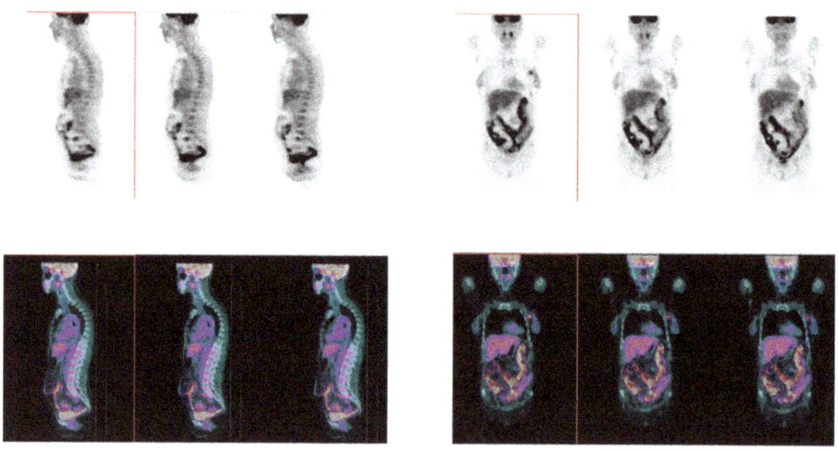

Figure 3. Sagittal (left panel) and coronal (right panel) images of whole-body ^{18}F-FDG PET/CT scan. PET (upper panels) and fused PET/CT (lower panels) images, showing diffuse nonhomogeneous ^{18}F-FDG uptake in colon and rectum in a BC patient who did not reach pCR.

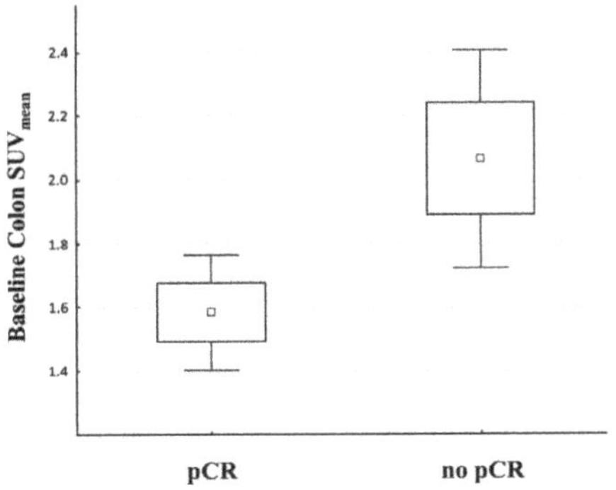

Figure 4. Distribution of baseline colon SUV_{mean} values in patients who subsequently experienced pCR after NAC compared to no pCR patients. The mean value for each group is indicated by a central square. Boxes indicate the Standard Error. Whiskers represent 1.96 × Standard Error. NAC, neoadjuvant chemotherapy; SUV_{mean}, mean standardized uptake value.

4. Discussion

Currently, it is well established that maintaining a healthy weight, being physically active, and following healthy eating patterns can reduce cancer risk and increase patients' outcome and quality of life [4,5,9,10]. Besides their impact on cancer management, dietary components have profound effects on inflammation. On one hand, human diet is a highly complex mixture of chemical compounds which make it difficult to clearly predict the final result of their overall effects. On the other hand, several dietary components have been demonstrated to modify cancer risk by modulating systemic inflammation. Some nutrients like omega-3 fatty acids and fiber can reduce inflammation, while others like refined carbohydrates, cholesterol, and saturated fatty acids have pro-inflammatory activities [29–35]. However, to our knowledge, no data has emerged, until now, concerning the influence of pro-inflammatory habits on response to NAC in BC patients. Here, we investigated this association, taking advantage of bowel ^{18}F-FDG uptake measured before NAC as a parameter for detecting bowel inflammation induced by unhealthy habits.

In agreement with previous data [28], in our study population at the time of diagnosis, the percentage of women following all recommendations included in an ideal healthy diet was very low (3.7%). Nevertheless, we noticed that most BC patients had a correct intake of fruits, vegetables, and cereals, did not drink alcohol or spirits, had an appropriate BMI, and regularly performed physical activity, just as suggested for cancer prevention and management [4,9]. On the other hand, in our cohort, few women adhered to recommendations concerning red and processed meat, which were classified by World Health Organization's International Agency for Research on Cancer as probably carcinogen and carcinogen, respectively [9]. Indeed, more than one third of the patients overemployed their consumption and about half of them abused cured meat.

Besides their effects on cancer, unhealthy habits may have an impact on bowel inflammation [29–34,36]. Thus, we assessed whether increased levels of bowel ^{18}F-FDG uptake correlated with well-known pro-inflammatory foods could be a marker of bowel inflammation. Drinks and foods were aggregated considering their well-known pro- or anti-inflammatory activities, since no data on the effects of individual foods were available. We observed that colon ^{18}F-FDG uptake positively correlated with the consumption of pro-inflammatory drinks and foods, whereas rectum ^{18}F-FDG uptake was inversely associated with anti-inflammatory food intake. Interestingly, colon SUV_{mean} showed a negative correlation trend with anti-inflammatory foods ($p = 0.05$). These findings reflect, at least in part, literature results. In fact, it has been proven that a diet rich in fats, processed meats, and sweet/salty snacks increased serum inflammatory markers [37]. Contrary, a decrease in inflammatory factors was associated with dietary patterns rich in fruits and vegetables [31–34]. Thus, we can speculate that bowel ^{18}F-FDG uptake may be affected by both pro- and anti-inflammatory foods and that we may use PET scan as an instrument for the identification of patients with increased levels of ^{18}F-FDG uptake suggestive of bowel inflammation. Furthermore, it could be useful to discuss the possible mechanisms influencing the absorption of ^{18}F-FDG in the small and large bowel. A low-carbohydrate, high-fat diet in the hours immediately before the PET examination reduced the ^{18}F-FDG uptake in the descending colon and small bowel when compared to a routine diet [38], possibly exerting its effect acting on the Randle cycle as previously demonstrated [39]. However, this condition should have marginally influenced our case since all patients were on the same dietary limitations (low-carbohydrate diet since lunch on the day preceding the PET scan). Surprisingly, we have shown here, for the first time, that colon mean ^{18}F-FDG uptake was also inversely correlated with pCR, thus suggesting a role for colonic inflammation and possibly its causative unhealthy foods and beverages in NAC response. Furthermore, since pCR is considered a surrogate endpoint for long-term outcome [40], we could speculate that unhealthy foods that trigger colonic inflammation may have an impact on long-term outcome in BC patients. These results have also been confirmed by the multivariate analysis carried out with DFA that highlighted that pCR was more strongly influenced by colon ^{18}F-FDG uptake and cured meat; these were the variables that most influenced the discrimination between pCR and not-pCR cases.

The influence of dietary patterns is widely recognized in inflammatory bowel disease. An unhealthy diet, rich in processed meat and low in fibers, has been associated with alterations in the gut microbiome and barrier function of the colonic epithelium [41]. In particular, some evidence indicated that fiber is more effective than the Mediterranean diet on influencing the gut microbiota composition [42–46]. In vivo studies have shown that disruption of the healthy gut microbiota has direct effects on the immune system by triggering a pro-inflammatory environment controlled by specific subpopulations of the immune system (e.g., natural killer cells) [47]. Therefore, a healthy diet including high-fiber foods, such as fruits, vegetables, and whole grains, could effectively reduce the risk of several metabolic diseases, including colorectal and breast cancer [48–50]. In addition, previous studies have demonstrated an association between gut microbiota and physiologic bowel ^{18}F-FDG activity both in healthy subjects and BC patients [51,52]. In healthy subjects, different levels of bowel uptake were associated with specific microbial taxa, thus suggesting that an increased ^{18}F-FDG uptake might be caused by an increment in intestinal permeability and might reflect impaired intestinal barrier function [51]. On the other hand, Yoon and colleagues found that changes in intestinal bacteria abundance in BC patients were associated with physiological intestinal ^{18}F-FDG and that the latter was associated with pro-inflammatory Tumor Necrosis Factor-α, thus further supporting the link between mucosal inflammation and physiologic intestinal ^{18}F-FDG uptake [52].

In the present study, no association was observed between the rectum SUV_{mean} and the pCR. However, compared to the rectum, the colon performs most of the large bowel functions. In addition to systemic immune control and microbiota function, colon physiology is determined by the role of the various epithelial cells that form its mucosa and are responsible for water and electrolyte absorption [53]. In addition, probiotic bacteria in the colon flora, such as *Lactobacillus* and *Bifidobacterium*, regulate micronutrient levels such as vitamins (e.g., folate-producing strains) and exert an immunomodulatory effect [54]. Therefore, the impact of inflammation at the colon level is far more compelling, due to its multiple functions that could interfere with response to NAC, than the possible inflammatory alterations of the rectum, which is primarily responsible as a reservoir for fecal content [55].

Different studies have investigated the effect of exercise in counteracting inflammation [56], and in obtaining benefits for patients with cancer [4,12]. However, in this population, no statistical significance has been reached between physical activity and bowel ^{18}F-FDG uptake, despite correlation analyses suggested an inverse association. This could be due to the limited sample size or to the lack of accurate information about type, intensity, and duration of exercise. Moreover, we could not exclude that physical activity may have an impact on inflammation and/or NAC response without affecting bowel tracts.

Similarly, despite the well-recognized role of excess body weight on BC risk [9] and inflammation [36], in this study, we did not find a correlation between bowel ^{18}F-FDG uptake and BMI. Nevertheless, we have to point out that our population was mainly composed by normal weight women (only the 3.7% of patients were obese), thus possibly mitigating the effect of this parameter on inflammation. Likewise, the lack of correlation between smoking and ^{18}F-FDG bowel uptake may be due to the small proportion of smokers in our population (i.e., 18.3%).

Some limitations of the present study should be mentioned. First of all, the monocentric design of the study impacted on the sample size. Due to the pilot nature of this study, we did not calculate an a priori formal sample size estimation, so the conclusions drawn by such a small group of patients should be taken with caution. However, this study could provide us the effect size needed to plan a larger observational study to confirm and validate our findings. Furthermore, we recognize that the present study lacks comparison of dietary diversity as a consequence of the small sample size and the omnivorous habits of most patients. On the other hand, we recruited more than 80 patients, which is on average, more than usual for PET studies [51,57–66]. Moreover, a single center study allows a better uniformity in patients' recruitment, data collection, and PET scanning procedures. This allowed us to exclude potential interferences in the results. In fact, the observed increased bowel ^{18}F-FDG uptake could be due to interfering factors different from eating habits. However, all women followed PET preparation guidelines, including at least 6 h fasting and avoiding the consumption of carbohydrates on the evening before in order to minimize variability due to the last meal before PET. In addition, during the analyses, we took into consideration all comorbidities of our population, without finding differences in bowel ^{18}F-FDG uptake between patients with a history of gastrointestinal diseases and women without. Finally, none of our patients assumed the oral hypoglycemic treatment Metformin that is known to affect intestinal ^{18}F-FDG uptake in diabetic patients [67]. Only one of the enrolled women was diabetic and she treated it through an insulin pen.

5. Conclusions

In conclusion, albeit the pilot nature of the study, the most striking result of our study is to have pinpointed an association between NAC not-complete response and increased levels of colon ^{18}F-FDG uptake, which are affected by BC patients' pro-inflammatory eating habits (i.e., consumption of unhealthy foods/drinks), for the first time. Additional investigations in enlarged cohorts are needed to confirm and validate our proof-of-principle study and to deeply investigate whether ^{18}F-FDG PET/CT could be an easy instrument for identifying BC patients who could be referred for nutritional counseling. Moreover, ongoing studies on transcriptome profiling will enhance our understanding of the interaction between bowel inflammation and NAC response in BC.

Supplementary Materials: The following supporting information can be downloaded at: https://www.mdpi.com/article/10.3390/nu15010211/s1, Figure S1: Flowchart of the recruitment and follow-up process; Figure S2: The survey fulfilled by the enrolled BC patients before NAC; Table S1: Discriminant Function Analysis (DFA) Summary.

Author Contributions: Conceptualization, L.A., A.C., A.S. and R.D.S.; Formal analysis, P.T., A.V., E.F. and R.D.S.; Funding acquisition, A.C., A.S. and R.D.S.; Investigation, P.T., L.A., M.G., A.V., M.P., C.M., F.J., C.B., E.F. and R.D.S.; Methodology, P.T., L.A., M.G., A.V., M.P., C.M., F.J., C.B., E.F. and R.D.S.; Project administration, A.C., A.S. and R.D.S.; Supervision, A.C., A.S. and R.D.S.; Writing—original draft, P.T., L.A. and R.D.S. All authors have read and agreed to the published version of the manuscript.

Funding: This work was supported by funds obtained through an Italian law that allows taxpayers to allocate the "5 × 1000" share of their payments to support a research institution of their choice (for the present study, funding donated to IRCCS Humanitas Research Hospital). Fundings for this study 018-04. The funders had no role in study design, data collection and analysis, decision to publish, or preparation of the paper.

Institutional Review Board Statement: The study was approved by the IRCCS Humanitas Research Hospital Ethics Committee (Protocol identifying number ONC/OSS-02/2019) on 19 February 2019.

Informed Consent Statement: Informed consent was obtained from all subjects involved in the study, in accordance with the Declaration of Helsinki.

Data Availability Statement: The data presented in this study are available on request from the corresponding author. The data are not publicly available due to privacy issue.

Acknowledgments: We are grateful to all BC patients enrolled in this study for their valuable collaboration.

Conflicts of Interest: The authors declare the following financial interests/personal relationships which may be considered as potential competing interests: Armando Santoro: Advisory Board: Bristol-Myers-Squibb (BMS), Servier, Gilead, Pfizer, Eisai, Bayer, Merck Sharp & Dohme (MSD). Consultancy: Arqule, Sanofi, Incyte. Speaker's Bureau: Takeda, BMS, Roche, Abb-Vie, Amgen, Celgene, Servier, Gilead, Astrazeneca, Pfizer, Arqule, Lilly, Sandoz, Eisai, Novartis, Bayer, MSD (all outside the submitted work); Rita De Sanctis: honoraria for advisory board consultancy from Novartis, Istituto Clinico Gentili, Amgen, EISAI, Lilly and Gilead (all outside the present work).

References

1. Sung, H.; Ferlay, J.; Siegel, R.L.; Laversanne, M.; Soerjomataram, I.; Jemal, A.; Bray, F. Global Cancer Statistics 2020: GLOBOCAN Estimates of Incidence and Mortality Worldwide for 36 Cancers in 185 Countries. *CA Cancer J. Clin.* **2021**, *71*, 209–249. [CrossRef] [PubMed]
2. Caprara, G.; Tieri, M.; Fabi, A.; Guarneri, V.; Falci, C.; Dieci, M.V.; Turazza, M.; Ballardini, B.; Bin, A.; Cinieri, S.; et al. Results of the ECHO (Eating habits CHanges in Oncologic patients) Survey: An Italian Cross-Sectional Multicentric Study to Explore Dietary Changes and Dietary Supplement Use; in Breast Cancer Survivors. *Front. Oncol.* **2021**, *11*, 705927. [CrossRef] [PubMed]
3. Chlebowski, R.T.; Aragaki, A.K.; Anderson, G.L.; Pan, K.; Neuhouser, M.L.; Manson, J.E.; Thomson, C.A.; Mossavar-Rahmani, Y.; Lane, D.S.; Johnson, K.C.; et al. Dietary Modification and Breast Cancer Mortality: Long-Term Follow-Up of the Women's Health Initiative Randomized Trial. *J. Clin. Oncol.* **2020**, *38*, 1419–1428. [CrossRef] [PubMed]
4. Ligibel, J.A.; Bohlke, K.; May, A.M.; Clinton, S.K.; Demark-Wahnefried, W.; Gilchrist, S.C.; Irwin, M.L.; Late, M.; Mansfield, S.; Marshall, T.F.; et al. Exercise; Diet; and Weight Management During Cancer Treatment: ASCO Guideline. *J. Clin. Oncol.* **2020**, *18*, JCO2200687. [CrossRef]
5. Mili, N.; Paschou, S.A.; Goulis, D.G.; Dimopoulos, M.A.; Lambrinoudaki, I.; Psaltopoulou, T. Obesity; metabolic syndrome; and cancer: Pathophysiological and therapeutic associations. *Endocrine* **2021**, *74*, 478–497. [CrossRef]
6. Pedersini, R.; di Mauro, P.; Bosio, S.; Zanini, B.; Zanini, A.; Amoroso, V.; Turla, A.; Vassalli, L.; Ardine, M.; Monteverdi, S.; et al. Changes in eating habits and food preferences in breast cancer patients undergoing adjuvant chemotherapy. *Sci. Rep.* **2021**, *11*, 12975. [CrossRef]
7. Steck, S.E.; Murphy, E.A. Dietary patterns and cancer risk. *Nat. Rev. Cancer.* **2020**, *20*, 125–138. [CrossRef]
8. Toledo, E.; Salas-Salvadó, J.; Donat-Vargas, C.; Buil-Cosiales, P.; Estruch, R.; Ros, E.; Corella, D.; Fitó, M.; Hu, F.B.; Arós, F.; et al. Mediterranean Diet and Invasive Breast Cancer Risk Among Women at High Cardiovascular Risk in the PREDIMED Trial: A Randomized Clinical Trial. *JAMA Intern. Med.* **2015**, *175*, 1752–1760. [CrossRef]
9. Rock, C.L.; Thomson, C.; Gansler, T.; Gapstur, S.M.; McCullough, M.L.; Patel, A.V.; Andrews, K.S.; Bandera, E.V.; Spees, C.K.; Robien, K.; et al. American Cancer Society guideline for diet and physical activity for cancer prevention. *CA Cancer J. Clin.* **2020**, *70*, 245–271. [CrossRef]
10. Thomson, C.A. Diet and breast cancer: Understanding risks and benefits. *Nutr. Clin. Pract.* **2012**, *27*, 636–650. [CrossRef]
11. Arends, J.; Bachmann, P.; Baracos, V.; Barthelemy, N.; Bertz, H.; Bozzetti, F.; Fearon, K.; Hütterer, E.; Isenring, E.; Kaasa, S.; et al. ESPEN guidelines on nutrition in cancer patients. *Clin. Nutr.* **2017**, *36*, 11–48. [CrossRef] [PubMed]
12. Carayol, M.; Ninot, G.; Senesse, P.; Bleuse, J.P.; Gourgou, S.; Sancho-Garnier, H.; Sari, C.; Romieu, I.; Romieu, G.; Jacot, W. Short- and long-term impact of adapted physical activity and diet counseling during adjuvant breast cancer therapy: The "APAD1" randomized controlled trial. *BMC Cancer.* **2019**, *19*, 737. [CrossRef] [PubMed]
13. De Cicco, P.; Catani, M.V.; Gasperi, V.; Sibilano, M.; Quaglietta, M.; Savini, I. Nutrition and Breast Cancer: A Literature Review on Prevention; Treatment and Recurrence. *Nutrients.* **2019**, *11*, 1514. [CrossRef] [PubMed]
14. Kwan, M.L.; Weltzien, E.; Kushi, L.H.; Castillo, A.; Slattery, M.L.; Caan, B.J. Dietary patterns and breast cancer recurrence and survival among women with early-stage breast cancer. *J. Clin. Oncol.* **2009**, *27*, 919–926. [CrossRef] [PubMed]

15. McTiernan, A. Diet and Prognosis in Women with Breast Cancer. *Cancer Epidemiol. Biomark. Prev.* **2021**, *30*, 252–254. [CrossRef] [PubMed]
16. Herrmann, K.; Benz, M.R.; Krause, B.J.; Pomykala, K.L.; Buck, A.K.; Czernin, J. (18)F-FDG-PET/CT in evaluating response to therapy in solid tumors: Where we are and where we can go. *Q. J. Nucl. Med. Mol. Imaging.* **2011**, *55*, 620–632.
17. Kubota, K. From tumor biology to clinical Pet: A review of positron emission tomography (PET) in oncology. *Ann. Nucl. Med.* **2001**, *15*, 471–486. [CrossRef]
18. Paydary, K.; Seraj, S.M.; Zadeh, M.Z.; Emamzadehfard, S.; Shamchi, S.P.; Gholami, S.; Werner, T.J.; Alavi, A. The Evolving Role of FDG-PET/CT in the Diagnosis; Staging; and Treatment of Breast Cancer. *Mol. Imaging Biol.* **2019**, *21*, 1–10. [CrossRef] [PubMed]
19. Holtmann, M.H.; Uenzen, M.; Helisch, A.; Dahmen, A.; Mudter, J.; Goetz, M.; Schreckenberger, M.; Galle, P.R.; Bartenstein, P.; Neurath, M.F. 18F-Fluorodeoxyglucose positron-emission tomography (PET) can be used to assess inflammation non-invasively in Crohn's disease. *Dig. Dis. Sci.* **2012**, *57*, 2658–2668. [CrossRef]
20. Rubin, D.T.; Surma, B.L.; Gavzy, S.J.; Schnell, K.M.; Bunnag, A.P.; Huo, D.; Appelbaum, D.E. Positron emission tomography (PET) used to image subclinical inflammation associated with ulcerative colitis (UC) in remission. *Inflamm. Bowel Dis.* **2009**, *15*, 750–755. [CrossRef]
21. Sena, Y.; Matsumoto, S.; Silman, C.; Otsuka, K.; Kiyota, T. Physiological 18F-FDG uptake in the normal adult anal canal: Evaluation by PET/CT. *Ann. Nucl. Med.* **2020**, *34*, 538–544. [CrossRef] [PubMed]
22. Boellaard, R.; Delgado-Bolton, R.; Oyen, W.J.; Giammarile, F.; Tatsch, K.; Eschner, W.; Verzijlbergen, F.J.; Barrington, S.F.; Pike, L.C.; Weber, W.A.; et al. FDG PET/CT: EANM procedure guidelines for tumour imaging: Version 2.0. *Eur. J. Nucl. Med. Mol. Imaging* **2015**, *42*, 328–354. [CrossRef] [PubMed]
23. CREA Centro di Ricerca Alimenti e Nutrizione. *Linee Guida Per Una Sana Alimentazione*; Centro di Ricerca Alimenti e Nutrizione: Rome, Italy, 2018; ISBN 978-88-96597-01-9. Available online: https://www.crea.gov.it/web/alimenti-e-nutrizione/-/linee-guidaper-una-sana-alimentazione-2018 (accessed on 1 September 2022).
24. De Sanctis, R.; Agostinetto, E.; Masci, G.; Ferraro, E.; Losurdo, A.; Viganò, A.; Antunovic, L.; Zuradelli, M.; Torrisi, R.M.C.; Santoro, A. Predictive Factors of Eribulin Activity in Metastatic Breast Cancer Patients. *Oncology* **2018**, *94*, 19–28. [CrossRef] [PubMed]
25. Viganò, A.; Savastano, E.; Petolicchio, B.; Toscano, M.; De Sanctis, R.; Maestrini, I.; Di Piero, V. A Study of Clinical Features and Risk Factors of Self-Referring Emergency Department Headache Patients: A Comparison with Headache Center Outpatients. *Eur. Neurol.* **2020**, *83*, 34–40. [CrossRef] [PubMed]
26. De Sanctis, R.; Viganò, A.; Giuliani, A.; Gronchi, A.; De Paoli, A.; Navarria, P.; Quagliuolo, V.; Santoro, A.; Colosimo, A. Unsupervised versus Supervised Identification of Prognostic Factors in Patients with Localized Retroperitoneal Sarcoma: A Data Clustering and Mahalanobis Distance Approach. *Biomed. Res. Int.* **2018**, *2018*, 2786163. [CrossRef]
27. Alessiani, M.; Petolicchio, B.; De Sanctis, R.; Squitieri, M.; Di Giambattista, R.; Puma, M.; Franzese, C.; Toscano, M.; Derchi, C.C.; Gilliéron, E.; et al. A Propensity Score Matching Study on the Effect of OnabotulinumtoxinA Alone versus Short-Term Psychodynamic Psychotherapy Plus Drug-of-Choice as Preventive Therapy in Chronic Migraine: Effects and Predictive Factors. *Eur. Neurol.* **2022**, *85*, 453–459. [CrossRef]
28. Clotas, C.; Serral, G.; Vidal Garcia, E.; Puigpinós-Riera, R.; DAMA Cohort Group. Dietary changes and food habits: Social and clinical determinants in a cohort of women diagnosed with breast cancer in Barcelona (DAMA cohort). *Cancer Causes Control* **2021**, *32*, 1355–1364. [CrossRef]
29. Ding, S.; Chi, M.M.; Scull, B.P.; Rigby, R.; Schwerbrock, N.M.; Magness, S.; Jobin, C.; Lund, P.K. High-fat diet: Bacteria interactions promote intestinal inflammation which precedes and correlates with obesity and insulin resistance in mouse. *PLoS ONE* **2010**, *5*, e12191. [CrossRef]
30. Malesza, I.J.; Malesza, M.; Walkowiak, J.; Mussin, N.; Walkowiak, D.; Aringazina, R.; Bartkowiak-Wieczorek, J.; Mądry, E. High-Fat; Western-Style Diet; Systemic Inflammation; and Gut Microbiota: A Narrative Review. *Cells* **2021**, *10*, 3164. [CrossRef]
31. Beam, A.; Clinger, E.; Hao, L. Effect of Diet and Dietary Components on the Composition of the Gut Microbiota. *Nutrients.* **2021**, *13*, 2795. [CrossRef]
32. Esmaillzadeh, A.; Kimiagar, M.; Mehrabi, Y.; Azadbakht, L.; Hu, F.B.; Willett, W.C. Dietary patterns and markers of systemic inflammation among Iranian women. *J. Nutr.* **2007**, *137*, 992–998. [CrossRef] [PubMed]
33. Holt, E.M.; Steffen, L.M.; Moran, A.; Basu, S.; Steinberger, J.; Ross, J.A.; Hong, C.P.; Sinaiko, A.R. Fruit and vegetable consumption and its relation to markers of inflammation and oxidative stress in adolescents. *J. Am. Diet Assoc.* **2009**, *109*, 414–421. [CrossRef] [PubMed]
34. Nanri, A.; Moore, M.A.; Kono, S. Impact of C-reactive protein on disease risk and its relation to dietary factors. *Asian Pac. J. Cancer Prev.* **2007**, *8*, 167–177. [PubMed]
35. Watzl, B.; Kulling, S.E.; Möseneder, J.; Barth, S.W.; Bub, A. A 4-wk intervention with high intake of carotenoid-rich vegetables and fruit reduces plasma C-reactive protein in healthy; nonsmoking men. *Am. J. Clin. Nutr.* **2005**, *82*, 1052–1058. [CrossRef] [PubMed]
36. Galland, L. Diet and inflammation. *Nutr. Clin. Pract.* **2010**, *25*, 634–640. [CrossRef]
37. Nettleton, J.A.; Steffen, L.M.; Mayer-Davis, E.J.; Jenny, N.S.; Jiang, R.; Herrington, D.M.; Jacobs, D.R., Jr. Dietary patterns are associated with biochemical markers of inflammation and endothelial activation in the Multi-Ethnic Study of Atherosclerosis (MESA). *Am. J. Clin. Nutr.* **2006**, *83*, 1369–1379. [CrossRef]

38. Moasses-Ghafari, B.; Fallahi, B.; Esfehani, A.F.; Eftekhari, M.; Rahmani, K.; Eftekhari, A.; Geramifar, P. Effect of Diet on Physiologic Bowel 18F-FDG Uptake. *J. Nucl. Med. Technol.* **2021**, *49*, 241–245. [CrossRef]
39. Fallahi, B.; Moasses-Ghafari, B.; Fard-Esfahani, A.; Geramifar, P.; Beiki, D.; Emami-Ardekani, A.; Eftekhari, M. Factors influencing the pattern and intensity of myocardial 18F-FDG uptake in oncologic PET-CT imaging. *Iran J. Nucl. Med.* **2017**, *25*, 52–61.
40. Cortazar, P.; Zhang, L.; Untch, M.; Mehta, K.; Costantino, J.P.; Wolmark, N.; Bonnefoi, H.; Cameron, D.; Gianni, L.; Valagussa, P.; et al. Pathological complete response and long-term clinical benefit in breast cancer: The CTNeoBC pooled analysis. *Lancet* **2014**, *384*, 164–172. [CrossRef]
41. Khalili, H.; Chan, S.S.M.; Lochhead, P.; Ananthakrishnan, A.N.; Hart, A.R.; Chan, A.T. The role of diet in the aetiopathogenesis of inflammatory bowel disease. *Nat. Rev. Gastroenterol. Hepatol.* **2018**, *15*, 525–535. [CrossRef]
42. Marsh, A.; Radford-Smith, G.; Banks, M.; Lord, A.; Chachay, V. Dietary intake of patients with inflammatory bowel disease aligns poorly with traditional Mediterranean diet principles. *Nutr. Diet.* **2022**, *79*, 229–237. [CrossRef]
43. Ioniță-Mîndrican, C.B.; Ziani, K.; Mititelu, M.; Oprea, E.; Neacșu, S.M.; Moroșan, E.; Dumitrescu, D.E.; Roșca, A.C.; Drăgănescu, D.; Negrei, C. Therapeutic Benefits and Dietary Restrictions of Fiber Intake: A State of the Art Review. *Nutrients* **2022**, *14*, 2641. [CrossRef]
44. Illescas, O.; Rodríguez-Sosa, M.; Gariboldi, M. Mediterranean Diet to Prevent the Development of Colon Diseases: A Meta-Analysis of Gut Microbiota Studies. *Nutrients* **2021**, *13*, 2234. [CrossRef] [PubMed]
45. Vernia, F.; Longo, S.; Stefanelli, G.; Viscido, A.; Latella, G. Dietary Factors Modulating Colorectal Carcinogenesis. *Nutrients* **2021**, *13*, 143. [CrossRef] [PubMed]
46. Campaniello, D.; Corbo, M.R.; Sinigaglia, M.; Speranza, B.; Racioppo, A.; Altieri, C.; Bevilacqua, A. How Diet and Physical Activity Modulate Gut Microbiota: Evidence, and Perspectives. *Nutrients* **2022**, *14*, 2456. [CrossRef]
47. Lozupone, C.A.; Stombaugh, J.I.; Gordon, J.I.; Jansson, J.K.; Knight, R. Diversity; stability and resilience of the human gut microbiota. *Nature* **2012**, *489*, 220–230. [CrossRef]
48. Abedpoor, N.; Taghian, F.; Hajibabaie, F. Physical activity ameliorates the function of organs via adipose tissue in metabolic diseases. *Acta Histochem.* **2022**, *124*, 151844. [CrossRef] [PubMed]
49. Hajibabaie, F.; Abedpoor, N.; Assareh, N.; Tabatabaiefar, M.A.; Shariati, L.; Zarrabi, A. The Importance of SNPs at miRNA Binding Sites as Biomarkers of Gastric and Colorectal Cancers: A Systematic Review. *J. Pers. Med.* **2022**, *12*, 456. [CrossRef]
50. Abedpoor, N.; Taghian, F.; Hajibabaie, F. Cross Brain-Gut Analysis Highlighted Hub Genes and LncRNA Networks Differentially Modified During Leucine Consumption and Endurance Exercise in Mice with Depression-Like Behaviors. *Mol. Neurobiol.* **2022**, *59*, 4106–4123. [CrossRef]
51. Kang, J.Y.; Kim, H.N.; Chang, Y.; Yun, Y.; Ryu, S.; Shin, H.; Kim, H.L. Gut microbiota and physiologic bowel 18F-FDG uptake. *EJNMMI Res.* **2017**, *7*, 72. [CrossRef] [PubMed]
52. Yoon, H.J.; Kim, H.N.; Bang, J.I.; Lim, W.; Moon, B.I.; Paik, N.S.; Kim, B.S.; Kim, H.L. Physiologic intestinal 18F-FDG uptake is associated with alteration of gut microbiota and proinflammatory cytokine levels in breast cancer. *Sci. Rep.* **2019**, *9*, 18273. [CrossRef] [PubMed]
53. Kumral, D.; Zfass, A.M. Gut Movements: A Review of the Physiology of Gastrointestinal Transit. *Dig. Dis. Sci.* **2018**, *63*, 2500–2506. [CrossRef] [PubMed]
54. Barone, M.; D'Amico, F.; Brigidi, P.; Turroni, S. Gut microbiome-micronutrient interaction: The key to controlling the bioavailability of minerals and vitamins? *Biofactors* **2022**, *48*, 307–314. [CrossRef]
55. Barleben, A.; Mills, S. Anorectal anatomy and physiology. *Surg. Clin. N. Am.* **2010**, *90*, 1–15. [CrossRef] [PubMed]
56. Angulo, J.; El Assar, M.; Álvarez-Bustos, A.; Rodríguez-Mañas, L. Physical activity and exercise: Strategies to manage frailty. *Redox Biol.* **2020**, *35*, 101513. [CrossRef]
57. Basu, S.; Chen, W.; Tchou, J.; Mavi, A.; Cermik, T.; Czerniecki, B.; Schnall, M.; Alavi, A. Comparison of triple-negative and estrogen receptor-positive/progesterone receptor-positive/HER2-negative breast carcinoma using quantitative fluorine-18 fluorodeoxyglucose/positron emission tomography imaging parameters: A potentially useful method for disease characterization. *Cancer* **2008**, *112*, 995–1000. [CrossRef]
58. Williams, J.M.; Rani, S.D.; Li, X.; Arlinghaus, L.R.; Lee, T.C.; MacDonald, L.R.; Partridge, S.C.; Kang, H.; Whisenant, J.G.; Abramson, R.G.; et al. Comparison of prone versus supine 18F-FDG-PET of locally advanced breast cancer: Phantom and preliminary clinical studies. *Med. Phys.* **2015**, *42*, 3801–3813. [CrossRef]
59. Jeong, Y.; Baek, S.; Park, J.W.; Joo, J.H.; Kim, J.S.; Lee, S.W. Lymph node standardized uptake values at pre-treatment 18F-fluorodeoxyglucose positron emission tomography as a valuable prognostic factor for distant metastasis in nasopharyngeal carcinoma. *Br. J. Radiol.* **2017**, *90*, 20160239. [CrossRef]
60. Husi, K.; Pinczés, L.I.; Fejes, Z.; Nagy, B., Jr.; Illés, Á.; Miltényi, Z. Combined prognostic role of TARC and interim 18F-FDG PET/CT in patients with Hodgkin lymphoma-real world observational study. *Hell. J. Nucl. Med.* **2022**, *25*, 125–131. [CrossRef]
61. Guzmán Ortiz, S.; Mucientes Rasilla, J.; Vargas Núñez, J.A.; Royuela, A.; Rodríguez Carrillo, J.L.; Dotor de Lama, A.; Navarro Matilla, M.B.; Mitjavila Casanovas, M. Evaluation of the prognostic value of the metabolic volumetric parameters calculated with 18F-FDG PET/CT and its value added to the molecular characteristics in patients with diffuse large B-cell lymphoma. *Rev. Esp. Med. Nucl. Imagen. Mol.* **2022**, *41*, 215–222. [CrossRef]

62. Wu, J.; Deng, H.; Zhong, H.; Wang, T.; Rao, Z.; Wang, Y.; Chen, Y.; Zhang, C. Comparison of 68Ga-FAPI and 18F-FDG PET/CT in the Evaluation of Patients With Newly Diagnosed Non-Small Cell Lung Cancer. *Front. Oncol.* **2022**, *12*, 924223. [CrossRef] [PubMed]
63. Guo, R.; Xu, P.; Cheng, S.; Lin, M.; Zhong, H.; Li, W.; Huang, H.; Ouyang, B.; Yi, H.; Chen, J.; et al. Comparison of Nasopharyngeal MR; 18 F-FDG PET/CT; and 18 F-FDG PET/MR for Local Detection of Natural Killer/T-Cell Lymphoma; Nasal Type. *Front. Oncol.* **2020**, *10*, 576409. [CrossRef] [PubMed]
64. Kampe, K.K.; Rotermund, R.; Tienken, M.; Thomalla, G.; Regier, M.; Klutmann, S.; Kluge, S. Diagnostic Value of Positron Emission Tomography Combined with Computed Tomography for Evaluating Critically Ill Neurological Patients. *Front. Neurol.* **2017**, *8*, 33. [CrossRef] [PubMed]
65. Bailly, C.; Eugène, T.; Couec, M.L.; Strullu, M.; Frampas, E.; Campion, L.; Kraeber-Bodéré, F.; Bodet-Milin, C. Prognostic Value and Clinical Impact of (18)FDG-PET in the Management of Children with Burkitt Lymphoma after Induction Chemotherapy. *Front. Med.* **2014**, *1*, 54. [CrossRef] [PubMed]
66. Lu, X.R.; Qu, M.M.; Zhai, Y.N.; Feng, W.; Gao, Y.; Lei, J.Q. Diagnostic role of 18F-FDG PET/MRI in the TNM staging of breast cancer: A systematic review and meta-analysis. *Ann. Palliat. Med.* **2021**, *10*, 4328–4337. [CrossRef]
67. Yoon, H.J.; Kim, H.N.; Yun, Y.; Kim, Y.; Ha, A.N.; Kim, H.L.; Kim, B.S. Background Intestinal 18F-FDG Uptake Is Related to Serum Lipid Profile and Obesity in Breast Cancer Patients. *PLoS ONE* **2015**, *10*, e0141473. [CrossRef]

nutrients
MDPI

Article

Healthy Diet, Polygenic Risk Score, and Upper Gastrointestinal Cancer Risk: A Prospective Study from UK Biobank

Wenmin Liu [1,†], Tianpei Wang [1,2,3,†], Meng Zhu [1,2] and Guangfu Jin [1,2,*]

1 Department of Epidemiology, School of Public Health, Nanjing Medical University, Nanjing 211166, China
2 Jiangsu Key Lab of Cancer Biomarkers, Prevention and Treatment, Collaborative Innovation Center for Cancer Personalized Medicine and China International Cooperation Center for Environment and Human Health, Nanjing Medical University, Nanjing 211166, China
3 Public Health Institute of Gusu School, The Affiliated Suzhou Hospital of Nanjing Medical University, Suzhou 215000, China
* Correspondence: guangfujin@njmu.edu.cn
† These authors contributed equally to this work.

Abstract: Dietary and genetic factors are considered to be associated with UGI cancer risk. However, examinations of the effect of healthy diet on UGI cancer risk and the extent to which healthy diet modifies the impact of genetic susceptibility on UGI cancer remains limited. Associations were analyzed through Cox regression of the UK Biobank data (n = 415,589). Healthy diet, based on "healthy diet score," was determined according to fruit, vegetables, grains, fish, and meat consumption. We compared adherence to healthy diet and the risk of UGI cancer. We also constructed a UGI polygenic risk score (UGI-PRS) to assess the combined effect of genetic risk and healthy diet. For the results high adherence to healthy diet reduced 24% UGI cancer risk ($HR_{\text{high-quality diet}}$: 0.76 (0.62–0.93), p = 0.009). A combined effect of high genetic risk and unhealthy diet on UGI cancer risk was observed, with HR reaching 1.60 (1.20–2.13, p = 0.001). Among participants with high genetic risk, the absolute five-year incidence risk of UGI cancer was significantly reduced, from 0.16% to 0.10%, by having a healthy diet. In summary, healthy diet decreased UGI cancer risk, and individuals with high genetic risk can attenuate UGI cancer risk by adopting a healthy diet.

Keywords: UGI cancer; dietary pattern; polygenic risk score; prospective cohort; UK Biobank

Citation: Liu, W.; Wang, T.; Zhu, M.; Jin, G. Healthy Diet, Polygenic Risk Score, and Upper Gastrointestinal Cancer Risk: A Prospective Study from UK Biobank. *Nutrients* **2023**, *15*, 1344. https://doi.org/10.3390/nu15061344

Academic Editors: Andriana Kaliora, Chara Tzavara and Charalampia Amerikanou

Received: 17 February 2023
Revised: 4 March 2023
Accepted: 7 March 2023
Published: 10 March 2023

1. Introduction

Upper gastrointestinal (UGI) cancer, including esophageal cancer (ESC) and gastric cancer (GC), account for 1.7 million new cancer cases and 1.3 million deaths each year worldwide [1]. Previous studies have identified several common environmental risk factors for UGI cancer, including tobacco [2] and alcohol consumption [3], obesity [4], physical activity [5], and dietary factors [6]. Dietary components have received an increasing amount of attention as a potentially modifiable factor [7,8].

It was estimated that 5.1–5.9% of cancer cases each year worldwide can be attributed directly to poor diet [9]. As recently reported by the World Cancer Research Fund International/American Institute for Cancer Research, the role of individual dietary components on UGI cancer risk remains controversial and limited [10]. Rather than individual dietary components, people consume diverse foods together, and the resulting complex combination of dietary components is likely to have interactive or synergistic effects [11]. In this context, dietary pattern analysis has been recommended as an approach because it considers the complexity of overall diet and can potentially facilitate public health interventions [12]. In recent years, cancer prevention guidelines have shifted from reductionist or nutrition-centric approaches to more holistic dietary concepts characterized by dietary patterns. Holistic dietary concepts emphasize how food as a whole can prevent chronic disease, associating nutrients, foods or food groups with health rather than studying the role played by nutrient/food interactions in health [13–15]. Adherence to a dietary pattern can be assessed using a priori method, which is constructed on the basis of a predefined set of criteria (generally based on guidelines) to measure diet quality in a given population [16], which would be easier to make comparisons between different studies and populations. A meta-analysis of the association of GC risk with dietary patterns indicated that Western dietary patterns (generally considered unhealthy, characterized by an increased consumption of meat, high-fat dairy products, sweets, and starchy foods) were associated with a higher GC risk, while prudent dietary patterns (generally considered healthy, characterized by higher intake of vegetables and fruits) played a protective factor [17]. A case-control study suggested that adherence to a healthy dietary pattern represented by high loadings of vegetables and fruits was associated with a lower risk of GC [18]. However, there is no large-scale prospective cohort study that systematically investigates the association between dietary patterns and UGI cancer risk.

Accumulating evidence has shown that genetic factors have major roles in the development of UGI cancer [19,20]. Recent genome-wide association studies (GWAS) have identified dozens of genetic variants associated with UGI cancer risk [21,22]. The PRSs, gathering genetic contribution and effects of all UGI cancer-associated genetic variants, have been proven to effectively predict incident cases of ESC and GC [23,24]. Both dietary factors and genetic risk play an essential role in the development of the disease. A Gene-Diet Interaction Study from the UK Biobank showed that, compared with those in the lowest intraocular pressure (IOP) polygenic risk score (PRS) quartile who consumed no caffeine, those in the highest IOP PRS quartile who consumed $\geq$321 mg/day showed a 3.90-fold higher glaucoma prevalence [25]. Moreover, one current study suggested that genetic factors modified the association between diet and cardiovascular disease (CVD) [26]. However, previous studies have typically focused on the separate effects of dietary factors and genetic factors on UGI cancer risk. Few studies provided insight into the combined effect of dietary factors and genetic factors on UGI cancer risk. It is unclear whether there is a gene-diet combined effect or interaction in the risk of UGI cancer development, as well as the extent to which participants with a high genetic risk of UGI cancer can offset that risk by adhering to a healthy diet.

In this study, we conducted dietary pattern analysis based on examining the adherence to healthy diet and investigated the association of adherence to healthy diet with UGI cancer risk using UK Biobank data. We also tested the hypothesis that dietary factors and genetic factors jointly contribute to incident UGI cancer and that adopting a healthy diet can attenuate UGI cancer risk for individuals at high genetic risk.

2. Materials and Methods

2.1. Study Design and Participants

UK Biobank is a large, population-based prospective study with genetic and phenotypic data. Between 2006 and 2010, UK Biobank recruited over 500,000 participants from the general population who were aged 40–69 years. Participants were recruited at 22 assessment centers located throughout England, Wales, and Scotland [27]. Participants completed a touch-screen questionnaire, took physical measurements, and provided biological samples at assessment centers. The basic collection details are described elsewhere [28,29]. We excluded participants with prevalent cancer (n = 46,531), those who were missing any dietary information data (n = 40,132), and individuals who had withdrawn consent for future linkage (n = 157), leaving 415,589 participants (193,083 men and 222,506 women) included in the study. First, we examined the association between the degree of adherence to healthy diet defined by healthy diet score and UGI cancer risk. Then, we compared the combined effect and interactions of healthy diet and genetic risk categories on UGI cancer risk across genetic risk groups. Last, we compared the benefit of adherence to a healthy diet within genetic risk groups (Figure 1).

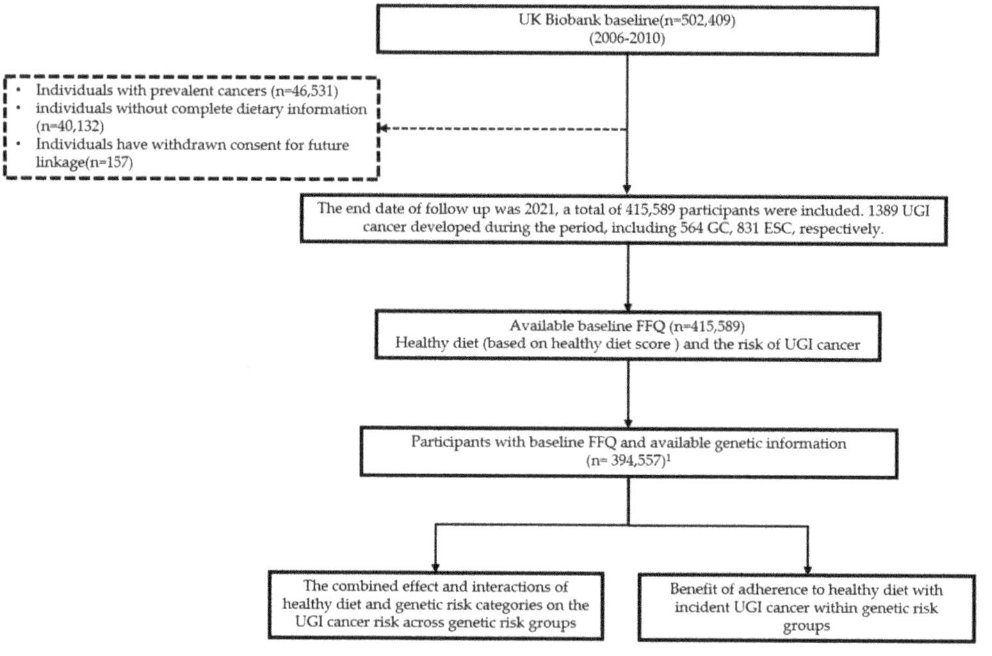

Figure 1. Study design and workflow. [1] For healthy diet and genetic risk on UGI cancer risk across and within genetic risk group analysis, participants without available genetic information were excluded (n = 21,032).

2.2. *Exposure Measurement*

2.2.1. Dietary Intake Assessment

The touch-screen questionnaire, self-completed at baseline, was used to collect the frequency of consumption of the following 12 food items over the previous year with FFQ: beef, lamb, pork, processed meat, oily fish, non-oily fish, fresh fruit, dried fruit, raw vegetables, cooked vegetables, cereal, and bread. We also created new data fields based on food items: (1) Red meat intake, (2) Total fish intake, (3) Total vegetables intake, (4) Total fruit intake, (5) Whole grains intake, and (6) Refined grains intake. We summed beef, lamb and pork intake to create red meat intake. We also summed oily fish and non-oily fish intake to generate total fish intake. To calculate total vegetables and fruit consumption respectively, we aggregated cooked vegetables and salad/raw vegetable intake as total vegetables intake, and fresh fruit and dried fruit as total fruit intake. We divided grains into whole grains and refined grains according to the type of bread and cereal mainly consumed. We defined wholemeal or wholegrain bread, bran cereal, oat cereal, and muesli as whole grains; white bread, brown bread, other bread, biscuit cereal, and other cereals as refined grains. We categorized the 12 food items into 7 food groups, including red meat, processed meat, total fish, total fruit, total vegetables, whole grains and refined grains. We also defined serving size for each baseline food items. For bread and cereal, data were provided for weekly consumption, which were converted into daily consumption. Detailed serving size and coding for each food item/food group are shown in Table S1.

2.2.2. Healthy Diet Score Estimation

We adopted seven dietary factors and cut-offs according to recommendations for dietary priorities on cardiometabolic health [30], that is, increasing fruit, vegetables, whole grains, and fish consumption, and decreasing red meat, processed meat, and refined grains intake. The healthy diet score was calculated using the seven dietary components: Total fruit

$\geq$ 4 servings/day; Total vegetables $\geq$ 4 servings/day; Total fish $\geq$ 2 servings/week; Processed meat $\leq$ 1 serving/week; Red meat $\leq$ 1.5 servings/week; Whole grains $\geq$ 3 servings/day; Refined grains $\leq$ 1.5 servings/day. Each favorable dietary factor was given one point (Table S2). The score ranged from 0 to 7; we defined score 0–1 as low-quality diet, 2–4 as intermediate-quality diet, and 5–7 as high-quality diet, according to data distribution characteristics. Next, we categorized the scores into unfavorable diet (healthy diet score < 4) and favorable diet (healthy diet score $\geq$ 4).

2.3. PRS Calculation and UGI-PRS Construction

Genotyping process and single nucleotide polymorphisms (SNPs) used in the UKB research have been described elsewhere in detail [31,32]. We extracted variants with $p < 5 \times 10^{-8}$ and minor allele frequency (MAF) $\geq$0.01 from GWAS with the largest sample size in European ancestry [23,33]. For variants that were not available in the UKB genotyping data, their strong correlated SNPs ($r^2 > 0.8$) were included in the present study. If more than one variant correlated in the same locus were reported, the SNPs with the smallest reported *p*-value were selected by using the linkage disequilibrium clumping procedure (at $r^2 < 0.2$) in PLINK. We excluded SNPs with allele mismatches or MAF differences > 0.10, compared with those in the European population of 1000 Genomes, and palindromic SNPs (A/T, G/C) with an MAF $\geq$0.45. Finally, we estimated site-specific PRS based on 13 SNPs and 3 SNPs for ESC and GC, respectively (Table S3). No SNPs were shared or in high LD ($r^2 > 0.6$) with each other in more than one site-specific PRS. Firstly, site-specific PRS was created following an additive model [34], generated by multiplying the genotype dosage of each risk allele by its respective effect size, summing all alleles together. Then, we built a UGI-PRS to assess UGI cancer risk by summing site-specific PRSs weighted by ESC and GC age-standardized incidence rate in the UK population [35]. Cancer site-specific PRS has been proven to effectively identify individuals with high risk of overall cancers and gastrointestinal cancer risk [36,37]. The UGI-PRS was divided into three levels of genetic risk: low (lowest quintile), moderate (quintiles 2–4), and high (top quintile).

2.4. Outcome Assessment

The outcomes in the study were first primary incident events due to UGI cancer (ESC and GC), which is identified through the national cancer registries of England, Wales, and Scotland, coded by the 10th revision of the International Classification of Diseases (ICD-10), as (C15) and (C16) for ESC and GC, respectively. After four years of baseline recruitment (2006–2010), UGI cancer risk in participants was assessed from baseline up to the UGI cancer diagnosis, death, completion of follow-up, or loss to follow-up, whichever occurred first. The time of risk was calculated according to date the participant attended the assessment center (Data Field: 53), date of cancer diagnosis (Data Field: 40005) and the end date of follow-up. The end date of follow-up was updated to September 2018 for Scotland and to June 2021 for England and Wales. For participants who developed a UGI cancer, time at risk was the interval between the date of cancer diagnosis and the date of attending assessment. For participants without UGI cancer, time at risk was calculated by the end date of follow-up minus date of attending assessment center.

2.5. Statistical Analysis

Cox proportional hazard models were used to investigate the associations between healthy diet and UGI cancer risk and to estimate hazards ratios (HRs) and 95% confidence intervals (CIs) with the time of follow-up used as the timeline variable. The proportional hazard assumptions were checked using Schoenfeld residuals. We determined UGI cancer risk for participants among healthy diet score categories (low-quality diet, intermediate-quality diet, and high-quality diet group). We also compared the UGI cancer risk for per two-point increase in healthy diet score. Furthermore, we investigated the combined effect and interactions of dietary and genetic factors on UGI cancer risk according to healthy diet and genetic risk categories to explore the extent to which healthy diet modified the associations between genetic susceptibility and UGI cancer risk across genetic risk groups. We examined the results for potential additive and multiplicative interaction between healthy diet and genetic risk [38]. The additive interaction was evaluated using two indexes: the relative excess risk due to the interaction (RERI) and the attributable proportion due to the interaction (AP) [39]. The 95% CIs of the RERI and AP were generated by drawing 5000 bootstrap samples from the estimation data set [40]. If there was no additive interaction, the CIs of the RERI and AP would include 0. In addition, we used RHR (ratio of HR) to evaluate the gene–diet multiplicative interactions by setting variable cross-product terms of the healthy diet with the genetic risk in the models. The 95% CIs of RHR would contain 1 if there was no multiplicative interaction. We also calculated the absolute risk as the percentage of incident UGI cancer cases occurring in each genetic risk group to compare the benefit of adherence to a healthy diet with incident UGI cancer within genetic risk groups. The absolute risk reduction was calculated according to the given groups UGI cancer incidences difference, and then the difference in five-year event rates was extrapolated among given groups. The calculation of 95% CIs for the absolute risk reduction were calculated by drawing 1000 bootstrap samples from the estimation dataset.

Two models were applied in our analyses: minimally adjusted model, adjusted for age at recruitment, sex, Townsend deprivation index, assessment center (10 regions) and ethnic background; fully adjusted model, additionally adjusted for BMI (kg/m^2, <25, 25–29.9, $\geq$30), glycosylated hemoglobin (HbA1c, mmol/mol, quintiles), smoking status (never, former, current, unknown), alcohol intake frequency (never/rare, twice or less per week, at least three times per week, unknown), education (college or university degree, no degree, unknown), multimorbidity (None, $\geq$1, unknown), physical activity (<600 MET minutes/week, 600–3000 MET minutes/week, >3000 MET minutes/week) [41] and family cancer history (yes, no, unknown) (Table S4). We additionally adjusted the top 10 genetic principal components of ancestry in the analysis including genetic risk. Missing data were coded as missing proxies (unknown) for categorical variables, while those for continuous variables were imputed with sex-specific median values.

We performed the following sensitivity analysis to further investigate the robustness of our results: (1) excluded participants who reported that they had made a major change in their diet in the past 5 years due to illness in the past 5 years (n = 41,292); (2) excluded participants followed up for less than two years (n = 1648); (3) excluded non-white participants (n = 21,680).

All statistical analyses were performed with R software for version 4.2.0 (R Core Team, Auckland, CA, USA). All p values were two-sided and $p < 0.05$ was considered statistically significant.

3. Results

3.1. Participants and Characteristics

A total of 415,589 participants (53.54% women) had available dietary data of this study. The median follow-up period was 12.12 (interquartile range: 11.32–12.84) years for UGI cancer incidence. A total of 1389 UGI cancer developed during the period, including 564 GC and 831 ESC. The baseline characteristics of participants are shown in Table 1. For 1389 UK Biobank participants (mean [SD] age, 61.21 [6.29] years; 27.93% women) with incidents of UGI cancer had a mean (SD) BMI of 28.61 (5.19) kg/m^2. Of all participants, the 16.99% with UGI cancer were current smokers, and 23.18% UGI cancer participants consumed alcohol at least three times per week. The 414,200 participants (mean [SD] age, 56.17 [8.09] years; 53.63% women) had a mean (SD) BMI of 27.39 (4.75) kg/m^2 without UGI cancer. A total of 10.25% participants with UGI cancer were current smokers, and 18.29% UGI cancer participants consumed alcohol at least three times per week.

Table 1. Baseline characteristics of participants in UK Biobank [1].

	Participants	
	With Incident UGI Cancer (n = 1389)	**Without UGI Cancer (n = 414,200)**
Age at baseline, y	61.21 ± 6.29	56.17 ± 8.09
Female	388 (27.93)	222,118 (53.63)
Townsend deprivation index, means ± SD	−1.04 ± 3.23	−1.40 ± 3.03
BMI, means ± SD	28.61 ± 5.19	27.39 ± 4.75
HbA1c, mmol/mol, means ± SD	38.12 ± 7.98	35.94 ± 6.47
Physical activity, MET minutes/week		
<600	259 (18.65)	63,772 (15.4)
600–3000	774 (55.72)	244,531 (59.04)
>3000	356 (25.63)	105,897 (25.57)
Ethnicity		
White	1346 (96.9)	392,733 (94.82)
Nonwhite	38 (2.74)	20,197 (4.88)
Unknown	5 (0.36)	1270 (0.31)
Education		
College or university degree	352 (25.34)	139,657 (33.72)
No degree	1023 (73.65)	271,190 (65.47)
Unknown	14 (1.01)	3353 (0.81)
Smoking status		
Never	506 (36.43)	228,680 (55.21)
Former	640 (46.08)	141,909 (34.26)
Current	236 (16.99)	42,454 (10.25)
Unknown	7 (0.5)	1157 (0.28)
Alcohol intake frequency		
Never/rare	621 (44.71)	184,431 (44.53)
Twice or less per week	445 (32.04)	153,836 (37.14)
At least three times per week	322 (23.18)	75,752 (18.29)
Unknown	1 (0.07)	181 (0.04)
Health status		
Multimorbidity, n (%)		
None	216 (15.55)	107,012 (25.84)
≥1	1172 (84.38)	306,847 (74.08)
Unknown	1 (0.07)	341 (0.08)
Family cancer history		
yes	832 (59.9)	257,969 (62.28)
no	380 (27.36)	109,695 (26.48)
Unknown	177 (12.74)	46,536 (11.24)

[1] Values are presented as mean ± SD or n (%) unless otherwise indicated.

3.2. Healthy Diet and the Risk of UGI Cancer

The association between adherence to healthy diet and UGI cancer risk was shown in Table 2. Individuals with a high-quality diet that included high intake of fruit, vegetables, fish and whole grains and reduced amount of red meat, processed meat and refined grains had a lower risk of UGI cancer incidents compared with those in low-quality diet group, with HR of 0.76 (95% CI: 0.62–0.93, p = 0.009). Having a two-point increase in healthy diet score was associated with a higher UGI cancer risk, with HR of 0.90 (95% CI: 0.83–0.97, p = 0.006). Similar results were noted in a series of sensitivity analyses (Table S5).

Table 2. Associations between healthy diet score and the risk of UGI cancer.

Healthy Diet	Total No. (Cases)	Minimally Adjusted Model [1]		Fully Adjusted Model [2]	
		HR (95% CI)	p Value	HR (95% CI)	p Value
Healthy diet score [3]					
Low-quality diet (0–1)	64,171 (304)	1.00 (ref)		1.00 (ref)	
Intermediate-quality diet (2–4)	297,417 (943)	0.81 (0.71, 0.92)	0.001	0.87 (0.77, 1.00)	0.047

Table 2. *Cont.*

Healthy Diet	Total No. (Cases)	Minimally Adjusted Model [1]		Fully Adjusted Model [2]	
		HR (95% CI)	*p* Value	HR (95% CI)	*p* Value
High-quality diet (5–7)	54,001 (142)	0.66 (0.54, 0.81)	<0.001	0.76 (0.62, 0.93)	0.009
Per two-point score increase	415,589 (1389)	0.84 (0.78, 0.91)	<0.001	0.90 (0.83, 0.97)	0.006
p for trend			<0.001		0.007

Definition of abbreviations: HR, hazard ratio; 95% CI: 95% confidence interval; ref, reference. 1 Minimally adjusted model: adjusted for age at recruitment, sex, assessment center (10 regions), Townsend deprivation index and ethnicity. 2 Fully adjusted model: minimally adjusted model additionally adjusted for education, BMI, glycosylated hemoglobin (HbAlc), smoking status, alcohol intake frequency, physical activity, multimorbidity and family history of cancer. 3 Healthy diet score: using available data from UK Biobank Food Frequency Questionnaire at baseline; Health diet score ranged from 0 to 7. Fruits: ≥4 servings/day; Vegetables: ≥4 servings/day; Fish: ≥2 servings/week; Processed meats: ≤1 serving/week; Unprocessed red meats: ≤1.5 servings/week; Whole grains: ≥3 servings/day; Refined grains: ≤1.5 servings/day.

3.3. Combined Effect and Interactions of Healthy Diet and Genetic Risk on UGI Cancer Risk

We determined that participants who had an unhealthy diet and were in a high genetic risk group had an approximately 1.60-fold risk of UGI cancer risk, with HR reaching 1.60 (95% CI: 1.20–2.13, $p = 0.001$), when compared with participants with a healthy diet and low genetic risk (Figure 2). The results of the sensitivity analysis did not change materially (Figure S1A–C). The RERI, AP, and RHR were not significant, which indicated no additive and multiplicative interactions of healthy diet and genetic risk on the risk of UGI cancer (Table 3).

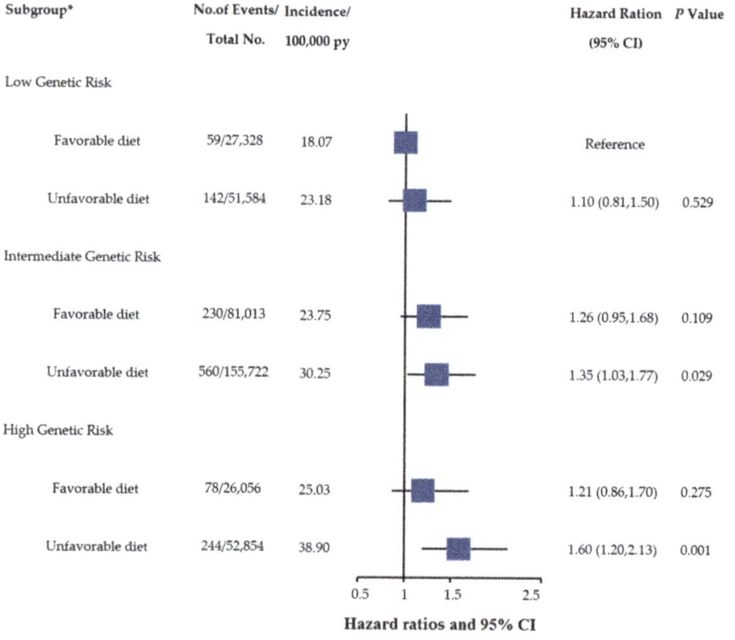

Figure 2. Risk of incident UGI cancer according to healthy diet and genetic risk categories in the UKB cohort. The HRs were estimated using Cox proportional hazard models with adjustment for age at recruitment, sex, assessment center (10 regions), ethnicity, Townsend deprivation index, education, BMI, glycosylated hemoglobin (HbA1c), smoking status, alcohol intake frequency, physical activity, multimorbidity, family history of cancer, and the first 10 principal components of ancestry. * For healthy diet and genetic risk on UGI cancer risk across and within genetic risk group analysis, participants without available genetic information were excluded (n = 21,032). Unfavorable diet (healthy diet score < 4) and Favorable diet (healthy diet score ≥ 4).

Table 3. Interaction between diet and genetic risk [1].

	PRS *	
	Intermediate	**High**
RERI (95% CI)	−0.01 (−0.47–0.31)	0.28 (−0.23–0.67)
AP (95% CI)	−0.01 (−0.29–0.26)	0.18 (−0.13–0.45)
RHR (95% CI)	1.03 (0.73–1.45)	0.84 (0.56–1.24)

Definition of abbreviations: RERI = relative excess risk due to the interaction; AP = attributable proportion due to the interaction; RHR = ratio of hazard ratio. * Defined by PRS: low (lowest quintile), intermediate (quintiles 2–4), and high (quintile 5). [1] Cox proportional hazards regression is adjusted for age at recruitment, sex, assessment center (10 regions), Townsend deprivation index, ethnicity, education, BMI, glycosylated hemoglobin (HbAlc), smoking status, alcohol intake frequency, physical activity, multimorbidity, and family history of cancer.

3.4. Benefits of Adherence to a Healthy Diet with UGI Cancer Risk

In further stratification analyses with an unhealthy dietary pattern as the reference group according to genetic risk categories, we found that in the intermediate and high genetic risk groups, similar risk reduction for UGI cancer were observed in those who adhered to a healthy dietary pattern compared to those who adhered to an unhealthy dietary pattern. Among participants with an intermediate genetic risk, the absolute five-year incidence risk of UGI cancer were 0.13 for participants with an unhealthy dietary pattern versus 0.11 for those with a healthy dietary pattern. Similarly, for individuals with high genetic risk, the absolute five-year incidence risk of UGI cancer decreased from 0.16 for participants with an unhealthy dietary pattern to 0.10 for those with a healthy dietary pattern (Table 4). The results of sensitivity analyses were similarly (Table S6).

Table 4. UGI cancer risk associated with healthy diet by genetic risk level in the UKB cohort [1].

	Low Genetic Risk		Intermediate Genetic Risk		High Genetic Risk	
Dietary Pattern	**Unfavorable**	**Favorable**	**Unfavorable**	**Favorable**	**Unfavorable**	**Favorable**
No. of cases/Person-years	142/61,2672	59/326,606	560/185,1465	230/968,645	244/627,221	78/311,605
HR (95% CI)	Ref.	0.85 (0.63–1.17)	Ref.	0.94 (0.80–1.10)	Ref.	0.78 (0.60–1.01)
p value		0.323		0.417		0.057
Absolute risk (%)-5 years (95% CI)	0.10 (0.07–0.12)	0.08 (0.05–0.10)	0.13 (0.12–0.15)	0.11 (0.09–0.12)	0.16 (0.13–0.19)	0.10 (0.07–0.13)
Absolute risk reduction (%)-5 years (95% CI)	Ref.	0.02 (−0.06–0.49)	Ref.	0.03 (0.01–0.05)	Ref.	0.06 (0.02–0.09)

[1] Cox proportional hazards regression is adjusted for age at recruitment, sex, assessment center (10 regions), Townsend deprivation index, ethnicity, education, BMI, glycosylated hemoglobin (HbAlc), smoking status, alcohol intake frequency, physical activity, multimorbidity and family history of cancer. Unfavorable dietary pattern (healthy diet score < 4) and Favorable dietary pattern (healthy diet score $\geq$ 4).

4. Discussion

In this large, prospective study using UK Biobank, we investigated dietary pattern analyses based on healthy diet and UGI cancer risk. We found that improving the quality of healthy diet was associated with a lower risk of UGI cancer. Across genetic risk groups, analysis further showed that individuals with high genetic risk and an unhealthy dietary pattern were at a greater risk of UGI cancer compared to those with low genetic risk and a healthy dietary pattern. Within genetic risk groups, analysis indicated that adherence to a healthy dietary pattern was consistently associated with a decreased absolute five-year incidence risk of UGI cancer in intermediate and high genetic risk groups.

Current studies suggested that dietary patterns analyses are regarded as good ways to explore diet and cancer risk. A systematic review and meta-analysis from prospective cohort studies supported an association between healthy dietary patterns and decreased risks of colon and breast cancer [42]. One study that focused on nutrition and breast cancer showed that adherence to a healthy dietary pattern might improve overall survival after diagnosis of breast cancer [43]. We performed dietary pattern analyses based on healthy diet score and the risk of UGI cancer. A systematic review and meta-analysis on dietary patterns and gastric cancer risk indicated that there is an approximately two-fold difference in GC risk between a 'prudent/healthy' diet, and a 'Western/unhealthy' diet [17]. A population-based case-control study suggested that a diet high in fruit and vegetables may decrease the risk of ESC cancer [44]. Another systematic review and meta-analysis suggested that a healthy dietary pattern was significantly associated with a decreased risk of ESC [45]. Our study also found similar results, i.e., that adherence to a healthy diet reduced the UGI cancer risk. We also compared the benefit of adherence to a healthy dietary pattern within genetic risk groups based on the calculation of absolute five-year incidence risk of UGI cancer. We found that individuals with intermediate and high genetic risk who adopted a healthy diet had a decreased risk of developing UGI cancer. For participants with high genetic risk, the absolute five-year incidence risk of UGI cancer was significantly reduced from 0.16% to 0.10% by having a healthy diet. Taken together, our findings along with previous evidence not only demonstrated the significance of adherence to healthy diet, but also provided collective support for public health interventions to promote a healthy dietary pattern for everyone, especially people with intermediate or high genetic risks, which will ultimately lead to a reduction of UGI cancer burden.

It has been estimated that ESC and GC could be prevented in 54% and 59% of patients in the UK, respectively [46]. It is important to understand the contribution of modifiable risk factors to UGI cancer and how they affect or add to the inherited genetic factors. At present, several studies have summarized the association between diet and nutrition and the UGI cancer risk; however, reported meta-analytic estimates from observational studies may not represent causality. Instead, they may result from common biases across studies, such as exposure measurement error, residual confounding, and publication bias, and thereby weaken the strength of the scientific evidence [47–49]. In addition, few studies have focused on the combined effect and interactions of gene–diet on the risk of UGI cancer. We systematically and comprehensively investigated the association between modifiable dietary factors with UGI cancer risk and tested the hypothesis that UGI cancer risk can be modified or reduced by adopting a healthy diet in a large prospective cohort study.

UK Biobank is a large, general population-based prospective cohort, which provides health outcomes and a wide range of potential confounders, including diet. One of the inevitable problems with large sample studies is that p values are more likely to be statistically different. In detail, a statistical p value is the distance between the data and the null hypothesis measured by an estimate of the parameter of interest. This distance is usually measured in terms of the standard deviation (standard error). The standard error shrinks as the sample size increases; in a very large sample, the standard error becomes very small, which leads to a statistically significant distance between the estimate and the null hypothesis that may be negligible. Therefore, to reduce type I errors, the null hypothesis cannot be rejected by the p-value alone in a large sample study. These problems can be solved by additionally reporting effect sizes and 95% confidence intervals (CI) [50]. In our study, we provided 95% CI as well as p values to more cautiously infer the association between healthy diet and UGI cancer.

The present study has several limitations. First, participants in the UK Biobank are of European descent; therefore, the summary statistics should be generalized to the general population with caution. Secondly, the use of self-reported recall of FFQ could introduce some level of recall bias. Third, it is generally accepted that associations between nutrients and disease should only be considered primary if the effects are independent of energy intake [51]. We were not able to adjust for total energy intake because the baseline touchscreen brief FFQ only covered some commonly consumed foods. Therefore, our findings may be biased by the differences in body size, physical activity, and metabolic efficiency resulting from energy intake. Last, covariates were evaluated only once at baseline, and changes during the follow-up or competitive risk of other illnesses may have an effect on risk estimates.

5. Conclusions

Our findings confirm and broaden the results from previous studies. Healthy diet was associated with a lower risk of UGI cancer. Dietary factors and genetic risk had a combined effect on risk of UGI cancer. Individuals with high genetic risk can attenuate UGI cancer risk by adopting a healthy dietary pattern.

Supplementary Materials: The following supporting information can be downloaded at: https://www.mdpi.com/article/10.3390/nu15061344/s1, Figure S1A: Risk of incident UGI cancer according to healthy diet and genetic risk categories in the UKB cohort after excluding participants who report changing their diet in the last 5 years due to illness; Figure S1B: Risk of incident UGI cancer according to genetic and healthy diet and genetic risk categories in the UKB cohort after excluding participants who report less than 2 years of follow-up; Figure S1C: Risk of incident UGI cancer according to healthy diet and genetic risk categories in the UKB cohort after excluding non-white participants; Table S1: Serving size and coding of intake for each touchscreen food items/food groups; Table S2: Healthy diet score factors definition; Table S3: Single nucleotide polymorphisms utilized to build the polygenic risk scores for UGI cancer; Table S4: Definition of covariates; Table S5: Associations between healthy diet and the risk of UGI cancer after excluding participants who report changing their diet due to in the last 5 years due to illness or after excluding participants who report less than 2 years of follow-up or after excluding non-white participants; Table S6: UGI cancer risk associated with healthy diet by genetic risk level after excluding participants who report changing their diet in the last 5 years due to illness or after excluding participants who report less than 2 years of follow-up or after excluding non-white participants.

Author Contributions: G.J.: Conceptualization; Resources; Supervision; Funding acquisition. W.L.: Methodology; Formal analysis; Writing—original draft; Writing—review & editing. T.W.: Writing—review & editing; Data acquisition. M.Z.: Data curation. All authors have read and agreed to the published version of the manuscript.

Funding: This work was supported by grants from the National Natural Science Foundation of China (81872702, 82125033).

Institutional Review Board Statement: The UK Biobank study was approved by the multicenter Research Ethics Committee, the National Information Governance Board for Health and Social Care in England and Wales, and the Community Health Index Advisory Group in Scotland as a Research Tissue Bank (RTB) approval (21/NW/0157) in 2011 and renewed in 2016 and 2021. http://www.ukbiobank.ac.uk/ethics (accessed on 11 November 2022).

Informed Consent Statement: Informed consent was obtained from all subjects involved in the study. Each eligible participant received an extensive baseline questionnaire, interview, and physical measurements to offer lifestyle and other potentially health-related information. In parallel, a blood sample was also collected for genotyping. Written informed consent has been obtained from the patients to publish this paper.

Data Availability Statement: Since the UK Biobank has proprietary rights to the data, data used for this analysis and the code book included in the manuscript are not publicly available. External researchers can apply to use the UK Biobank data set by registering and applying at the website: http://www.ukbiobank.ac.uk/register-apply (accessed on 11 November 2022).

Acknowledgments: The authors thank the investigators and participants of UK Biobank for their contributions to this study. This research was conducted using the UK Biobank Resource (Application Number: 60169). UK Biobank has received ethics approval from the Research Ethics Committee.

Conflicts of Interest: The authors declare no conflict of interest.

References

1. Sung, H.; Ferlay, J.; Siegel, R.L.; Laversanne, M.; Soerjomataram, I.; Jemal, A.; Bray, F. Global Cancer Statistics 2020: GLOBOCAN Estimates of Incidence and Mortality Worldwide for 36 Cancers in 185 Countries. *Cancer J. Clin.* **2021**, *71*, 209–249. [CrossRef] [PubMed]
2. Wang, S.M.; Katki, H.A.; Graubard, B.I.; Kahle, L.L.; Chaturvedi, A.; Matthews, C.E.; Freedman, N.D.; Abnet, C.C. Population Attributable Risks of Subtypes of Esophageal and Gastric Cancers in the United States. *Am. J. Gastroenterol.* **2021**, *116*, 1844–1852. [CrossRef]
3. Lu, L.; Mullins, C.S.; Schafmayer, C.; Zeißig, S.; Linnebacher, M. A global assessment of recent trends in gastrointestinal cancer and lifestyle-associated risk factors. *Cancer Commun.* **2021**, *41*, 1137–1151. [CrossRef] [PubMed]
4. Avgerinos, K.I.; Spyrou, N.; Mantzoros, C.S.; Dalamaga, M. Obesity and cancer risk: Emerging biological mechanisms and perspectives. *Metab. Clin. Exp.* **2019**, *92*, 121–135. [CrossRef] [PubMed]
5. Behrens, G.; Jochem, C.; Keimling, M.; Ricci, C.; Schmid, D.; Leitzmann, M.F. The association between physical activity and gastroesophageal cancer: Systematic review and meta-analysis. *Eur. J. Epidemiol.* **2014**, *29*, 151–170. [CrossRef]
6. Navarro Silvera, S.A.; Mayne, S.T.; Risch, H.A.; Gammon, M.D.; Vaughan, T.; Chow, W.H.; Dubin, J.A.; Dubrow, R.; Schoenberg, J.; Stanford, J.L.; et al. Principal component analysis of dietary and lifestyle patterns in relation to risk of subtypes of esophageal and gastric cancer. *Ann. Epidemiol.* **2011**, *21*, 543–550. [CrossRef]
7. Gonzalez, C.A.; Riboli, E. Diet and cancer prevention: Contributions from the European Prospective Investigation into Cancer and Nutrition (EPIC) study. *Eur. J. Cancer* **2010**, *46*, 2555–2562. [CrossRef]
8. Abnet, C.C.; Corley, D.A.; Freedman, N.D.; Kamangar, F. Diet and upper gastrointestinal malignancies. *Gastroenterology* **2015**, *148*, 1234–1243.e4. [CrossRef]
9. GBD 2019 Cancer Risk Factors Collaborators. The global burden of cancer attributable to risk factors, 2010–2019: A systematic analysis for the Global Burden of Disease Study 2019. *Lancet* **2022**, *400*, 563–591. [CrossRef]
10. World Cancer Research Fund; American Institute for Cancer Research. Diet, Nutrition, Physical Activity and Cancer: A Global Perspective. Continuous Update Project Expert Report 2018. Available online: http://dietandcancerreport.org (accessed on 11 March 2019).
11. Hu, F.B. Dietary pattern analysis: A new direction in nutritional epidemiology. *Curr. Opin. Lipidol.* **2002**, *13*, 3–9. [CrossRef]
12. Sacks, F.M.; Obarzanek, E.; Windhauser, M.M.; Svetkey, L.P.; Vollmer, W.M.; McCullough, M.; Karanja, N.; Lin, P.H.; Steele, P.; Proschan, M.A.; et al. Rationale and design of the Dietary Approaches to Stop Hypertension trial (DASH). A multicenter controlled-feeding study of dietary patterns to lower blood pressure. *Ann. Epidemiol.* **1995**, *5*, 108–118. [CrossRef]
13. Fardet, A.; Rock, E. Toward a new philosophy of preventive nutrition: From a reductionist to a holistic paradigm to improve nutritional recommendations. *Adv. Nutr.* **2014**, *5*, 430–446. [CrossRef]
14. Fardet, A.; Rock, E. Perspective: Reductionist Nutrition Research Has Meaning Only within the Framework of Holistic and Ethical Thinking. *Adv. Nutr.* **2018**, *9*, 655–670. [CrossRef]
15. Fardet, A.; Rock, E. Exclusive reductionism, chronic diseases and nutritional confusion: The degree of processing as a lever for improving public health. *Crit. Rev. Food Sci. Nutr.* **2022**, *62*, 2784–2799. [CrossRef]
16. Steck, S.E.; Murphy, E.A. Dietary patterns and cancer risk. *Nat. Rev. Cancer* **2020**, *20*, 125–138. [CrossRef]
17. Bertuccio, P.; Rosato, V.; Andreano, A.; Ferraroni, M.; Decarli, A.; Edefonti, V.; La Vecchia, C. Dietary patterns and gastric cancer risk: A systematic review and meta-analysis. *Ann. Oncol. Off. J. Eur. Soc. Med. Oncol.* **2013**, *24*, 1450–1458. [CrossRef]
18. Kim, J.H.; Lee, J.; Choi, I.J.; Kim, Y.I.; Kim, J. Dietary patterns and gastric cancer risk in a Korean population: A case-control study. *Eur. J. Nutr.* **2021**, *60*, 389–397. [CrossRef]
19. Mucci, L.A.; Hjelmborg, J.B.; Harris, J.R.; Czene, K.; Havelick, D.J.; Scheike, T.; Graff, R.E.; Holst, K.; Möller, S.; Unger, R.H.; et al. Familial Risk and Heritability of Cancer Among Twins in Nordic Countries. *JAMA* **2016**, *315*, 68–76. [CrossRef]
20. Sun, W.Y.; Yang, H.; Wang, X.K.; Fan, J.H.; Qiao, Y.L.; Taylor, P.R. The Association Between Family History of Upper Gastrointestinal Cancer and the Risk of Death from Upper Gastrointestinal Cancer-based on Linxian Dysplasia Nutrition Intervention Trial (NIT) Cohort. *Front. Oncol.* **2022**, *12*, 897534. [CrossRef]
21. Gharahkhani, P.; Fitzgerald, R.C.; Vaughan, T.L.; Palles, C.; Gockel, I.; Tomlinson, I.; Buas, M.F.; May, A.; Gerges, C.; Anders, M.; et al. Genome-wide association studies in oesophageal adenocarcinoma and Barrett's oesophagus: A large-scale meta-analysis. *Lancet Oncol.* **2016**, *17*, 1363–1373. [CrossRef]
22. Yan, C.; Zhu, M.; Ding, Y.; Yang, M.; Wang, M.; Li, G.; Ren, C.; Huang, T.; Yang, W.; He, B.; et al. Meta-analysis of genome-wide association studies and functional assays decipher susceptibility genes for gastric cancer in Chinese populations. *Gut* **2020**, *69*, 641–651. [CrossRef] [PubMed]
23. Kunzmann, A.T.; Cañadas Garre, M.; Thrift, A.P.; McMenamin, Ú.C.; Johnston, B.T.; Cardwell, C.R.; Anderson, L.A.; Spence, A.D.; Lagergren, J.; Xie, S.H.; et al. Information on Genetic Variants Does Not Increase Identification of Individuals at Risk of Esophageal Adenocarcinoma Compared to Clinical Risk Factors. *Gastroenterology* **2019**, *156*, 43–45. [CrossRef] [PubMed]
24. Jin, G.; Lv, J.; Yang, M.; Wang, M.; Zhu, M.; Wang, T.; Yan, C.; Yu, C.; Ding, Y.; Li, G.; et al. Genetic risk, incident gastric cancer, and healthy lifestyle: A meta-analysis of genome-wide association studies and prospective cohort study. *Lancet Oncol.* **2020**, *21*, 1378–1386. [CrossRef] [PubMed]
25. Kim, J.; Aschard, H.; Kang, J.H.; Lentjes, M.A.H.; Do, R.; Wiggs, J.L.; Khawaja, A.P.; Pasquale, L.R. Intraocular Pressure, Glaucoma, and Dietary Caffeine Consumption: A Gene-Diet Interaction Study from the UK Biobank. *Ophthalmology* **2021**, *128*, 866–876. [CrossRef]

26. Zhang, H.; Zeng, Y.; Yang, H.; Hu, Y.; Hu, Y.; Chen, W.; Ying, Z.; Sun, Y.; Qu, Y.; Li, Q.; et al. Familial factors, diet, and risk of cardiovascular disease: A cohort analysis of the UK Biobank. *Am. J. Clin. Nutr.* **2021**, *114*, 1837–1846. [CrossRef]
27. Collins, R. What makes UK Biobank special? *Lancet* **2012**, *379*, 1173–1174. [CrossRef]
28. Palmer, L.J. UK Biobank: Bank on it. *Lancet* **2007**, *369*, 1980–1982. [CrossRef]
29. Sudlow, C.; Gallacher, J.; Allen, N.; Beral, V.; Burton, P.; Danesh, J.; Downey, P.; Elliott, P.; Green, J.; Landray, M.; et al. UK biobank: An open access resource for identifying the causes of a wide range of complex diseases of middle and old age. *PLoS Med.* **2015**, *12*, e1001779. [CrossRef]
30. Lourida, I.; Hannon, E.; Littlejohns, T.J.; Langa, K.M.; Hyppönen, E.; Kuzma, E.; Llewellyn, D.J. Association of Lifestyle and Genetic Risk With Incidence of Dementia. *JAMA* **2019**, *322*, 430–437. [CrossRef]
31. Bycroft, C.; Freeman, C.; Petkova, D.; Band, G.; Elliott, L.T.; Sharp, K.; Motyer, A.; Vukcevic, D.; Delaneau, O.; O'Connell, J.; et al. The UK Biobank resource with deep phenotyping and genomic data. *Nature* **2018**, *562*, 203–209. [CrossRef]
32. Choi, S.H.; Weng, L.C.; Roselli, C.; Lin, H.; Haggerty, C.M.; Shoemaker, M.B.; Barnard, J.; Arking, D.E.; Chasman, D.I.; Albert, C.M.; et al. Association Between Titin Loss-of-Function Variants and Early-Onset Atrial Fibrillation. *JAMA* **2018**, *320*, 2354–2364. [CrossRef]
33. Helgason, H.; Rafnar, T.; Olafsdottir, H.S.; Jonasson, J.G.; Sigurdsson, A.; Stacey, S.N.; Jonasdottir, A.; Tryggvadottir, L.; Alexiusdottir, K.; Haraldsson, A.; et al. Loss-of-function variants in ATM confer risk of gastric cancer. *Nat. Genet.* **2015**, *47*, 906–910. [CrossRef]
34. Dai, J.; Lv, J.; Zhu, M.; Wang, Y.; Qin, N.; Ma, H.; He, Y.Q.; Zhang, R.; Tan, W.; Fan, J.; et al. Identification of risk loci and a polygenic risk score for lung cancer: A large-scale prospective cohort study in Chinese populations. *Lancet Respir. Med.* **2019**, *7*, 881–891. [CrossRef]
35. Office for National Statistics. Cancer Registration Statistics, England. Available online: https://www.ons.gov.uk/peoplepopulationandcommunity/healthandsocialcare/conditionsanddiseases/datasets/cancerregistrationstatisticscancerregistrationstatisticsengland (accessed on 26 April 2019).
36. Zhu, M.; Wang, T.; Huang, Y.; Zhao, X.; Ding, Y.; Zhu, M.; Ji, M.; Wang, C.; Dai, J.; Yin, R.; et al. Genetic Risk for Overall Cancer and the Benefit of Adherence to a Healthy Lifestyle. *Cancer Res.* **2021**, *81*, 4618–4627. [CrossRef]
37. Liu, Y.; Yan, C.; Yin, S.; Wang, T.; Zhu, M.; Liu, L.; Jin, G. Genetic risk, metabolic syndrome, and gastrointestinal cancer risk: A prospective cohort study. *Cancer Med.* **2022**, *12*, 597–605. [CrossRef]
38. Knol, M.J.; VanderWeele, T.J. Recommendations for presenting analyses of effect modification and interaction. *Int. J. Epidemiol.* **2012**, *41*, 514–520. [CrossRef]
39. Li, R.; Chambless, L. Test for additive interaction in proportional hazards models. *Ann. Epidemiol.* **2007**, *17*, 227–236. [CrossRef]
40. Assmann, S.F.; Hosmer, D.W.; Lemeshow, S.; Mundt, K.A. Confidence intervals for measures of interaction. *Epidemiology* **1996**, *7*, 286–290. [CrossRef]
41. Arthur, R.S.; Wang, T.; Xue, X.; Kamensky, V.; Rohan, T.E. Genetic Factors, Adherence to Healthy Lifestyle Behavior, and Risk of Invasive Breast Cancer Among Women in the UK Biobank. *J. Natl. Cancer Inst.* **2020**, *112*, 893–901. [CrossRef]
42. Grosso, G.; Bella, F.; Godos, J.; Sciacca, S.; Del Rio, D.; Ray, S.; Galvano, F.; Giovannucci, E.L. Possible role of diet in cancer: Systematic review and multiple meta-analyses of dietary patterns, lifestyle factors, and cancer risk. *Nutr. Rev.* **2017**, *75*, 405–419. [CrossRef]
43. De Cicco, P.; Catani, M.V.; Gasperi, V.; Sibilano, M.; Quaglietta, M.; Savini, I. Nutrition and Breast Cancer: A Literature Review on Prevention, Treatment and Recurrence. *Nutrients* **2019**, *11*, 1514. [CrossRef] [PubMed]
44. Chen, H.; Ward, M.H.; Graubard, B.I.; Heineman, E.F.; Markin, R.M.; Potischman, N.A.; Russell, R.M.; Weisenburger, D.D.; Tucker, K.L. Dietary patterns and adenocarcinoma of the esophagus and distal stomach. *Am. J. Clin. Nutr.* **2002**, *75*, 137–144. [CrossRef] [PubMed]
45. Liu, X.; Wang, X.; Lin, S.; Yuan, J.; Yu, I.T. Dietary patterns and oesophageal squamous cell carcinoma: A systematic review and meta-analysis. *Br. J. Cancer* **2014**, *110*, 2785–2795. [CrossRef] [PubMed]
46. Cancer Research UK. Cancer Statistics for the UK. Available online: https://www.cancerresearchuk.org/health-professional/cancer-statistics-for-the-uk (accessed on 5 April 2019).
47. Ioannidis, J.P. Why most discovered true associations are inflated. *Epidemiology* **2008**, *19*, 640–648. [CrossRef]
48. Ioannidis, J.P. Why most published research findings are false. *PLoS Med.* **2005**, *2*, e124. [CrossRef]
49. Dwan, K.; Gamble, C.; Williamson, P.R.; Kirkham, J.J. Systematic review of the empirical evidence of study publication bias and outcome reporting bias—An updated review. *PLoS ONE* **2013**, *8*, e66844. [CrossRef]
50. Lin, M.; Lucas, H.C.; Shmueli, G. Research Commentary—Too Big to Fail: Large Samples and the p-Value Problem. *Inf. Syst. Res.* **2013**, *24*, 906–917. [CrossRef]
51. Willett, W.C. *Nutritional Epidemiology;* Oxford University Press: New York, NY, USA, 1998.

nutrients

MDPI

Article

Updated Meal Patterns in the Child and Adult Care Food Program and Changes in Quality of Food and Beverages Served: A Natural Experimental Study

Tatiana Andreyeva [1,*], Rebecca S. Mozaffarian [2] and Erica L. Kenney [3]

[1] Department of Agricultural and Resource Economics, Rudd Center for Food Policy and Health, University of Connecticut, One Constitution Plaza, Hartford, CT 06103, USA
[2] Department of Nutrition, Harvard T.H. Chan School of Public Health, 655 Huntington Ave, Boston, MA 02115, USA
[3] Department of Social and Behavioral Sciences, Harvard T.H. Chan School of Public Health, 655 Huntington Ave, Boston, MA 02115, USA
* Correspondence: tatiana.andreyeva@uconn.edu

Abstract: With diet-related chronic diseases being the largest contributors to U.S. morbidity and mortality, identifying population-level strategies to promote healthier diets is essential. Intervention during early childhood may be particularly important. The Child and Adult Care Food Program (CACFP), a federal nutrition assistance program in the U.S. that supports serving meals and snacks in child care settings, reaches millions of U.S. children. Recent 2017 updates to CACFP's meal patterns were meant to improve the nutritional quality of food served through CACFP by providing more whole grains, fruit, and vegetables. In this study, we used a natural experimental, longitudinal study of child care centers participating in CACFP compared to nonparticipating centers to assess whether the quality of food and beverages served (per menu analysis) improved following the CACFP meal pattern changes. While we found that CACFP centers were more likely to meet several key nutrition standards in comparison to non-CACFP centers overall, there were no differences in menu quality from before to after the 2017 standards change between CACFP and non-CACFP centers. Nutrition standards for CACFP may need to be further strengthened with adequate financial and technical support given to child care programs for effective implementation.

Keywords: child nutrition; CACFP; menu analysis; preschools

Citation: Andreyeva, T.; Mozaffarian, R.S.; Kenney, E.L. Updated Meal Patterns in the Child and Adult Care Food Program and Changes in Quality of Food and Beverages Served: A Natural Experimental Study. *Nutrients* **2022**, *14*, 3786. https://doi.org/10.3390/nu14183786

Academic Editor: Anna Gramza-Michałowska

Received: 12 August 2022
Accepted: 8 September 2022
Published: 14 September 2022

1. Introduction

Poor diet quality contributes more to global morbidity than any other behavioral, environmental, occupational, or metabolic risk factor [1]. Identifying how to shift dietary patterns towards choices that promote health and prevent chronic disease is a critical public health challenge. Ensuring that young children develop healthy eating habits may be a particularly effective strategy, given that habits formed during this developmental stage can persist throughout the life course [2]. However, in the U.S., young children currently consume foods high in sugar, sodium, and saturated fat and low in fiber, so diets of many children fall short of dietary recommendations [3,4].

Child care settings are important to focus on because they have the potential to help reshape food choices for the majority of American children who attend regular non-parental care, including center-based child care [5]. Children, especially those attending all day programs, could obtain a significant fraction of daily calories from meals and snacks served and have more choices in terms of a variety and types of foods served outside of home. Child care programs can influence children's dietary intake by providing nutritious foods and beverages and implementing feeding practices that encourage healthy choices [6–8]. A large role in supporting nutrition in child care settings belongs to the USDA Child and

Adult Care Food Program (CACFP), which serves 4.6 million children per year, targeting benefits to children from households with low incomes [9]. Foods provided to children in CACFP programs must meet specific nutrition standards in order to be reimbursed with federal funds. These standards can thus help ensure that the meals and snacks served to children in child care promote healthy eating habits and nutrition.

Until recently, the nutrition standards for CACFP had not been updated to be in line with dietary science [10]. While there was some evidence that CACFP meals and snacks had some nutritional benefits over those served in non-CACFP participating programs, such as serving more fruit and vegetables and fewer sugary beverages, these benefits tended to be small and inconsistent across studies [11–16]. The standards themselves were originally designed before the onset of the childhood obesity epidemic and thus did not focus on nutrition for healthy child weight and chronic disease prevention [10]. In 2017, as a result of the Healthy, Hunger-Free Kids Act of 2010, the standards were updated for the first time since 1968 to be more in line with what dietary science has found promotes health and reduces risk of chronic disease [3]. The updated guidelines increase foods such as whole grain, fruit, and vegetable offerings and decrease added sugars [17]. Such changes could help promote healthier eating for the millions of mostly low-income children who attend CACFP-participating programs [9].

Emerging research suggests that the updated standards have been widely implemented [18] and may have resulted in some improvements in young children's dietary intake [19,20]. However, studies to date have only examined changes within CACFP-participating programs or were limited to survey data only [21]. Without a comparison group of non-participating programs, it is difficult to assess whether any improvements in child care menu and/or meal quality are truly due to the updated meal pattern standards in CACFP or whether improvements may be due to natural time trends or some other influence on child care meals. Finally, prior research assessing menu quality in child care settings has been limited to cross-sectional studies and often lack comparison groups of non-participating programs [22,23].

This study aims to address this evidence gap by leveraging pre-update and post-update menu data on the reported meals and snacks served in a sample of both CACFP-participating and -nonparticipating child care centers. Using a longitudinal, difference-in-difference approach, this study evaluates the extent to which the 2017 CACFP meal pattern changes were associated with improvements in the quality of meals and snacks served. We hypothesized that: (1) CACFP participation would be associated with better menu quality as compared to nonparticipation both before and after the updates and (2) that menu quality would improve in CACFP-participating programs from before and after the updates while staying the same in nonparticipating programs.

2. Materials and Methods

2.1. Study Sample and Design

The sample consisted of licensed child care centers in the state of Connecticut (CT) that served meals and/or snacks to children 0–5 years of age (not including school-age after-school programs). To identify eligible centers, administrative records of licensed child care centers in the state were obtained in 2016 (prior to the implementation of the updated CACFP standards) and again in 2019 from the CT Office of Early Childhood. This list was compared to the State Department of Education's records on CACFP participation. All CACFP-participating centers were assumed to serve meals and/or snacks by design. For non-participating centers, which could either opt to serve their own food or have parents/guardians send in meals and snacks, researchers verified via telephone whether the center served meals and/or snacks. The study invited all CACFP-participating centers (n = 176) and a sample of non-CACFP centers serving children ages 0–5 (n = 391 randomly selected from 733 non-CACFP centers serving young children and known to provide food) to participate. Non-CACFP centers located in low-income communities were oversampled to provide a stronger comparison group with CACFP centers given that CACFP participation is associated with serving low-income communities.

To recruit these centers, center directors were contacted by email to participate in an online survey about food service practices affecting young children. This initial e-mail requested that the survey be completed by the person most familiar with the food service at the child care center. As part of the survey, participants were asked to submit a copy of their current menu to show food selections of children in their care. All participants consenting into the study and completing the survey received a monetary incentive (a USD 20 gift card). The survey data were reported elsewhere [13,21,24].

We collected data from 237 child care centers in 2016 as part of what was initially planned to be a cross-sectional study and then conducted a second data collection with a sample of 201 centers in 2019, many of whom had participated in the original data collection, after the updated CACFP standards had been implemented. Of these centers, directors provided weekly menus for 92 centers at both baseline and follow-up, which resulted in a final sample for longitudinal analysis of 55 non-CACFP participating centers and 37 centers participating in CACFP.

2.2. Measures

2.2.1. Child Care Center-Level Characteristics

Child care providers completed surveys about center characteristics, including: CACFP participation status; whether the center was accredited by the National Association for the Education of Young Children (NAEYC); whether the center participates in Connecticut's School Readiness program (which is an indicator of how many low-income children are served by the center); and non-profit/for-profit status. Providers also reported whether they had received CACFP training and what types of kitchen facilities their center had. Center capacity, i.e., the maximum number of enrolled children allowed per center, was determined from administrative licensing data, and data on household income, racial/ethnic composition, and poverty status for the U.S. Census block in which the center was located were assessed by linking the center's address with data from the 2013–2017 American Community Survey [25].

2.2.2. Menu Quality

For each participating child care center, the research team extracted one week's worth of menu data. Each food and beverage item listed on the menu for each meal on the sampled week was classified into one or more food and beverage categories relevant to CACFP nutrition standards using a coding protocol from prior studies, which is available upon request [13,24,26]. We categorized milks as low-fat (1%) or skim, reduced fat (2%), whole fat, and flavored (for any flavor and percent fat). We classified other beverages as 100% fruit juice; fruit drinks/other sugar-sweetened beverages (SSBs, such as sodas, lemonades, non-100% juice drinks); or water (tap or bottled, unsweetened). We classified foods as follows: fruit of any type (excluding juice); vegetables of any type; dark green vegetables (e.g., spinach, broccoli, mixed greens); red and orange vegetables (e.g., carrots, squash, red peppers); starchy vegetables (e.g., potatoes, corn, green peas); legumes; grain products of all types; whole-grain products (first ingredient on nutrition label is a whole grain); refined grain products (first ingredient on package label is not a whole grain); grain products of unknown whole-grain content; grain-based desserts (cookies, pastries, granola bars); meat/meat alternates of any type (includes both meats and vegetarian protein sources); lean meats (poultry); red or processed meats (beef, lamb, goat, ham, sausage, hot dogs); pre-fried meats (chicken nuggets, chicken fingers, fish sticks); nuts/nut butter; eggs; natural cheese; processed cheese (American cheese, spray cheese, Velveeta); yogurt; tofu or other soy products; sweets (non-grain-based: ice cream, candy, pudding); and other foods. Cereals were evaluated for compliance with CACFP standards on sugar content by checking manufacturer's website for nutrition information.

With these coded data, we first assessed the extent to which center menus met basic CACFP standards for serving all required meal/snack components. Centers were coded as meeting basic component standards if they reported serving the following to preschool-age children: three required components for breakfast (fruits/vegetables, grain/meat/meat alternate, low-fat/skim milk); five required components for lunch (fruit, vegetable, grain, low-fat/skim milk, meat/meat alternate), and two components for snack (any of two of the five lunch components above) [9].

We then assessed if the food and beverage items on the menus met each of five of the 2017 updated daily minimum CACFP nutritional requirements for meals [17] served for each day in the sampled menu week. The five standards assessed were: (1) only unflavored low-fat/skim milk served to children ages 2–5 years old; (2) at least one serving of whole grains per day; (3) both fruit and vegetable served at lunch; (4) 100% fruit/vegetable juice limited to one serving per day; and (5) no grain-based desserts served as grains. We chose these five standards, as we were able to assess them with menu data; some other standards, such as eliminating on-site frying or limiting the sugar content of cereals and yogurts, were not possible to evaluate with menus at both time points. The minimum standard for serving fruit and vegetable at lunch was assessed per meal, and the other four minimum CACFP standards were assessed as meeting or not meeting the standards per day.

Lastly, we evaluated if menus met the following voluntary CACFP "best-practice" standards, which are encouraged but not required by CACFP: (1) serve fruit or vegetable as one of the two components at every snack; (2) serve whole fruit more often than juice; (3) serve dark green vegetables at least weekly; (4) serve red and orange vegetables at least weekly; (5) serve beans and legumes at least weekly; (6) serve starchy vegetables least weekly; (7) serve other vegetables at least weekly; (8) serve at least two servings of whole grains daily; (9) serve lean meats, nuts, and legumes only; (10) limit to one serving or less of processed meats weekly; (11) limit cheese to natural cheese only; (12) limit to one serving or less of pre-fried meat weekly; and (13) provide no non-creditable foods with added sugars (e.g., candy, sugary drinks). The best-practice standard for fruit or vegetable at snack was assessed per snack; whole fruit served more often than juice and at least two daily servings of whole grains were assessed per day; the remaining best-practice standards were assessed on a weekly basis.

2.3. Statistical Analysis

We calculated the proportion of centers meeting each of the accreditations, type of preschool, staff CACFP training, meal preparation methods, access to food service facilities and equipment, and the mean (±SD) for descriptive and demographic data. Data were stratified by CACFP participation status and baseline (2016) versus follow-up (2019). Because many centers in our sample provided snacks but not lunches (these were household-provided) and thus were not comparable to centers providing lunches on several of the CACFP outcomes, we also stratified our analysis by centers that provided snacks only and centers that provided meals only or meals and snacks.

We calculated the proportion of centers by CACFP status, year, and food service type that met each CACFP meal component requirement, minimum nutrition standard, and best-practice standard. We also calculated the mean (±SD) total number of minimum standards met per day and best-practice standards met each week.

To test whether the updated minimum CACFP standards increased the nutritional quality of meals served in CACFP centers (based on menus), we used a difference-in-difference approach. Generalized estimating equations (GEE) logistic regression models were used to calculate the odds of meeting each of the CACFP minimum standards and voluntary best practices for menus, accounting for the clustering of menu observation days within centers. Models included a term for (1) time comparing 2019 to 2016; (2) CACFP status comparing CACFP to non-CACFP centers, and (3) an interaction term between time and CACFP status to estimate whether CACFP centers experienced additional changes in menu quality from 2016–2019 beyond the overall time effect (the difference-in-difference estimator and the CACFP policy change effect). To select demographic and center-characteristic covariates for the model, we used a backward selection process; the only covariate that meaningfully altered parameter estimates and remained significant was center capacity; thus, all models also adjusted for center capacity. Two-sided tests of significance were conducted in these models; alpha was set at $p < 0.05$.

The 2016 data collection was approved by the University of Connecticut Institutional Review Board in June 2015, and the 2019 study was approved in May 2018. Analyses were conducted on SAS 9.4. (Cary, NC, USA).

3. Results

The administrative and sociodemographic characteristics of the centers are presented in Table 1, showing important differences between CACFP- and non-CACFP participating programs. Of the 37 CACFP-participating centers in the final sample, 100% served meals and/or snacks, while only 18 (32.7%) of the 55 non-CACFP participating centers did; the remaining 37 non-CACFP centers served snacks only. All CACFP centers in the sample were accredited by NAEYC, while only 54.6% of non-CACFP centers were; similarly, participation in CT's School Readiness program was much more common among CACFP centers. Capacity was higher on average in CACFP centers, and tuition was lower. CACFP centers were more likely to have kitchens on site or to heat up foods on site delivered from a vendor than non-CACFP centers, who were more likely to have partial kitchens or microwaves in classrooms. CACFP centers were located in areas that tended to have lower median household incomes and lower proportions of the population identifying as non-Hispanic White or as having a college degree and higher levels of poverty.

The frequencies with which centers serving meals and snacks ($n = 37$ CACFP centers, $n = 18$ non-CACFP centers) met each of the minimum CACFP meal component and nutrition standards as well as the voluntary best-practice standards, are shown in Table 2; frequencies for centers only serving snacks are shown in Appendix A Table A1. At both time points, CACFP centers were significantly more likely to meet the minimum meal component requirements for breakfasts, lunches, and snacks than non-CACFP centers. They were also more likely to meet the 2017 updated minimum nutrition standards for serving both fruits and vegetables at lunch and serving at least one whole grain per day at both time points; both CACFP and non-CACFP centers were overwhelmingly likely, before and after the updates, to serve low-fat milk to 2–5-year-olds and to limit the serving of 100% juice to once per day. Meanwhile, CACFP and non-CACFP centers had similar frequencies of not serving grain-based desserts in 2016, and these frequencies both increased in 2019. In general, compliance with the voluntary best-practice standards was less prevalent among both CACFP and non-CACFP centers, with some exceptions. Particularly poor compliance was for serving only lean meats, nuts, and legumes for meat/meat alternates and serving natural cheese only. Although rates of serving at least two servings of whole grains improved over time, only one-third of non-CACFP centers and about one-half of CACFP centers managed to satisfy this standard in 2019.

Results from the difference-in-difference GEE models are shown in Table 3 for centers serving meals and snacks (results for centers serving snacks only are shown in Appendix A Table A2). In evaluating the role of overall time trends and controlling for CACFP status, centers were more likely to limit grain-based desserts and serve whole fruits more often than juice in 2019 as compared to 2016 (aOR for grain-based dessert standard, 2019 vs. 2016: 4.40, 95% CI: 1.2, 16.2; aOR for whole fruit best-practice standard, 2019 vs. 2016: 4.23, 95% CI 1.5, 11.6); however, no other significant differences from 2016 to 2019 were observed. There were, however, several significant differences in the likelihood of meeting CACFP standards between non-CACFP and CACFP centers at baseline. In 2016, CACFP centers were more likely to serve both fruits and vegetables at lunch (aOR = 4.42, 95% CI 1.25, 15.5), serve whole grains at least once per day (aOR = 2.72, 95% CI: 1.3, 5.9), and not serve foods with added sugars (aOR = 3.50, 95% CI: 1.1, 11.3). However, when evaluating whether the change in CACFP standards in 2017 was associated with additional increases in meeting standards for CACFP centers, no significant changes at $p < 0.05$ were found.

Table 1. Demographics characteristics of $n = 92$ Connecticut-licensed child care centers in 2019 [1].

	Non-CACFP Centers ($n = 55$)	CACFP Centers ($n = 37$)	p-Value [3]
Meal and/or snack served, n (%):			<0.001
Meal only or meal and snack served	18 (32.7)	37 (100)	
Snack only served	37 (67.3)	0 (0.0)	
Accreditations, n (%):			
Center accredited by the National Association for the Education of Young Children (NAEYC)	30 (54.6)	37 (100)	<0.001
Center has School Readiness program slots available	8 (15.1)	30 (81.1)	<0.001
Center has a sponsoring organization, n (%) [1]	12 (21.8)	13 (39.4)	0.08
Center capacity, mean (SD)	88.4 (60.1)	117.9 (94.3)	0.10
Weekly tuition, mean (SD) [2]	USD 262.4 (78.7)	USD 210.2 (57.9)	0.003
Meal preparation, n (%):			<0.001
Kitchen on site from scratch	11 (20.0)	19 (51.4)	
Kitchen of another center and delivered	0 (100)	0 (100)	
Purchased from a vendor and heat up on site	5 (9.1)	5 (13.5)	
Purchased from a vendor and do not heat up	8 (14.6)	7 (18.9)	
Other	31 (56.4)	6 (16.2)	
Kitchen and equipment, n (%):			0.003
Full kitchen on site	25 (45.5)	24 (62.2)	
Partial kitchen on site (fridge/freezer alone)	21 (38.2)	10 (27.0)	
Microwaves in the classroom	15 (27.3)	1 (2.7)	
No cooking equipment on site	4 (7.3)	1 (2.7)	
Sociodemographic characteristics of center's Census tract, mean (SD):			
Median household income	USD 98,717 (USD 43,852)	USD 52,255 (USD 24,791)	<0.001
Percent of population with college degree	47.1 (16.4)	28.6 (21.4)	<0.001
Percent of households below poverty level	6.4 (6.2)	21.2 (13.7)	<0.001
Percent of population that is non-Hispanic White	78.8 (12.9)	47.2 (28.9)	<0.001

Notes: [1] Center characteristics were also measured in 2016; as these variables did not differ significantly in 2019, we present 2019 values only for clarity, but 2016 data are available on request; [2] measured in 2016 data collection; [3] p-values are from chi-square or Fisher's exact tests for categorical variables and t-tests for continuous variables.

Table 2. Frequencies of menus meeting CACFP minimum standards and voluntary best practices among non-CACFP and CACFP centers providing meals and snacks in Connecticut, 2016–2019.

	2016 (Old CACFP Meal Pattern)		2019 (New CACFP Meal Pattern)	
	Non-CACFP Centers	CACFP Centers	Non-CACFP Centers	CACFP Centers
N menu days	90	179	89	185
N meals and snacks	246	492	232	544
Breakfast, n (%)	35 (14.2)	159 (32.3)	54 (23.3)	175 (32.2)
Lunch, n (%)	86 (35.0)	174 (35.4)	80 (34.5)	180 (33.1)
Snack (a.m./p.m.), n (%)	125 (50.8)	159 (32.3)	98 (42.2)	189 (34.7)
Meeting CACFP meal component requirements, n (%) of meals [1]				
Breakfasts	30 (85.7)	153 (96.2)	40 (74.1)	159 (90.9)
Lunches	57 (66.3)	149 (85.6)	50 (62.5)	163 (90.6)
Snacks	112 (89.6)	154 (96.8)	81 (82.7)	184 (97.4)

Table 2. *Cont.*

Meeting CACFP minimum nutrition standards, n (%) of days				
Fruit and vegetables as 2 components at lunch	66 (76.7)	161 (92.5)	64 (80.0)	164 (91.1)
Unflavored skim/low-fat milk to 2–5-year-old children	84 (93.3)	178 (99.4)	84 (94.4)	183 (98.9)
≥1 serving of whole grains per day	35 (38.9)	110 (61.5)	49 (55.1)	160 (86.5)
Limit 100% fruit/vegetable juice to 1 serving per day	84 (93.3)	173 (96.7)	89 (100)	184 (99.5)
No grain-based desserts served	78 (86.7)	150 (83.8)	86 (96.6)	177 (95.7)
Mean (SD) total number of minimum standards met per day	4.1 (0.8)	4.4 (0.7)	4.4 (0.7)	4.9 (0.6)
Following CACFP voluntary best practices, n (%) of snacks, days, and weeks				
Meal-level best practices				
Fruit or vegetable as 1 of 2 components at snack	93 (74.4)	95 (61.6)	61 (62.2)	107 (56.6)
Daily best practices				
Whole fruit served more often than juice	59 (65.6)	156 (87.2)	78 (87.6)	168 (90.8)
≥2 servings of whole grains per day	13 (14.4)	50 (27.9)	27 (30.3)	101 (54.6)
Weekly best practices				
Dark green vegetables ≥ 1 time/week	11 (61.1)	25 (69.4)	10 (55.6)	27 (73.0)
Red and orange vegetables ≥ 1 time/week	14 (77.8)	33 (91.7)	15 (83.3)	32 (86.5)
Bean and peas ≥ 1 time/week	11 (61.1)	13 (36.1)	10 (55.6)	20 (54.1)
Starchy vegetables ≥ 1 time/week	13 (72.2)	27 (75.0)	14 (77.8)	29 (78.4)
Other vegetables ≥ 1 time/week	18 (100)	34 (94.4)	17 (94.4)	36 (97.3)
Processed meats ≤ 1 time/week	15 (83.3)	28 (77.8)	15 (83.3)	37 (100)
Pre-fried meats ≤ 1 time/week	18 (100)	28 (77.8)	15 (83.3)	36 (97.3)
Natural cheese only	0 (0.0)	0 (0.0)	1 (5.9)	3 (8.6)
Only lean meats, nuts, and legumes served for meat/meat alternates	4 (22.2)	3 (8.3)	3 (17.7)	3 (8.3)
No non-creditable foods with added sugars served	7 (38.9)	25 (69.4)	10 (55.6)	23 (62.2)
Mean (SD) total number of best practices followed per week	7.3 (1.6)	6.9 (1.5)	7.2 (1.5)	7.6 (1.5)

Notes: [1] Meal component requirements for CACFP are as follows: Breakfast (three components): milk, vegetables, fruit, or both and grains; Lunch (five components): milk, meat/meat alternates, vegetables*, fruit, and grains; Snacks (two components from milk, meat/meat alternates, vegetables, fruit, grains). *2 servings of vegetables were not considered due to low availability of this practice.

Table 3. Difference-in-difference model results: centers servings meals or meals and snacks.

	2019 (Post-Update) vs. 2016 (Pre-Update)			**CACFP vs. Non CACFP**			**Interaction between 2019 to 2016 and CACFP Status**		
	Odds Ratio	**95% CI**	***p*-Value**	**Odds Ratio**	**95% CI**	***p*-Value**	**Odds Ratio**	**95% CI**	***p*-Value**
CACFP minimum nutrition standards									
Fruit and vegetables as 2 components at lunch	1.09	0.38, 3.1	0.88	4.42	1.25, 15.5	0.02	0.78	0.20, 3.04	0.72
Unflavored skim/low fat milk to 2–5-year-old children	0.76	0.13, 4.3	0.75	13.0	0.98 172.2	0.05	0.66	0.12, 3.71	0.64
At least one serving of whole grains per day	1.97	0.98, 4.0	0.06	2.72	1.3, 5.9	0.01	2.02	0.75, 5.4	0.17

Table 3. *Cont.*

	2019 (Post-Update) vs. 2016 (Pre-Update)			CACFP vs. Non CACFP			Interaction between 2019 to 2016 and CACFP Status		
	Odds Ratio	95% CI	*p*-Value	Odds Ratio	95% CI	*p*-Value	Odds Ratio	95% CI	*p*-Value
No grain-based dessert	4.40	1.2, 16.2	0.03	0.80	0.42, 1.5	0.49	0.97	0.22, 4.4	0.97
Best Practice Standards									
Fruit or vegetable as 1 of 2 components at snack	0.41	0.12, 1.46	0.17	0.19	0.08, 0.50	<0.001	2.27	0.59, 8.82	0.23
Whole fruit served more often than juice	4.23	1.5, 11.6	0.005	4.82	1.4, 16.2	0.01	0.34	0.10, 1.15	0.08
At least 2 servings of whole grains per day	2.77	1.0, 7.7	0.05	2.53	0.80, 8.0	0.11	1.14	0.35, 3.71	0.82
Dark green vegetables at least once per week	0.79	0.27, 2.3	0.66	1.58	0.47, 5.27	0.46	1.5	0.38, 6.0	0.56
Red and orange vegetables at least once per week	1.50	0.29, 7.9	0.63	4.33	0.82, 22.8	0.08	0.97	0.04, 3.29	0.38
Bean and peas at least once per week	0.77	0.32, 1.83	0.55	0.36	0.11, 1.2	0.10	2.97	0.92, 9.6	0.07
Starchy vegetables at least once per week	1.50	0.64, 3.5	0.35	1.26	0.36, 4.4	0.72	0.80	0.21, 3.0	0.73
Lean meats, nuts, and legumes only	0.71	0.20, 2.5	0.60	0.31	0.06, 1.6	0.16	1.41	0.27, 7.2	0.68
No non-creditable foods with added sugars, including candy, sugary drinks	1.99	0.61, 6.5	0.26	3.50	1.1, 11.3	0.04	0.36	0.09, 1.4	0.14

Notes: All regressions adjusted for center capacity. The CACFP 2017 minimum nutrition standard of limiting 100% fruit/vegetable juice to one serving per day as well as the CACFP best-practice standards of serving other vegetables at least once per week, serving processed meats once a week or less, serving pre-fried meats once a week or less, and serving only natural cheese were not modeled due to lack of non-positive cases in certain cells.

4. Discussion

In this longitudinal, natural experimental study of whether the 2017 CACFP nutrition standard changes were associated with better adherence to best-practice standards for child care menus, we found that CACFP-participating centers had a higher likelihood of meeting several nutrition standards compared to non-participating centers. CACFP centers were overall more likely than non-CACFP centers to serve both a fruit and a vegetable at lunch, to serve whole grains at least once per day, to serve fruit more often than juice, and to refrain from serving foods with added sugars. However, at the same time, we also found that there was no evidence of an extra improvement in nutrition standard adherence for the CACFP centers related to the 2017 CACFP standard changes when comparing these centers to non-CACFP centers over time.

Our findings regarding the higher likelihood of CACFP centers meeting several key nutrition standards (although not all) are similar to prior investigations comparing CACFP to non-CACFP centers using cross-sectional study designs [19,21]. Previous studies have also found that CACFP centers are more likely to serve whole grains and limit foods with added sugars such as sugary drinks and candy when compared to non-CACFP centers while also finding few consistent differences between CACFP and non-CACFP centers when it comes to the serving of fruit and vegetables and the types of meats/meat alternates served [11,13–15,21,24,27]. Menus from both CACFP participating and non-participating centers were highly adherent with the CACFP nutrition standards. The centers in this sample, regardless of CACFP participation, overwhelmingly served low-fat milk to 2–5-year-olds, offered fruits and vegetables daily, and limited serving juice and grain-based desserts, making it difficult to impossible for there to be a significant difference for CACFP centers.

Previous evaluations of the 2017 CACFP nutrition standards change have examined changes among CACFP-participating programs only [18–20,26]. One study found significant increases in CACFP-participating center directors reporting not serving sugary cereals or flavored milk, serving 100% whole grain products, and serving processed meats less than once a week from before to after the change in standards [18]. Another found increases in the likelihood of meeting the whole-grain standard but no other significant changes [26], alongside findings that children increased intake of whole grains and fruit but did not increase intake of vegetables or milk [19]. Sisson et al. found that children's intake of fiber increased and sugar decreased at CACFP centers from before to after the standards change, but that there were no significant changes in adherence to CACFP requirements and best practices [20]. Taken together with our study, which is the first to use a comparison group not exposed to CACFP to control for time effects independent of the standards change, these studies suggest that while there may have been some small improvements in what was served to and consumed by children in CACFP centers, this may not have been due to the change in standards themselves.

The minimal impact of the standards change for CACFP can be contextualized by comparing with the impact of the standards changes for the National School Lunch and Breakfast Programs that were also required by the Healthy, Hunger-Free Kids Act of 2010 [28]. The USDA requested recommendations based on the best available science for both programs from the National Academies of Medicine (NAM) [10,29]. The recommendations for the school meals programs were nearly adopted in full by the USDA [30], and Congress authorized increases in school meal reimbursements to account for the higher food costs associated with compliance with these new standards as well as funding to support implementation [31]. In contrast, the final standards issued by USDA for CACFP were necessarily less of a change than what had been recommended by NAM because no corresponding increase in CACFP meal reimbursements was authorized, and no grants were set aside to support child care programs in updating kitchen equipment or training staff [17]. The differing strength in standards for school meals versus CACFP as well as the differing investment of resource appears to have resulted in differing impacts; while several studies have found the changes to school meals have significantly improved students' nutritional quality [32,33] and reduced obesity risk [34,35], our study, alongside the prior evaluations within CACFP centers [19,20,26], suggests very limited impact. Although following the stronger voluntary best-practice standards for CACFP could potentially lead to larger impacts, there appears to be little indication that participating programs are aware of them. Training by CACFP state agencies to support adoption of the best practices may help. Resources for providing and encouraging such training by state CACFP and licensing agencies are needed.

Strengths and Limitations

Strengths of this study include the use of a longitudinal study design with a comparison group of centers not participating in CACFP. Nearly all studies of CACFP to date have used cross-sectional designs or have not leveraged a comparison group of nonparticipating centers, making it difficult to test whether any changes observed in CACFP menus over time are due to CACFP's nutrition standards changes or whether they might be due to an underlying trend affecting all centers regardless of CACFP participation. An additional strength was the use of centers' menus for analyzing nutrition standard adherence rather than relying on self-report measures, which may not be accurate.

A limitation of this study is the unbalanced covariates between the CACFP and non-CACFP centers. Because this study could not use a randomized, controlled trial design, we made every effort to draw a comparison sample of non-participating (i.e., unexposed to the nutrition standards changes) centers that would be as similar to CACFP centers as possible, limiting the sample to only those providing meals and attempting to oversample from low-income areas. Despite these efforts, the nonparticipating centers were substantially different from the CACFP centers. They were less likely to be accredited, to serve children from households with low incomes, and to have kitchen facilities while having higher tuition. Such differences may be impossible to avoid, given that CACFP is uniquely focused on centers that serve children from households with low income and centers make a choice to join the program. Despite these differences, our analyses indicated that these covariates were not associated with menu quality, suggesting the differences were unlikely to be confounders. However, the possibility of confounding cannot be ruled out. Another important limitation of the longitudinal menu assessment was a lower response rate than in cross-sectional surveys [21,24].

This study also was not able to evaluate whether there were differences in children's actual consumption of foods and beverages served over time. Our study does not evaluate what foods and beverages were served to children or consumed by children. While evaluating menus and their compliance with nutrition standards is important, data on meals served and consumed are needed and should be prioritized in future studies. Such studies can answer a critical question of nutrient density, which menu assessments cannot. Finally, our analysis was limited to one state (Connecticut), which could affect generalizability of our findings, particularly with respect to states with much higher CACFP participation rates. Future research should prioritize using national longitudinal samples to understand impacts of the CACFP meal patterns across diverse areas and child care programs. While nationally representative studies examining CACFP exist [16], they are not currently longitudinal and do not involve examination of meals for unexposed children in non-CACFP programs.

5. Conclusions

Participation in CACFP is associated with increased likelihood of serving whole grains, fruits and vegetables instead of juice, and refraining from serving foods and beverages with added sugars. However, the updates to CACFP's nutrition standards via the Healthy, Hunger-Free Kids Act of 2010 were not associated with improved likelihood of centers adhering to nutrition standards when accounting for changes that may have been occurring over time for centers regardless of CACFP participation. To effect more beneficial changes to the foods and beverages served to the millions of children who receive subsidized meals through CACFP, stronger nutrition standards for the program, with appropriate accompanying financial and technical support for implementation, may be needed.

Author Contributions: Conceptualization, T.A. and E.L.K.; methodology, T.A.; formal analysis, R.S.M., E.L.K., and T.A.; data curation, R.S.M. and T.A.; writing—original draft preparation, E.L.K., T.A., and R.S.M.; writing—review and editing, E.L.K., T.A., and R.S.M.; supervision, T.A. and E.L.K.; funding acquisition, T.A. All authors have read and agreed to the published version of the manuscript.

Funding: This study was supported by Healthy Eating Research, a national program of the Robert Wood Johnson Foundation (grant number 77232). Data collection for 2016 was additionally supported by the National Institute of Food and Agriculture, U.S. Department of Agriculture, award number 2015-69001-23243, and for 2019 by a grant administered by the Child Health and Development Institute (CHDI) of Connecticut and funded by the Children's Fund of Connecticut, Connecticut Health Foundation, and Newman's Own Foundation.

Institutional Review Board Statement: The study was conducted in accordance with the Declaration of Helsinki, and approved by the Institutional Review Board of the University of Connecticut (protocol codes H14-339 of June 10, 2015 and H18-029 of 30 May 2018).

Informed Consent Statement: Informed consent was obtained from all subjects involved in the study.

Acknowledgments: The authors wish to thank all child care centers that provided data to this study. The authors are grateful to research assistants of the UConn Rudd Center for help with data collection and entry.

Conflicts of Interest: The authors declare no conflict of interest.

Appendix A

Table A1. Frequencies of meeting CACFP minimum standards and voluntary best practices among non-CACFP and CACFP centers providing snacks only, Connecticut, 2016–2019.

	2016 (Pre-Update)		2019 (Post-Update)	
	Non-CACFP Centers	CACFP Centers	Non-CACFP Centers	CACFP Centers
N days	254	163	247	174
N snacks	449	169	408	189
Morning snack, n (%)	234 (52.1)	11 (6.5)	227 (55.6)	15 (7.9)
Afternoon snack, n (%)	215 (47.9)	158 (93.5)	181 (44.4)	174 (92.1)
N weeks	51	33	50	35
CACFP meal component requirements, n (%)				
Morning snack	198 (84.6)	10 (90.9)	202 (89.0)	15 (100)
Afternoon snack	175 (81.4)	153 (96.8)	161 (89.0)	169 (97.1)
CACFP minimum nutrition standards, n (%)				
N days	254	163	247	174
Unflavored skim/low-fat milk to 2–5-year-old children	253 (99.6)	162 (99.4)	242 (98.0)	172 (98.9)
At least one serving of whole grains per day	76 (29.9)	27 (16.6)	67 (27.3)	58 (33.3)
Limit 100% fruit/vegetable juice to one serving per day	237 (93.3)	162 (99.4)	233 (94.3)	174 (100)
No grain-based dessert	220 (86.6)	133 (81.6)	224 (90.7)	166 (95.4)
Mean total number of minimum standards met per day, SD	3.1 (0.7)	3.0 (0.6)	3.1 (0.7)	3.3 (0.6)
CACFP best practices, n (%)				
Morning snacks, n	234	11	227	15
Fruit or vegetable as 1 of 2 components, n (%)	139 (59.4)	7 (63.6)	140 (61.7)	11 (73.3)
Afternoon snacks, n	215	158	181	174
Fruit or vegetable as 1 of 2 components, n (%)	149 (69.3)	96 (60.8)	120 (66.3)	96 (55.2)
Day service, n	254	163	247	174
Whole fruit served more often than juice, n (%)	106 (41.7)	59 (36.2)	118 (47.8)	72 (41.4)
≥2 servings of whole grains per day, n (%)	7 (2.8)	2 (1.2)	6 (2.4)	1 (0.6)
Weeks, n	51	33	50	35
Dark green vegetables at least once per week, n (%)	3 (5.9)	1 (3.0)	3 (6.0)	1 (2.9)
Red and orange, n (%) vegetables at least once per week, n (%)	19 (37.3)	4 (12.1)	13 (26.0)	5 (14.3)
Bean and peas at least once per week, n (%)	17 (33.3)	5 (15.2)	17 (34.0)	6 (17.1)
Starchy vegetables at least once per week, n (%)	3 (5.9)	0 (0.0)	1 (2.0)	1 (2.9)
Other vegetables at least once per week, n (%)	20 (39.2)	10 (30.3)	25 (50.0)	14 (40.0)
Processed meats one serving per week or less, n (%)	51 (100)	33 (100)	50 (100)	35 (100)
Pre-fried meats one serving per week or less, n (%)	51 (100)	33 (100)	50 (100)	35 (100)
Natural cheese only, n (%)	4 (11.8)	5 (21.7)	13 (44.8)	13 (50.0)
Lean meats, nuts, and legumes only, n (%)	21 (95.5)	11 (100)	21 (95.5)	10 (100)
No non-creditable foods with added sugars, including candy, sugary drinks, n (%)	30 (58.8)	32 (97.0)	38 (76.0)	33 (94.3)
Mean total number of best practices met per week, SD	4.7 (2.1)	4.3 (1.5)	5.1 (1.6)	4.5 (1.4)

Notes: Meal component requirements for CACFP are as follows: Snacks (two components from milk, meat/meat alternates, vegetables, fruit, grains).

Table A2. Difference-in-difference model results: centers servings snacks only.

	n	2016 (Pre-Update) vs. 2019 (Post-Update)			CACFP vs. Non CACFP			Interaction between 2019 to 2016 and CACFP Status		
		Odds Ratio	95% CI	*p*-Value	Odds Ratio	95% CI	*p*-Value	Odds Ratio	95% CI	*p*-Value
CACFP minimum nutrition standards										
Unflavored skim/low-fat milk to 2–5-year-old children	838	0.16	0.005, 5.2	0.30	2.40	0.0005, 11,895	0.84	3.06	0.11, 86.5	0.51
At least one serving of whole grains per day	838	0.87	0.58, 1.3	0.52	0.49	0.29, 0.85	0.01	2.62	1.2, 5.5	0.01
No grain-based dessert	838	1.47	0.87, 2.5	0.15	0.69	0.43, 1.1	0.13	3.12	1.2, 7.9	0.02
Best practice standards										
Morning snack: Fruit or vegetable as 1 of 2 components	487	1.11	0.72, 1.70	0.65	1.02	0.19, 5.6	0.98	1.39	0.16, 12.1	0.76
Afternoon snack: Fruit or vegetable as 1 of 2 components	728	0.86	0.58, 1.3	0.48	0.70	0.40, 1.2	0.20	0.91	0.50, 1.6	0.75
Whole fruit served more often than juice	838	1.24	0.84, 1.8	0.28	0.81	0.46, 1.4	0.46	0.91	0.49, 1.7	0.77
At least 2 servings of whole grains per day	838	0.87	0.29, 2.6	0.80	0.38	0.07, 2.0	0.26	0.54	0.03, 8.9	0.66
Dark green vegetables at least once per week	169	1.01	0.23, 4.5	0.99	0.40	0.02, 7.4	0.54	0.97	0.22, 4.3	0.94
Red and orange vegetables at least once per week	169	0.59	0.27, 1.3	0.18	0.23	0.07, 0.77	0.02	2.00	0.43, 9.2	0.38
Bean and peas at least once per week	169	0.98	0.51, 1.9	0.96	0.30	0.10, 0.94	0.04	1.23	0.36, 4.2	0.74
Other vegetables at least once per week	169	1.51	0.90, 2.5	0.12	0.68	0.26, 1.7	0.42	0.93	0.29, 3.0	0.90
Natural cheese only	169	6.10	1.6, 22.6	0.007	2.07	0.49, 8.8	0.32	0.60	0.10, 3.6	0.58
No non-creditable foods with added sugars, including candy, sugary drinks	169	2.13	1.0, 4.3	0.04	30.5	2.6, 352.9	0.006	0.23	0.01, 3.9	0.31

Notes: All regressions adjusted for center capacity. The CACFP 2017 minimum nutrition standard of limiting 100% fruit/vegetable juice to one serving per day as well as the CACFP best practice standards of serving starchy vegetables at least once per week, serving processed meats once a week or less, serving pre-fried meats once a week or less, and serving lean meats, nuts, and legumes only were not modeled due to lack of non-positive cases in certain cells.

References

1. GBD 2015 Risk Factors Collaborators. Global, regional, and national comparative risk assessment of 79 behavioural, environmental and occupational, and metabolic risks or clusters of risks, 1990–2015: A systematic analysis for the Global Burden of Disease Study 2015. *Lancet* **2016**, *388*, 1659–1724. [CrossRef]
2. Savage, J.S.; Fisher, J.O.; Birch, L.L. Parental Influence on Eating Behavior: Conception to Adolescence. *J. Law Med. Ethic.* **2007**, *35*, 22–34. [CrossRef] [PubMed]
3. U.S. Department of Health and Human Services. 2015–2020 Dietary Guidelines for Americans. 2015. Available online: https://health.gov/our-work/food-nutrition/previous-dietary-guidelines/2015 (accessed on 1 August 2022).
4. Welker, E.B.; Jacquier, E.F.; Catellier, D.J.; Anater, A.S.; Story, M.T. Room for Improvement Remains in Food Consumption Patterns of Young Children Aged 2–4 Years. *J. Nutr.* **2018**, *148*, 1536S–1546S. [CrossRef] [PubMed]
5. de Brey, C.; Snyder, T.D.; Zhang, A.; Dillow, S.A. *Digest of Education Statistics 2019*; NCES 2021-009; National Center for Education Statistics: Washington, DC, USA, 2021.
6. American Academy of Pedatrics; American Public Health Association; National Resource Center for Health and Safety in Child Care and Early Education. Preventing Childhood Obesity in Early Care and Education: Selected Standards from Caring for Our Children: National Health and Safety Performance Standards; Guidelines for Early Care and Education Programs, 3rd Edition. 2012. Available online: https://nrckids.org/CFOC/Childhood_Obesity (accessed on 1 August 2022).
7. Kaphingst, K.M.; Story, M. Child care as an untapped setting for obesity prevention: State child care licensing regulations related to nutrition, physical activity, and media use for preschool-aged children in the United States. *Prev. Chronic. Dis.* **2009**, *6*, A11.
8. Reynolds, M.A.; Cotwright, C.J.; Polhamus, B.; Gertel-Rosenberg, A.; Chang, D. Obesity Prevention in the Early Care and Education Setting: Successful Initiatives across a Spectrum of Opportunities. *J. Law Med. Ethic.* **2013**, *41*, 8–18. [CrossRef]
9. United States Department of Agriculture, Food and Nutrition Service. Child and Adult Care Food Program (CACFP), Participation. 2018. Available online: https://fns-prod.azureedge.us/sites/default/files/resource-files/12ccfypart-9.pdf (accessed on 11 August 2022).
10. Murphy, S.P.; Yaktine, A.L.; Suitor, W.; Moats, S. *Child and Adult Care Food Program: Aligning Dietary Guidance for All*; National Academies Press: Washington, DC, USA, 2010.
11. Ritchie, L.D.; Boyle, M.; Chandran, K.; Spector, P.; Whaley, S.E.; James, P.; Samuels, S.; Hecht, K.; Crawford, P. Participation in the Child and Adult Care Food Program Is Associated with More Nutritious Foods and Beverages in Child Care. *Child. Obes.* **2012**, *8*, 224–229. [CrossRef]

12. Korenman, S.; Abner, K.S.; Kaestner, R.; Gordon, R.A. The Child and Adult Care Food Program and the nutrition of preschoolers. *Early Child. Res. Q.* **2013**, *28*, 325–336. [CrossRef]
13. Andreyeva, T.; Kenney, E.L.; O'Connell, M.; Sun, X.; Henderson, K.E. Predictors of Nutrition Quality in Early Child Education Settings in Connecticut. *J. Nutr. Educ. Behav.* **2018**, *50*, 458–467. [CrossRef]
14. Erinosho, T.; Vaughn, A.; Hales, D.; Mazzucca, S.; Gizlice, Z.; Ward, D. Participation in the Child and Adult Care Food Program Is Associated with Healthier Nutrition Environments at Family Child Care Homes in Mississippi. *J. Nutr. Educ. Behav.* **2018**, *50*, 441–450. [CrossRef]
15. Liu, S.T.; Graffagino, C.L.; Leser, K.A.; Trombetta, A.L.; Pirie, P.L. Obesity Prevention Practices and Policies in Child Care Settings Enrolled and Not Enrolled in the Child and Adult Care Food Program. *Matern. Child Health J.* **2016**, *20*, 1933–1939. [CrossRef]
16. Glenn, M.E.; Patlan, K.; Connor, P.; Stidsen, C.; Ball, S.; Peterson, K.E.; Olsho, L.E.; Gola, A.A.H.; Copeland, K.A. Dietary Intakes of Children Enrolled in US Early Child-Care Programs During Child-Care and Non-Child-Care Days. *J. Acad. Nutr. Diet.* **2022**, *122*, 1141–1157.e3. [CrossRef] [PubMed]
17. United States Department of Agriculture. *Child and Adult Care Food Program: Meal Pattern Revisions Related to the Healthy, Hunger-Free Kids Act of 2010*; United States Department of Agriculture: Washington, DC, USA, 2015.
18. Chriqui, J.F.; Leider, J.; Schermbeck, R.M.; Sanghera, A.; Pugach, O. Changes in Child and Adult Care Food Program (CACFP) Practices at Participating Childcare and Education Centers in the United States Following Updated National Standards, 2017–2019. *Nutrients* **2020**, *12*, 2818.
19. Kenney, E.L.; Poole, M.K.; Cory, H.; Cradock, A.L. Impact of changes to the Child and Adult Care Food Program on children's dietary intake in family child care homes. *Public Heal. Nutr.* **2020**, *23*, 2016–2023. [CrossRef] [PubMed]
20. Sisson, S.B.; Sleet, K.; Rickman, R.; Love, C.; Bledsoe, A.; Williams, M.; Jernigan, V.B.B. Impact of the 2017 Child and Adult Care Food Program Meal Pattern Requirement Change on Menu Quality in Tribal Early Care Environments: The Food Resource Equity and Sustainability for Health Study. *Curr. Dev. Nutr.* **2020**, *4*, 12–22. [CrossRef] [PubMed]
21. Andreyeva, T.; Sun, X.; Cannon, M.; Kenney, E.L. Implementation of Minimum Nutrition Standards and Best Practices in Childcare Centers. *J. Acad. Nutr. Diet.* **2021**, *121*, 2454–2463. [CrossRef]
22. Dave, J.M.; Cullen, K.W. Foods Served in Child Care Facilities Participating in the Child and Adult Care Food Program: Menu Match and Agreement with the New Meal Patterns and Best Practices. *J. Nutr. Educ. Behav.* **2018**, *50*, 582–588. [CrossRef]
23. Williams, B.D.; Sisson, S.B.; Stinner, E.L.; Hetrick, H.N.; Dunlap, M.; Graef-Downard, J.; Eliot, K.; Finnell, K.; Salvatore, A.L. Quality of Nutrition Environments, Menus and Foods Served, and Food Program Achievement in Oklahoma Family Child Care Homes. *Nutrients* **2021**, *13*, 4483. [CrossRef]
24. Andreyeva, T.; Henderson, K.E. Center-Reported Adherence to Nutrition Standards of the Child and Adult Care Food Program. *Child. Obes.* **2018**, *14*, 421–428. [CrossRef]
25. United States Census Bureau. 2013–2017 American Community Survey 5-Year Estimates. 2017. Available online: https://www.census.gov/programs-surveys/acs/technical-documentation/table-and-geography-changes/2017/5-year.html (accessed on 1 August 2022).
26. Poole, M.K.; Cradock, A.L.; Kenney, E.L. Changes in Foods Served and Meal Costs in Boston Family Child Care Homes after One Year of Implementing the New Child and Adult Care Food Program Nutrition Standards. *Nutrients* **2020**, *12*, 2817. [CrossRef]
27. Gurzo, K.; Lee, D.L.; Ritchie, K.; Yoshida, S.; Vitale, E.H.; Hecht, K.; Ritchie, L.D. Child Care Sites Participating in the Federal Child and Adult Care Food Program Provide More Nutritious Foods and Beverages. *J. Nutr. Educ. Behav.* **2020**, *52*, 697–704. [CrossRef]
28. Department of Agriculture, Food and Nutrition Service. National School Lunch Program and School Breakfast Program: Nutrition standards for all foods sold in school as required by the Healthy, Hunger-Free Kids Act of 2010. *Fed. Regist.* **2016**, *81*, 50131–50151.
29. Stallings, V.A.; Suitor, C.W.; Taylor, C.L. *School Meals: Building Blocks for Healthy Children*; National Academies Press: Washington, DC, USA, 2010.
30. Schwartz, C.; Wootan, M.G. How a Public Health Goal Became a National Law: The Healthy, Hunger-Free Kids Act of 2010. *Nutr. Today* **2019**, *54*, 67–77. [CrossRef] [PubMed]
31. *FACT SHEET: Healthy, Hunger-Free Kids Act School Meals Implementation*; United States Department of Agriculture Food and Nutrition Service: Washington, DC, USA, 2014.
32. Kinderknecht, K.; Harris, C.; Jones-Smith, J. Association of the Healthy, Hunger-Free Kids Act with Dietary Quality Among Children in the US National School Lunch Program. *JAMA* **2020**, *324*, 359–368. [CrossRef] [PubMed]
33. Johnson, D.B.; Podrabsky, M.; Rocha, A.; Otten, J.J. Effect of the Healthy Hunger-Free Kids Act on the Nutritional Quality of Meals Selected by Students and School Lunch Participation Rates. *JAMA Pediatr.* **2016**, *170*, e153918. [CrossRef]
34. Kenney, E.L.; Barrett, J.L.; Bleich, S.N.; Ward, Z.J.; Cradock, A.L.; Gortmaker, S.L. Impact of the Healthy, Hunger-Free Kids Act on Obesity Trends. *Health Aff.* **2020**, *39*, 1122–1129. [CrossRef] [PubMed]
35. Richardson, A.S.; Weden, M.M.; Cabreros, I.; Datar, A. Association of the Healthy, Hunger-Free Kids Act of 2010 With Body Mass Trajectories of Children in Low-Income Families. *JAMA Netw. Open* **2022**, *5*, e2210480. [CrossRef] [PubMed]

nutrients

MDPI

Article

Dietary Patterns, Socio-Demographic Predictors Thereof, and Associations of Dietary Patterns with Stunting and Overweight/Obesity in 1–<10-Year-Old Children in Two Economically Active Provinces in South Africa

Marjanne Senekal [1,*], Johanna H. Nel [2], Gabriel Eksteen [3] and Nelia P. Steyn [1]

1 Department of Human Biology, University of Cape Town, Cape Town 7925, South Africa; nelia.steyn@uct.ac.za
2 Department of Logistics, Stellenbosch University, Stellenbosch 7602, South Africa; jhnel@sun.ac.za
3 Clinical and Experimental Endocrinology, Department of Chronic Diseases and Metabolism, KU Leuven, 3000 Leuven, Belgium; gabrieljohannes.eksteen@kuleuven.be
* Correspondence: marjanne.senekal@uct.ac.za; Tel.: +27-(22)-423-8684

Abstract: A review of the literature showed that there were only a few studies that reported on the dietary patterns of children in South Africa. The aim of the present study was to characterise the dietary patterns of children aged 1–<10 years who were studied during the Provincial Dietary Intake Survey (PDIS) in 2018 and to investigate the socio-demographic predictors thereof, as well as the associations with stunting and overweight/obesity. Dietary pattern analysis was conducted within three age groups, namely 1–<3-year-olds, 3–<6-year-olds, and 6–<10-year-olds using iterated principal factor analysis with varimax rotation and 24 h recall data from the PDIS. The dietary patterns that emerged seem to be far from ideal. Energy-dense, nutrient-poor patterns were included in the top three strongest patterns in all three age groupings that were investigated. Few of the dietary patterns included vegetables other than starchy vegetables, fruit, dairy, quality proteins, and unrefined carbohydrates. There were no associations between any of the dietary patterns and stunting or overweight/obesity in the children. Key predictors of greater adherence to the mostly unhealthy patterns included indicators of a higher socio-economic status in all three age groups, as well as having an obese mother in the 6–<10-year-old group. Key predictors of greater adherence to the mostly healthy patterns were a higher wealth index and having an obese mother in the two younger groups, with no predictors in the 6–<10-year-old group. We conclude that the dietary patterns of children in the Western Cape contain strong elements of the energy-dense, nutrient-poor dietary patterns. Interventions to improve the dietary intake of children should be directed at both poorer and higher income communities.

Keywords: dietary patterns; stunting; overweight; obesity; children < 10 years; South Africa

Citation: Senekal, M.; Nel, J.H.; Eksteen, G.; Steyn, N.P. Dietary Patterns, Socio-Demographic Predictors Thereof, and Associations of Dietary Patterns with Stunting and Overweight/Obesity in 1–<10-Year-Old Children in Two Economically Active Provinces in South Africa. *Nutrients* **2023**, *15*, 4136. https://doi.org/10.3390/nu15194136

Academic Editors: Chara Tzavara, Charalampia Amerikanou and Andriana Kaliora

Received: 17 July 2023
Revised: 7 September 2023
Accepted: 15 September 2023
Published: 25 September 2023

1. Introduction

The Joint Malnutrition Estimates (JME) report that, globally, 149.2 million children under five years of age are stunted, 45.4 million are wasted, and 38.9 million are overweight. Overweight and obesity have emerged as global challenges, affecting low- and high-income countries alike. Their figures show that the share of both adults and children who are overweight or obese is increasing globally [1].

Optimal nutrition during infancy and childhood is essential for the growth and development of children. Malnutrition in children can result in a range of nutrition-related health problems, depending on the nature thereof, i.e., deficient and/or excessive intakes of energy, macro, and micronutrients. Childhood malnutrition phenotypes that are a concern in South Africa include stunting, overweight/obesity, iron anaemia (iron deficiency), and low vitamin A levels [2].

Malnutrition during childhood has been linked to poor cognitive development and, over time, the promotion of the development of non-communicable disease [3]. According to Liberali et al. [4], one of the most serious public health problems of the 21st century in both low-, middle-, and high-income countries is the fact that childhood obesity is a predictor of risk of obesity in adulthood. Not only is childhood obesity associated with adult obesity but also with increased risk for numerous non-communicable diseases (NCDs) such as diabetes, cardiovascular diseases (CVDs), and certain cancers [5,6].

In South Africa, as in many sub-Saharan African countries, a double burden of malnutrition is prevalent in children [2]. A recent study of children under five in two provinces in South Africa clearly illustrates the prevalence of the double burden of malnutrition in children per se. The prevalence of stunting in under five-year-olds was 21.6%, underweight 5.6%, overweight 10.3%, and obesity 7.0%. In the five- to younger than 10-year-old group, 6.7% were stunted, 6.8% underweight, 13.4% overweight, and 6.8% obese [7]. A comparison of anthropometric indicators in 1–9-year-olds in the Western Cape (WC) province between 1999 and 2018 shows no increase in stunting in the total group of children (14.9–13.7%), but an increase in overweight/obesity (13.5–21.8%) [8]. The 2022 Obesity Atlas projects that 28.2% of 5–9-year-old children in South Africa will be obese by 2030 [9].

While many NCDs may not present until later in life, dietary habits and patterns are generally formed in childhood and continue into adulthood [10]. A World Health Organization (WHO) systematic review concluded that a person's predisposition to developing obesity and other NCDs can be influenced as early as during foetal development and during childhood [11]. This is confirmed by researchers exploring what is known as the "life course of disease approach" to NCD prevention and control. This approach contends that all areas of life should be considered in relation to controlling NCD rates from preconception and prenatal care, to infancy, childhood, adolescence, adulthood, and the elderly [12]. The "life course of disease approach" thus advocates for the importance of educating and introducing intervention strategies from early in life.

The Global Burden of Disease Study (GBDS) noted that the highest rates of mortality and disability-adjusted life years (DALYs) related to diet were recorded in low- and middle-income countries [13]. The risk factors associated with the highest rate of mortality in these countries descending order were as follows: "a diet high in sodium, low in whole grains, low in fruits, low in nuts and seeds, low in vegetables, low in fish/seafood (omega-3-rich fatty acids), low in fibre, low in poly-unsaturated fats, low in legumes, high in trans fats, low in calcium, high in sugar-sweetened beverages, high in processed meat, low in milk and high in red meat" [13].

The nutrition transition documented in many LMICs is closely associated with the development of many NCDs [14,15]. This transition is described as a major change in diet from a traditional pattern which is largely composed of unrefined grains, legumes, fruit, and vegetables towards one of refined starches, added sugars, and animal products, fats and oils combined with a reduced intake of fruit, vegetables, legumes, nuts, and seeds [14,15], which has also been described as an obesogenic dietary pattern [4]. The obesogenic dietary pattern has been associated with several potential drivers, including economic growth, fast urbanization, and increase in the production and consumption of highly processed foods, as well as socio-economic and lifestyle characteristics [16]. Numerous studies have shown that these ultra-processed foods do not have the same nutritional benefits as unprocessed foods [17].

The analysis of dietary patterns for investigation of diet–disease interactions was introduced in the 1980s [18–21]. Dietary patterns provide a broader picture of food and nutrient consumption that reflects not only the effect of individual nutrients, but also the contribution of dietary variety and interactions between dietary components [22]. According to Malekua et al. [23], identifying dietary patterns that consider the overall eating habits, rather than focusing on individual foods or simple counts of consumed foods, better helps to understand the combined effects of dietary components. This information can be translated into suitable dietary guidelines for children to prevent the establishment of obesogenic dietary patterns, while promoting healthy dietary patterns [24].

Healthy dietary patterns have been described as a "diet [that] largely consists of vegetables, fruits, whole grains, legumes, nuts, and unsaturated oils, includes a low to moderate amount of seafood and poultry, and includes no or a low quantity of red meat, processed meat, added sugar, refined grains, and starchy vegetables"; thus, diverse and nutrient-dense [25]. As a result, many countries have adopted food-based dietary guidelines [26] to assist children and adults in making healthy food choices, including South Africa [27,28].

The aim of the present study was to characterise the dietary patterns of children aged 1–<10 years who were studied during the Provincial Dietary Intake Survey (PDIS) in 2018, and to investigate socio-demographic predictors of the identified dietary patterns, as well as associations between dietary patterns and stunting and overweight/obesity. In South Africa, little data are available regarding the diet of children and how the nutrition transition has affected their dietary intake. The only national survey in children was undertaken in 1999, the National Food Consumption Survey (NFCS), with no follow up for comparison to show trends and changes in diet [8]. The PDIS study was a follow-up of the NFCS study in two rapidly urbanizing and economically active provinces, Gauteng and the Western Cape, to investigate dietary intake and growth status in children aged 1–<10-years [8].

2. Materials and Methods

2.1. Study Area

The two provinces selected were Gauteng (GTG) and the Western Cape (WC), because they are the most rapidly urbanizing and wealthiest provinces with extensive migration from rural areas to cities in search of jobs and a better quality of life [29].

2.2. Structure of the Sample and the Sampling Procedure

The detail of sample structure and sampling procedure has been described elsewhere [7]. Briefly, six strata were identified, namely two provinces (GTG and WC), with each having three areas of residence: urban formal, urban informal, and rural areas. Enumerator areas (EAs) were identified in each stratum. A stratified two-stage sample design was used with a probability proportional to the size sampling of EAs at the first stage, and systematic sampling of households within the EAs at the second stage. The formula included below was used to determine the number of households per stratum, for six domains:

$$N = \text{Deft}^2 \times \{[(1/P) - 1]/a^2\}/(R_1 \times R_2 \times d)$$

where

N = the number of households per sampling stratum, taking non-response into account was calculated to be N (=175);
Deft (=1.3) is the design effect;
P (=0.21) is the estimated proportion of children classified as stunted;
a (=0.2) is the desired relative standard error;
R1 (=0.96) is the individual response rate;
R2 (=0.89) is the household gross response rate;
d (=1.06) is the number of eligible individuals per households [7].

The number of eligible individuals per households (d = 1.06) was calculated as the average number of children aged 1–<10 years per household. It was proposed to survey 175 × 6 strata, or 1050 households. Precision of estimates across regions (rural, urban informal, and urban formal) was ensured by including a minimum of 50 interviews per stratum [7]. Since the sample sizes of GTG rural, WC rural, and urban informal were less than 150, we increased sampling accordingly to ensure sufficient observations per cell in each age group, with the proposed sample size then being 1050 + 218 = 1268. A total of 84 EAs were selected from the six strata, 25 formal residential, 10 informal residential, and 11 rural EAs in GTG, as well as 18 formal residential, 10 informal residential, and 10 rural EAs in the WC.

2.3. Selection of Households and Children within Households

Maps of primary sampling units were generated and passed on to fieldwork teams. The total number of households (HHs) in each EA and a listing of eligible HHs was compiled for each EA, which served as the sampling frame for the selection of HHs. A maximum of 16 HHs were selected per EA based on a predetermined fixed interval (calculated to be specific to each EA) starting from a randomly determined point. A backup sampling frame was constructed in each EA by asking members of the 16 selected HHs to identify nearby HHs with women and children of the appropriate age of 1–<10 years old.

One child in each randomly selected HH was included in the survey. If more than one child in the prescribed age interval was present in the HH, then all eligible children in the HH were numbered in age order for random selection of one child using a "Random Number Table" designed for this purpose.

Inclusion criteria were as follows: children aged 1–<10 years (12–119 months) old; male or female; availability of a parent/primary caregiver to provide consent; and availability of a parent/primary caregiver to assist with completion of the research questionnaires. Children who were mentally or physically handicapped; who were on a prescribed diet for a childhood disease, e.g., Type 1 diabetes, phenylketonuria, and other conditions; who were ill at the time of the visit or were ill during the past 24 h; whose mothers/caregivers were unable to respond, or appeared to be incapable of responding or providing reliable information; whose mother/caregiver was under the influence of alcohol/drugs or was under 15 years old were not eligible for participation.

Sampling weights were calculated to adjust for the oversampling in the rural and urban informal areas and the number of children in the 1–<3, 3–<6, and 6–<10 year age groups, bearing in mind the survey design. The final post hoc stratification weighting reflects the census population of the Western Cape and Gauteng provinces [30]. The three age groups were demarcated to reflect children who are in the first 1000 days of life (children in their third year, but not yet three years old, were included in this group), older preschool children, and primary school-aged children.

2.4. Fieldwork Teams

Fieldwork in each province was led and managed by a registered dietitian (fieldwork coordinators). Research teams in the two provinces included a team leader and two pairs of experienced field workers. Team leaders and field workers received a week-long extensive training session using a manual developed for the purpose of the study. Training was facilitated by researchers experienced in administration of sociodemographic and dietary questionnaires, as well as the fieldwork coordinators. Training of the WC fieldwork teams took place in Cape Town and was attended by the GTG fieldwork coordinator. Subsequently, training of GTG fieldworkers took place in Johannesburg, which was co-facilitated by the WC fieldwork coordinator to contribute to data fidelity. After each training module, the field workers practiced using the questionnaires through role play sessions with each other. At the end of the week, the field workers completed a practical and written test based on case studies. Field workers who did not achieve a certain percentage were not selected.

2.5. Measures

2.5.1. Socio-Demographic Questionnaire

The socio-demographic questionnaire included questions (predictors) which could impact the dietary intake and health outcomes of children and were based on the child, family, household, and environment. Questions about the child included birth date, gender, primary caregiver, and whether they attended a creche or preschool facility. Questions about the family and household were head of household, marital status of mother, education and employment status of mother and father, type of house, availability of electricity or other energy devices, source of drinking water, type of toilet, and household density. These variables were selected as they were used in the National Food Consumption Survey (1999), and many were found to be significant predictors of nutritional status at the time [31].

A wealth index was calculated as indicated by the World Bank [32] and applied in the 2016 South Africa Demographic and Health Survey [33]. Iterated principal factor analysis was used to estimate relative wealth, and this estimation is based on items loading on the first factor. The first factor contributes to a wealth index that assigns a larger weight to assets that vary the most across households, so that an asset found in all households is given a weight of zero. The wealth index in this study was based on amenities available in the home and environment [7].

Hunger (food security) was measured using the Community Childhood Hunger Identification Project (CCHIP) questionnaire [34]. This questionnaire measures household, child, and individual-level food security. The scale comprises eight questions and a score of one is given for affirmative answers. A total score of 5–8 indicates the presence of food shortage (hunger) in the household, a score of 1–4 reflects risk of hunger, and a score of zero indicates that the house is food-secure (no hunger).

2.5.2. Dietary Intake

A 24 h recall was conducted with each participant to determine dietary intake using the multiple pass method [35] (details on the application of this method have been described elsewhere, Steyn et al. [36]). The literature indicates that the accuracy of reporting one's own dietary intake in younger children is not good, but that it improves between the ages of 8 and 12 years [37]. Consequently, in this study, all dietary interviews took place in the presence and with the input of the mother/primary caregiver. For 1–6-year-old children, the mother/caregiver reported on the intake of the child on the previous day with no input from the child. For 7–<10-year-old children, the mother/caregiver and child were interviewed together to record the dietary intake during the prior 24 h. If the child had been at a day care centre the previous day, the centre was visited by the fieldworker and the meals and portion sizes determined for the 24 h in question. All weekdays and Sundays were covered proportionally by each team to ensure that potential variation due to day of the week was captured.

Portion sizes were obtained using a booklet adapted from the Dietary Assessment and Education Kit (DAEK) [38]. The booklet comprises life-size sketches of generic household utensils and crockery (Figure 1) and life-size portions of actual foods, e.g., different slices of bread varying in size and thickness, to make estimations of portion size as accurately as possible. The sketches were validated in adolescents [39]. Generic three-dimensional food models made from flour were also used to assist in recording volume measures such as porridge and rice.

Breast milk consumption was quantified by asking mothers whether their child was still receiving breast milk and, if yes, the number of feeds the child received during the previous 24 h. Based on the study by Neville et al. [40], we used an estimate of 100 mL per feed to calculate the volume of breast milk consumed per day.

A common concern with a single 24 h recall is the day-to-day variation in the diet of free-living populations. The National Cancer Institute (NCI) method [41,42] that was developed to distinguish within-person from between-person variation accounts for extreme intakes, including zero intake, allows for adjustment for covariates and association analyses, and was applied in this study to estimate the usual dietary intake from repeated 24 h dietary recall assessments on a subsample of 148 (second recall) and 146 (third recall) children (details on the application of this method in this research have been described elsewhere) [36].

Figure 1. Life-size generic models used in the DAEK [38].

After completion of an EA, the questionnaires were verified by the two registered dietitians who managed the fieldwork in the two provinces for quality control purposes. The 24 h recalls were coded by these dietitians using the South African Food Composition Tables (SAFCTs) [43], each coding the recalls collected in their allocated province. Where necessary, codes were confirmed in consultation with the rest of the research team to ensure uniform decisions and coding of foods.

Food items were allocated to 30 food parameters based on the similarity of nutrient profiles in the allocated SAFCT food codes [43] (Table 1).

Table 1. Food parameters and allocated foods for dietary pattern analyses.

Food Parameters	Terminology in Result Tables	Allocated Foods
Infant food	Infant food	Breast milk, breast milk substitutes, infant cereals
White bread	Bread white	White bread or rolls
Brown bread	Bread brown	Brown and whole wheat bread or rolls
Unrefined cereals	UCs	Hi-fibre breakfast cereals, e.g., All-Bran, Weetbix
Refined cereals	RCs	Refined breakfast cereals, sweetened and unsweetened
Maize porridge	Maize pap	Soft, stiff, and crumbly
Other refined carbohydrates	Rcarb-other	Rice, pasta, samp, mabella, mageu
Cheese	Cheese	Cheddar, gouda
Dairy	Dairy	Milk, yoghurt, and maas (sour milk)
Poultry	Poultry	With or without skin, any preparation
Red meat	Red meat	Beef, mutton, lamb and organ meat, any preparation
Processed meat	Proc meat	Cold meats, sausages, canned meat, dried meat
Eggs	Eggs	Any preparation
Fish	Fish	Fresh, canned, any preparation
Legumes	Legumes	Beans, lentils—soup and other preparations, soy mince
Vegetables: starchy	Veg-starchy	Potatoes, sweet potato, corn, sweet corn
Vegetables: starchy + fat	Veg-starchy + fat	"Slap chips" [1], potato roasted in fat, candied sweet potato
Vegetables: non starchy	Veg-non starchy	All vegetables except for starchy vegetables
Fruit	Fruit	Any fresh, canned or dried fruit, juice
Fats and oils: saturated	Fats-oils-sat	Butter, lard, hard margarine, coconut oil, non-dairy creamer
Fats and oils: unsaturated	Fats-oils-unsat	Soft margarine, plant oils, avocado, nuts, salad dressing
Refined carbohydrate + fat	Rcarb + fat	Savoury snacks—crisps, crackers
Refined carbohydrate + fat + sugar	Rcarb + fat + sugar	Cake, tarts, doughnuts, ice-cream, chocolates
Refined carbohydrate +protein + fat	Rcarb + prot + fat	Samoosas, pies, vetkoek, pizza, pasta dishes, fish cake
Refined carbohydrates + sugar	Rcarb + sug	Sweets: boiled, jelly-like
Sugar-sweetened beverages	SSBs	Fizzy drinks, squash, sport drinks
Sugar or syrup	Sugar	Granulated sugar, syrup, jam
Tea-coffee	Tea-coffee	Rooibos tea, Ceylon tea, coffee (no milk/sugar added)
Soup-sauces	Soup-sauces	Commercial soups, tomato sauce, chutney
Miscellaneous	Misc.	Condiments, Marmite, Bovril, fish paste

Mageu = fermented maize drink; Maas = sour milk; Vetkoek = balls of dough fried in oil; [1] French fries.

2.6. Data Management and Analysis

All data were captured centrally by two experienced researchers. Data analyses were conducted using SAS Version 9.4, SAS for Windows (SAS Institute, Carry, NC, USA). Frequencies were tallied for the socio-demographic variables which were compared between the WC and GTG using the Rao–Scott chi-squared test, incorporating the complex survey design.

Dietary pattern analysis was conducted within each of the designated age groups, namely 1–<3-year-olds, 3–<6-year-olds, and 6–<10-year-olds using principal factor analysis with varimax rotation [44]. It is evident from other research that the dietary units of analysis used for dietary pattern identification include percentage of total daily energy contribution (kJ) of each food group/item [45], or daily amount consumed from each food group/item in grams [46], or daily frequency of consumption of a food group/item [47]. We conducted principal factor analyses using each of these options. After consideration of the Kaiser–Meyer–Olkin (KMO) statistic, which tests the appropriateness of applying principal factor analysis to the dataset (post hoc sampling adequacy) [48], as well as the percentage total variance explained by the identified patterns, it was evident that the frequency of consumption of food groups/items was the most appropriate dietary unit to use for our dataset.

As the frequencies of intake of the 30 food parameters calculated for this research (Table 1) were not normally distributed, the data were normalised using Blom's transformation [49]. Ricci et al. explain that this transformation is particularly suitable to normalise and standardise food or nutrient intakes before principal factor analysis is conducted [50]. The decision on the number of dietary patterns to be retained was based on the visual inspection of the scree plot, eigenvalues of >1.5, and interpretability of the pattern, as was conducted by Faber et al. [51]. No golden rule on the cut-off for the exclusion of food parameters in the principal factor analysis, using the pattern (factor) loadings, has yet been set. We retained food parameters with a loading of >0.3 and <−0.3 for interpretation and naming of each dietary pattern, which is in line with the cut-offs used by several other researchers [50–52]. Factor scores were generated for each food parameter using the loading of all 30 food parameters on each factor pattern. Higher factor scores reflect greater adherence of the food parameter to the specific pattern.

Predictors of dietary patterns were identified by constructing multiple regression models with backward elimination. As five patterns were retained in each age group, a total of 15 regression models were constructed. The dietary pattern scores were standardized-dependent variables with zero means and standard deviations of 1 unit. The socio-demographic variables outlined above, as well as province (the WC and GTG) were the independent variables in these analyses. The parameter estimate, standard error (SE), as well as the *p*-value for the independent variables that showed a significant association with the outcome variables that were retained in the final models are reported in the results. Variance inflation factors (VIFs) were calculated and were all less than 1.6, indicating the absence of multicollinearity.

The association between dietary patterns and anthropometric variables was investigated using logistic regression with (1) BAZ > 2SD and (2) HAZ < −2SD as dependent variables, with the pattern score as independent variable while controlling for age, gender, and province.

3. Results

3.1. Results for Sociodemographic Profile of HHs

Data on the socio-demographic profile of HHs included in the study are shown in Table 2. The sample comprised 49.3% of boys and 50.7% of girls. For 70.4% of the children, the primary caregiver was the mother, while the head of the household was mostly the father (39.7%) or the grandmother (24.0%). Fifty-three percent of mothers did not complete grade 12 compared with 29.1% of fathers. Significantly more mothers were employed in the WC compared with GTG (38.4% vs. 22.4%), while 65% of all fathers were employed.

In the WC, 68.0% of the sample were of mixed ancestry, while in GTG 97.8% were black African. The majority of the sample were urban formal residents (88.2%). Food insecurity was present in 20.7% of the households.

Table 2. Sociodemographic and other characteristics of the 1–<10-year-old children in the two provinces studied.

	Gauteng N = 733 % (95% CI)	Western Cape N = 593 % (95% CI)	Rao–Scott Chi-Sq Values	All N = 1326 % (95% CI)
Primary caregiver				
Mother	70.1 (65.6–74.6)	71.0 (64.7–77.2)	0.045 *	70.4 (66.8–74.0)
Father	6.6 (3.4–9.7)	1.8 (0.2–3.3)		5.0 (2.8–7.1)
Grandparent	16.7 (12.9–20.4)	21.0 (15.5–26.4)		18.1 (15.0–21.2)
Other (e.g., sibling, aunt)	6.7 (4.0–9.5)	6.3 (2.1–10.4)		6.6 (4.3–8.8)
Age in years				
1–<3 years	26.3 (22.1–30.6)	25.3 (19.4–31.2)	0.923	26.0 (22.6–29.4)
3–<6 years	35.4 (31.0–39.8)	35.1 (30.7–39.5)		35.3 (32.1–38.5)
6–<10 years	38.3 (34.1–42.4)	39.6 (33.1–46.1)		38.7 (35.2–42.2)
Gender				
Male	50.2 (45.5–54.9)	47.5 (43.1–51.9)	0.391	49.3 (45.9–52.7)
Female	49.8 (45.1–54.5)	52.5 (48.1–56.9)		50.7 (47.3–54.1)
Head of household				
Father	40.2 (33.8–46.6)	38.8 (34.6–43.0)	0.132	39.7 (35.3–44.1)
Mother	16.8 (13.8–19.9)	10.8 (7.0–14.5)		14.8 (12.5–17.2)
Grandmother	21.9 (15.5–28.3)	28.3 (21.8–34.9)		24.0 (19.3–28.8)
Grandfather	11.7 (8.3–15.1)	14.0 (10.0–18.0)		12.5 (9.9–15.0)
Other (e.g., aunt, uncle)	9.4 (5.7–13.1)	8.1 (4.9–11.4)		9.0 (6.3–11.7)
Marital status of mother				
Unmarried	41.1 (34.9–47.2)	34.8 (28.4–41.1)	<0.001 ***	39.0 (34.4–43.5)
Married	24.9 (20.5–29.4)	41.3 (33.3–49.2)		30.4 (26.4–34.3)
Divorced/widowed	4.8 (2.5–7.0)	2.4 (0.7–4.2)		4.0 (2.4–5.6)
Living together	27.8 (22.0–33.6)	20.8 (15.9–25.7)		25.5 (21.4–29.6)
Other	1.4 (0.2–2.6)	0.8 (0.0–1.8)		1.2 (0.3–2.1)
Mother's highest education				
Not completing Gr. 12	51.2 (44.9–57.4)	57.7 (47.1–68.3)	0.183	53.3 (47.9–58.7)
Completion of Gr. 12	33.9 (28.4–39.4)	24.7 (17.6–31.8)		30.8 (26.5–35.2)
Qualification after Gr.12	12.2 (8.7–15.7)	15.6 (7.6–23.6)		13.3 (9.9–16.8)
Do not know	2.8 (1.4–4.1)	2.0 (0.5–3.5)		2.5 (1.5–3.5)
Father's highest education				
Not completing Gr. 12	26.9 (22.0–31.7)	33.8 (29.0–38.5)	0.323	29.1 (25.6–32.7)
Completion of Gr. 12	32.6 (26.9–38.3)	30.4 (25.2–35.6)		31.9 (27.8–36.0)
Qualification after Gr.12	13.1 (9.4–16.9)	10.7 (5.7–15.7)		12.3 (9.4–15.3)
Do not know	27.4 (22.4–32.4)	25.2 (19.7–30.6)		26.7 (22.9–30.4)
Mother's employment status				
Yes	22.4 (17.8–26.9)	38.4 (31.0–45.9)	<0.001 **	27.7 (23.9–31.5)
No	74.6 (69.6–79.6)	60.2 (53.0–67.5)		69.8 (65.8–73.9)
Do not know/not applicable	3.0 (1.3–4.7)	1.3 (0.3–2.4)		2.5 (1.3–3.6)
Father's employment status				
Yes	64.8 (60.6–69.1)	65.3 (59.7–70.9)	0.953	65.0 (61.6–68.4)
No	21.4 (17.5–25.3)	20.5 (15.1–25.9)		21.1 (18.0–24.2)
Do not know/not applicable	13.8 (11.1–16.4)	14.1 (10.2–18.1)		13.9 (11.7–16.1)
Wealth index quintiles				
One	21.1 (14.6–27.6)	17.7 (10.7–24.7)	0.263	20.0 (15.1–24.8)
Two	17.8 (12.0–23.6)	24.3 (20.0–28.6)		20.0 (15.9–24.0)
Three	21.3 (17.0–25.7)	17.0 (12.6–21.4)		19.9 (16.7–23.1)
Four	21.5 (16.7–26.3)	17.5 (12.4–22.6)		20.2 (16.6–23.7)
Five	18.3 (11.6–25.0)	23.5 (14.5–32.5)		20.0 (14.7–25.3)

Table 2. *Cont.*

	Gauteng N = 733 % (95% CI)	Western Cape N = 593 % (95% CI)	Rao–Scott Chi-Sq Values	All N = 1326 % (95% CI)
Ethnicity				
Black African	97.8 (96.0–99.6)	27.6 (12.9–42.3)	<0.001 **	74.5 (69.5–79.4)
Mixed ancestry	2.2 (0.3–4.0)	68.0 (53.7–82.4)		24.1 (19.2–28.9)
Other	0.0 (0.0–0.1)	4.4 (0.6–8.2)		1.5 (0.3–2.7)
Type of residence				
Rural	2.4 (0.7–4.1)	6.6 (1.6–11.5)	0.194	3.8 (1.9–5.7)
Urban formal	88.9 (82.3–95.4)	86.8 (79.1–94.5)		88.2 (83.2–93.2)
Urban informal	8.7 (2.7–14.7)	6.6 (1.7–11.5)		8.0 (3.7–12.3)
Mother's BMI [39]				
Underweight/normal BMI = <18.5 and 18.5–24.9 kgm^2	33.3 (28.0–38.5)	29.1 (23.6–34.5)	0.002 **	32.0 (28.0–35.9)
Overweight BMI = 25–29.9 kgm^2	27.7 (23.6–31.8)	20.4 (16.5–24.3)		25.4 (22.4–28.5)
Obese BMI ≥ 30 kgm^2	39.1 (35.8–42.3)	50.6 (43.0–58.1)		42.6 (39.4–45.8)
Hunger scale [25]				
Total score = 0: No risk	57.9 (49.5–66.3)	48.8 (38.9–58.7)	0.1483	54.9 (48.5–61.3)
1–4: At risk of hunger	22.1 (17.2–27.0)	28.9 (23.0–34.9)		24.4 (20.6–28.2)
5–8: Food shortage in house	20.0 (14.8–25.1)	22.3 (16.5–28.0)		20.7 (16.8–24.6)

95% CI, 95% confidence intervals; * Significant relationship between the variable and province, chi-squared p-value < 0.05; ** p < 0.01; *** p < 0.001; N-values reflect actual number of cases, estimates are adjusted using relevant weighting.

3.2. Results for 1–<3-Year-Old Children

The five dietary patterns and significant socio-demographic predictors for 1–<3-year-olds are shown in Table 3. The total variance explained by the five patterns in this age group is 31.7%.

Maize porridge had a very high (PL = 0.84) positive pattern loading (PL) on the first dietary pattern. Soup/sauces also loaded positively (PL = 0.44), while dairy and refined carbohydrates loaded negatively. This pattern was labelled the "***Pap & sauce pattern***" Pap is the term commonly used for maize porridge which can be eaten soft, stiff, or crumbly as a starch. Predictors for greater adherence to this pattern were living in Gauteng and being at risk of or experiencing hunger. Predictors of lesser adherence were having an aunt or uncle as head of the household, a mother with a grade 12 qualification, a father who has a post grade 12 qualification, and a greater wealth index.

The two food parameters that had the highest positive loading on the second pattern were tea/coffee (PL = 0.74) and sugar (PL = 0.72) and this was labelled the ***Tea/coffee and sugar pattern***. Fats and oils high in saturated fat and legumes also loaded positively on this pattern (PL < 0.6). Predictors of greater adherence to this pattern were a higher wealth index and being at risk of hunger. Being looked after by a sibling or aunt predicted lesser adherence.

Foods that comprised a combination of refined carbohydrates and fat, e.g., crisps or savoury biscuits (PL = 0.52) and foods that comprise a combination of refined carbohydrates, sugar, and fat, e.g., cake, ice cream, and chocolate (PL = 0.5), brown bread (PL = 0.42), SSBs (PL = 0.41), and fruit (PL = 0.36) loaded positively on this pattern, and was labelled the "***Mostly unhealthy snack pattern***". Baby foods loaded negatively on this pattern. Predictors of greater adherence to this pattern were having a mother with a grade 12 qualification and a father with a post grade 12 qualification. Having a grandmother as primary caregiver predicted lesser adherence.

Table 3. Dietary patterns and significant socio-demographic predictors for 1–<3-year-old children in the Western Cape and Gauteng (n = 333).

Pap Soup/Sauce Pattern		Tea/Coffee & Sugar Pattern		Mostly Unhealthy Snack Pattern		White Bread & Topping Pattern		Healthy Pattern	
Food Parameters	**PL**	**Food Parameters**	**PL**	**Food Parameters**	**PL**	**Food Parameters**	**PL**	**Food Parameters**	**PL**
Maize pap	0.84	Tea and/or coffee	0.74	RC-Fat	0.52	Bread White	0.65	Fats-oils-Unsat	0.60
Soup-sauces	0.44	Sugar	0.72	RC-Fat-sugar	0.50	Processed meat	0.53	Veg non-starchy	0.41
Dairy	−0.39	Fats-oils-Sat	0.59	Bread Brown	0.42	Miscellaneous	0.36	Fish	0.31
RC-Other	−0.55	Legumes	0.33	SSB	0.41	Eggs	0.32	Poultry	−0.38
URC	−0.59			Fruit	0.36			RC-Sugar	−0.55
				Baby food	−0.52				
% Variance explained	2.16	% Variance explained	2.1	% Variance explained	2.0	% Variance explained	1.66	% Variance explained	1.6
Pattern Predictors [1]	**PE (SE) *p*-Value** [2]	**Pattern Predictors** [1]	**PE (SE) *p*-Value** [2]	**Pattern Predictors** [1]	**PE (SE) *p*-Value** [2]	**Pattern Predictors** [1]	**PE (SE) *p*-Value** [2]	**Pattern Predictors** [1]	**PE (SE) *p*-Value** [2]
HHH Other *-lesser adherence*	−0.29 (0.13) 0.034	Higher WI *-greater adherence*	0.04 (0.01) 0.013	PCG: Grandmother *-lesser adherence*	−0.38 (0.15) 0.015	Gauteng *-lesser adherence*	−0.27 (0.11) 0.015	HHH: Grandparent *-lesser adherence*	−0.29 (0.11) 0.009
Mother has Gr 12 *-lesser adherence*	−0.31 (0.09) <0.001	PCG: Other *-lesser adherence*	−0.4 (0.19) 0.035	Mother has Gr 12 *-greater adherence*	0.37 (0.12) 0.002	Mother obese *-lesser adherence*	−0.24 (0.11) 0.026	Gauteng *-greater adherence*	0.45 (0.11) <0.001
Father has Gr12+ *-less adherence*	−0.27 (0.14) 0.049	Hunger risk *-greater adherence*	0.35 (0.11) 0.002	Father has Gr12+ *- greater adherence*	0.4 (0.18) 0.028	PCG: Other *-lesser adherence*	−0.35 (0.18) 0.06	Mother overweight *-greater adherence*	0.31 (0.14) 0.023
Higher WI *-lesser adherence*	−0.03 (0.01) 0.016							Mother obese *-greater adherence*	0.33 (0.12) 0.007
Gauteng *-greater adherence*	1.23 (0.09) <0.001							Greater WI *-greater adherence*	0.03 (0.02) 0.04
Hunger risk *-greater adherence*	0.25 (0.1) 0.009								
Hunger present *-greater adherence*	0.33 (0.12) 0.008								

PL = pattern loading; RC = refined carbohydrates; sat = saturated; unsat = unsaturated; PE = parameter estimate; SE = standard error; WI = wealth index, PCG = Primary caregiver; PCG Other = sibling or aunt; HHH = head of household; HHH Other = aunt or uncle; Gr = grade; URCs = unrefined breakfast cereals, RC-Other = other refined carbohydrates, e.g., pasta, rice, and samp; RC-Fat-sugar = combination of refined carbohydrates, fat, and sugar, e.g., cake, tarts, doughnuts, ice cream, chocolates; RC-Fat = combination of refined carbohydrates and fat, e.g., crisps (any type) and salty biscuits; Miscellaneous = salty spreads and condiments; Soup-sauces = commercial powdered soup, tomato sauce, and chutney; RC- Sugar = sugar in the form of sweets; Veg non-starchy = all vegetables excluding starchy vegetables; Fats-oils-Sat = Butter, lard, hard margarine, coconut oil, non-dairy creamer; Fats-oils-Unsat = Soft margarine, plant oils, avocado, nuts, salad dressing; SSBs = sugar-sweetened beverages, e.g., fizzy drinks, squash, and sport drinks. [1] A positive parameter estimate indicates greater adherence and a negative parameter estimate lesser adherence to a dietary pattern. [2] Multiple regression model with backward elimination constructed for each pattern; only significant predictors that remained in the final model are included in the table. Dietary pattern scores were standardized with means of 0 and a unit standard deviation.

White bread had the highest positive loading on the fourth pattern (PL = 0.65), with processed meat (PL = 0.53), miscellaneous items, e.g., Marmite, Bovril, fish paste, and condiments (PL = 0.36), as well as eggs (PL = 0.32) also loading positively on this pattern, and was labelled the "***White bread & topping pattern***". Significant predictors of lesser adherence to this pattern were living in Gauteng, being looked after by a sibling or aunt, and having an obese mother. There were no predictors of greater adherence.

Unsaturated fats and oils had the highest positive loading (PL = 0.6) on the fifth pattern. Vegetables (all except starchy vegetables) (PL = 0.41) and fish (PL = 0.31) also loaded positively, while poultry (PL = −0.35) and sweets (PL= −0.55) loaded negatively on this pattern, and were labelled the "***Healthy pattern***". Predictors of greater adherence to this pattern were living in Gauteng, having an overweight mother or obese mother, and a higher wealth index. Having a grandparent as head of the household predicted lesser adherence.

Starchy vegetables combined with fat, e.g., "slap chips" (French fries) did not load on any dietary pattern in this age group.

There were no significant associations between dietary patterns and HAZ or BAZ variables (results not shown in a table).

3.3. Results for 3–<6-Year-Old Children

Dietary patterns and significant socio-demographic predictors for 3–<5-year-olds are shown in Table 4. The total variance explained by the five patterns in this age group is 30.5%.

Tea/coffee (PL = 0.85) and sugar (PL = 0.82) had very high positive loadings on the first dietary pattern. Fats and oils (saturated PL = 0.49 and unsaturated PL = 0.31 fats/oils), as well as brown bread (PL = 0.33) also loaded positively on this pattern, which was labelled the "***Tea/coffee, sugar & sandwich pattern***". There were no significant socio-demographic predictors of this pattern.

The second pattern was labelled the "***Unhealthy pattern***" as all food parameters that loaded positively on it were deemed to be unhealthy. These items are white bread (PL = 0.65); starchy vegetables combined with fat, e.g., "slap chips" (French fries) (PL = 0.55); foods which combine refined carbohydrates with animal protein and fat, e.g., pies, "vetkoek", pasta dishes and pizza (PL = 0.41); foods which combine refined carbohydrate with sugar and fat, e.g., cake, doughnuts, ice cream, and chocolates (PL = 0.41); and processed meat (PL = 0.36). Predictors of greater adherence to this pattern were having a grandparent as head of the household, a higher wealth index, and having a mother who is employed. Living in Gauteng was a predictor of lesser adherence.

The food parameter, RC-other, which includes rice and pasta, had the highest loading (PL = 0.71) on the third pattern. Starchy vegetables (PL = 0.48) and poultry (PL = 0.43) also loaded positively on this pattern, while maize porridge loaded negatively (−0.53). It was labelled the "***Starch & poultry pattern***". Being looked after by a sibling or aunt, living in Gauteng, and having an overweight mother were predictors of lesser adherence to this pattern. There was no significant predictor of greater adherence.

Four food items that could typically be consumed as part of breakfast loaded positively on the fourth pattern, including dairy (PL = 0.62), fruit (PL = 0.57), cheese (PL = 0.46), and refined breakfast cereal (PL = 0.46). This pattern was labelled the "***Breakfast pattern***". Predictors of greater adherence to this pattern were being a girl, a higher wealth index, and having a mother with a post grade 12 qualification, a father with a grade 12 or a post grade 12 qualification, an employed mother, and an obese mother. Predictors of lesser adherence were living in Gauteng, being looked after by a grandmother, and having an employed father.

Table 4. Dietary patterns and socio-demographic predictors for 3–<6-year-old children in the Western Cape and Gauteng (n = 514).

Tea/Coffee, Sugar, & Sandwich Pattern		Unhealthy Food & Snack Pattern		Starch & Poultry Pattern		Breakfast Item Pattern		Vegetable & Legume Pattern	
Food Parameters	**PL**	**Food Parameters**	**PL**	**Food Parameters**	**PL**	**Food Parameters**	**PL**	**Food Parameters**	**PL**
Tea/coffee	0.85	Bread-White	0.65	RC-Other	0.71	Dairy	0.62	Legumes	0.41
Sugar-syrup	0.82	Veg-Starchy-F	0.55	Veg starchy	0.48	Fruit	0.57	Veg non-starchy	0.41
Fats-oils-Unsat	0.49	RC- Prot-Fat	0.41	Poultry	0.43	Cheese	0.46	Miscellaneous	0.40
Bread Brown	0.33	RC-Fat-sugar	0.41	Maize pap	−0.53	RC-Fort-Cereal	0.46	URC	−0.62
Fats-oils-Sat	0.31	Processed meat	0.36						
% Variance explained	2.2	% Variance explained	1.81	% Variance explained	1.74	% Variance explained	1.72	% Variance explained	1.68
Pattern Predictors [1]	**PE (SE) *p*-Value [2]**	**Pattern Predictors [1]**	**PE (SE) *p*-Value [2]**	**Pattern Predictors [1]**	**PE (SE) *p*-Value [2]**	**Pattern Predictors [1]**	**PE (SE *p*-Value [2]**	**Pattern Predictors [1]**	**PE (SE) *p*-Value [2]**
None		HHH: Grandparent *-greater adherence*	0.27 (0.1) 0.008	PCG: Other *-lesser adherence*	−0.37 (0.14) 0.009	PCG: Grandmother *-lesser adherence*	−0.32 (0.12) 0.008	Gauteng *-greater adherence*	0.66 (0.1) <0.001
		Gauteng *-lesser adherence*	−0.63 (0.09) <0.001	Gauteng *-lesser adherence*	−0.74 (0.09) <0.001	Gender: Female *-greater adherence*	0.24 (0.09) 0.006	Hunger present *-greater adherence*	0.25 (0.11) 0.022
		Greater WI *-greater adherence*	0.03 (0.01) 0.016	Mother overweight *-lesser adherence*	−0.30 (0.09) 0.001	Mother has Gr12+ *-greater adherence*	0.28 (0.14) 0.045		
		Mother employed *-greater adherence*	0.22 (0.1) 0.026			Father has Gr12 *-greater adherence*	0.37 (0.1) <0.001		
						Father has Gr12+ *-greater adherence*	0.42 (0.15) 0.006		
						Mother employed *-greater adherence*	0.41 (0.1) <0.001		
						Father employed *-lesser adherence*	−0.25 (0.1) 0.009		
						Greater WI *-greater adherence*	0.05 (0.01) <0.001		
						Gauteng *-lesser adherence*	−0.24 (0.1) 0.012		
						Mother obese *-greater adherence*	0.25 (0.09) 0.005		

PL = pattern loading; RCs = refined carbohydrates; URCs = unrefined carbohydrates; sat = saturated; unsat = unsaturated; PE = parameter estimate; SE = standard error; WI = wealth index, Gr = grade, PCG = Primary caregiver; PCG Other = sibling or aunt; HHH = head of household; HHH Other = aunt or uncle; RC-Other = other refined carbohydrates, e.g., pasta and rice; RC-Fat-sugar = combination of refined carbohydrates, fat, and sugar, e.g., cake, tarts, doughnuts, ice cream, chocolates; RC-Protein-Fat = combination of refined carbohydrates, fat and animal protein, e.g., samosa, fat cakes, pie, pizza, and lasagna pasta dishes; Miscellaneous = salty spreads and condiments; RC Sugar = sugar in the form of sweets; Veg non-starchy = all vegetables excluding starchy vegetables; Veg-starchy-F = starchy vegetables combined with fat, e.g., "slap chips" (French fries); Fats-oils-Sat = Butter, lard, hard margarine, coconut oil, non-dairy creamer; Fats-Oils-Unsat = Soft margarine, plant oils, avocado, nuts, salad dressing; SSB = sugar-sweetened beverages, e.g., fizzy drinks, squash, and sport drinks. [1] A positive parameter estimate indicates greater adherence and a negative parameter estimate lesser adherence to a dietary pattern; [2] Multiple regression model with backward elimination constructed for each pattern; only significant predictors remaining in the final model are included in the table. Dietary pattern scores were standardized with means of 0 and a unit standard deviation.

Legumes (PL = 0.41), vegetables (all except starchy vegetables) (PL = 0.41), and miscellaneous items, e.g., Bovril, marmite, fish paste (PL = 0.40) loaded positively on the fifth pattern, while unrefined carbohydrates loaded negatively (PL= −0.64). This pattern was labelled the *"**Vegetable & legume pattern**"*. Predictors of greater adherence to this pattern were living in Gauteng and the presence of hunger in the household. There was no significant predictor of lesser adherence.

Soups/sauces, as well as foods that combine refined carbohydrates and fat, e.g., crisps (any type) and salty biscuits did not load on any pattern in this age group.

There were no significant associations between dietary patterns and HAZ or BAZ variables (results not shown in a table).

3.4. Results for 6–<10-Year-Old Children

Dietary patterns and significant socio-demographic predictors for 6–<10-year-olds are shown in Table 5. Total variance explained by the five patterns in this age group is 31.37%.

Foods that combined refined carbohydrates with fat, e.g., crisps and salty biscuits, had the highest loading on the first pattern (PL = 0.47), followed by SSBs (PL = 0.44), fruit (PL = 0.41), unrefined cereals (PL = 0.4), and sweets (PL = 0.36), while fish (PL = −0.33), legumes (PL = −0.39), and maize porridge (PL = −0.5) loaded negatively. As only two healthy food items loaded positively (versus three unhealthy items) and three healthy items loaded negatively on this pattern, it was deemed to be more reflective of unhealthy eating. It was labelled the *"**Mostly unhealthy pattern 1**"*. Predictors of greater adherence to this pattern were having a father with a grade 12 qualification, a higher wealth index, and having an obese mother. Predictors of lesser adherence to this pattern were living in Gauteng, being at risk of experiencing hunger, and having hunger present in the household.

Tea and/or coffee (PL = 0.85), sugar (PL = 0.82) and dairy loaded positively on the second pattern and this was labelled the *"**Tea/coffee, sugar and dairy pattern**"*. Predictors of lesser adherence to this pattern were the mother being the head of the household, being looked after by a sibling or aunt, being a girl, and living in Gauteng. There was no significant predictor of greater adherence to this pattern.

Food parameters that loaded positively on the third pattern were mostly unhealthy, including refined breakfast cereals (PL = 0.53); foods which combined refined carbohydrates with animal protein and fat, e.g., pies, "vetkoek", pasta dishes, and pizza (PL = 0.34); and foods which combined refined carbohydrates with sugar and fat, e.g., cake, doughnuts, ice cream, and chocolates (PL = 0.32), with red meat being the exception (PL = 0.44). As two healthy food parameters, non-starchy vegetables and poultry also loaded negatively on this pattern, which was labelled the *"**Mostly unhealthy pattern 2**"*. Predictors of greater adherence to this pattern were a higher wealth index and having a mother with a post grade 12 qualification. Predictors of lesser adherence were living in Gauteng and having hunger present in the household.

White bread had a high loading PL = (0.84) on the fourth pattern together with unsaturated fats and oils (PL = 0.48) and processed meat (PL = 0.42), while brown bread loaded negatively (PL = −0.51). This pattern was labelled the *"**White bread & processed meat pattern**"*. The only significant predictor for this pattern was living in Gauteng, which predicted lesser adherence.

The fifth pattern was labelled the *"**Non-maize pap or bread starch pattern,**"* as "other" refined carbohydrates, e.g., rice and pasta (PL = 0.68) and starchy vegetables (PL = 0.46) loaded positively on this pattern. Predictors of greater adherence to this pattern were having an aunt or uncle as head of the household, being looked after by a grandmother, being at risk of experiencing hunger, and having hunger in the household. Predictors of lesser adherence were having an employed father and living in Gauteng.

Soups/sauces, and starchy vegetables that were combined with fat, e.g., "slap chips" (French fries), did not load on any pattern in this age group.

There were no significant associations between dietary patterns and HAZ or BAZ variables (results not shown in a table).

Table 5. Dietary patterns and socio-demographic predictors for 6–<10-year-old children in the Western Cape and Gauteng (n = 479).

Mostly Unhealthy Pattern 1		Tea/Coffee, Sugar, & Milk Pattern		Mostly Unhealthy Pattern 2		White Bread & Topping Pattern		Starchy Pattern	
Food Parameters	**PL**	**Food Parameters**	**PL**	**Food Parameters**	**PL**	**Food Parameters**	**PL**	**Food Parameters**	**PL**
RC-Fat	0.47	Sugar	0.85	RC-BF cereal	0.53	Bread White	0.84	RC-Other	0.68
SSB	0.44	Tea or coffee	0.82	Red meat	0.44	Fats-oils-UnSat	0.48	Veg starchy	0.46
Fruit	0.41	Dairy	0.56	RC-Prot-Fat	0.34	Processed meat	0.42		
URC	0.40			RC-Fat-Sugar	0.32	Bread Brown	−0.51		
RC Sugar	0.36			Veg non-starchy	−0.32				
Fish	−0.33			Fats oils Sat	−0.33				
Legumes	−0.39			Poultry	−0.46				
Maize pap	−0.50								
Variance explained	2.2%	Variance explained	2.1%	Variance explained	1.83%	Variance explained	1.67%	Variance explained	1.61%
Pattern Predictors [1]	**PE (SE)** *p***-Value** [2]	**Pattern Predictors** [1]	**PE (SE)** *p***-Value** [2]	**Pattern Predictors** [1]	**PE (SE)** *p***-Value** [2]	**Pattern Predictors** [1]	**PE(SE)** *p***-Value** [2]	**Pattern predictors** [1]	**PE(SE)** *p***-Value** [2]
Father has Gr12 *-greater adherence*	0.19(0.09) 0.040	HHH: Mother *-less adherence*	−0.41(0.11) <0.001	Higher WI *-greater adherence*	0.05(0.01) <0.001	Gauteng *-less adherence*	−0.32(0.09) <0.001	HHH: Other *-greater adherence*	0.35(0.16) 0.3
Higher WI *-greater adherence*	0.03(0.01) 0.017	PCG: Other *-less adherence*	−0.28(0.12) 0.022	Hunger present *-less adherence*	−0.34(0.1) <0.001			PCG: Grandmother *-greater adherence*	0.41(0.11) <0.001
Gauteng *-less adherence*	−0.43(0.09) <0.001	Female *-less adherence*	−0.28(0.09) 0.001	Gauteng *-less adherence*	−0.34(0.09) <0.001			Father employed *-less adherence*	−0.25(0.09) 0.006
Mother obese *-greater adherence*	0.21(0.09) 0.018	Gauteng *-less adherence*	−0.34(0.09) <0.001	Mother has Gr12+ *-greater adherence*	0.21(0.13) 0.1			Gauteng *-less adherence*	−0.91(0.09) <0.001
Hunger risk *-less adherence*	−0.33(0.11) 0.002							Hunger risk *-greater adherence*	0.35(0.11) 0.001
Hunger present *-less adherence*	−0.7(0.11) <0.001							Hunger present *-greater adherence*	0.28(0.11) 0.01

PL = pattern loading; RCs = refined carbohydrates; sat = saturated; unsat = unsaturated; PE = parameter estimate; SE = standard error; WI = wealth index, PCG = Primary caregiver; PCG Other = sibling or aunt; HHH = head of household; HHH Other = aunt or uncle; Gr = grade; URCs = unrefined breakfast cereals; RC-Other = other refined carbohydrates, e.g., pasta, rice, and samp; RC-Fat = combination of refined carbohydrates and fat, e.g., crisps (any type) and salty biscuits; RC-Fat -sugar = combination of refined carbohydrates, fat, and sugar, e.g., cake, tarts, doughnuts, ice cream, chocolates; RC-Protein-Fat = combination of refined carbohydrates, fat, and animal protein, e.g., samosa, fat cakes, pie, pizza, and lasagna pasta dishes; RC Sugar = sugar in the form of sweets; Veg non-starchy = all vegetables excluding starchy vegetables; Fats-oils-Sat = Butter, lard, hard margarine, coconut oil, non-dairy creamer; Fats-Oils-Unsat = Soft margarine, plant oils, avocado, nuts, salad dressing; SSBs = sugar-sweetened beverages, e.g., fizzy drinks, squash, and sport drinks. [1] A positive parameter estimate indicates greater adherence, and a negative parameter estimate indicates lesser adherence to a dietary pattern. [2] Multiple regression model with backward elimination was constructed for each pattern; only significant predictors remaining in the final model are included in the table. Dietary pattern scores were standardized with means of 0 and a unit standard deviation.

4. Discussion

A review of the literature shows that there are not many studies that report on the dietary patterns in children. In the present study, we set out to examine dietary patterns and socio-demographic predictors thereof in 1–<10-year-old children in two economically active provinces in South Africa. We also investigated associations between identified patterns and anthropometric indicators in the children. The results show that the dietary patterns were far from ideal, with no associations with anthropometric indicators. Predictors of both healthy and unhealthy patterns related to having a higher socio-economic status and an obese mother.

Tea (mostly rooibos tea) with sugar, but not so much coffee, is a drink that has previously been found to load strongly on dietary patterns in children under 2 years of age in lower income areas in in KwaZulu Natal and the North West province in South Africa (using a single unadjusted 24 h recall for dietary pattern analysis) [51]. Moreover, these researchers found that, in their 18–24-month-old group, sugar also had a high loading on the tea pattern. This indicates that tea was taken with sugar and the researchers speculated that mothers were substituting breast milk/formula milk with tea as children grew older [51], despite the recommendation in the paediatric food-based dietary guidelines that tea, coffee, and sugary drinks should be avoided [27]. Our results confirm that a *tea/coffee–sugar* pattern seems to be common in young children in South Africa as a *Tea/coffee and sugar* pattern emerged as one of the two strongest patterns in each age group, with pattern loadings (PLs) for tea/coffee and sugar being >0.8 in the two older groups and >0.7 in the 1–<3-year-old group.

Dairy also loaded on the *Tea/coffee and sugar* pattern in the 6–<10-year-olds, but not in the two younger age groups. This may indicate that tea is not necessarily given with milk in the younger age groups, reducing the potential nutrient density of the pattern in terms of quality protein, calcium, and other micronutrients. The practice of feeding children sugar in combination with black tea was also reported for children in Kenya and Tanzania [52]. However, in the 4–<6-year-olds, brown bread and fats (saturated and unsaturated), loaded on the tea/coffee and sugar pattern, indicating that the drink may be accompanied by a sandwich made form healthy bread with butter/margarine as a spread, increasing nutrient density. Predictors of the *Tea and coffee pattern* were contradictory, with indicators of both a higher socio-economic and a hunger profile being contributors to greater adherence to the *Tea/coffee and sugar* pattern.

Patterns that reflect the energy-dense, nutrient-poor Western dietary pattern [14] were prominent in the study sample. Pattern 3 (*Mostly unhealthy snacks*) in 1–<3-year-olds comprised unhealthy snacks such as crisps, salty biscuits, cake, sweet biscuits, ice cream, chocolate, and SSBs (fruit also loaded on this pattern, but with a lower PL). Pattern 4 (*White bread and toppings*) in this age group comprised a sandwich on white bread with mostly unhealthy toppings (processed meat, salty spreads). Pattern 2 (*Unhealthy foods and snacks*) in the 4–<6-year-old age group was deemed to be unhealthy and comprised white bread, "slap chips," items such as pies, "vetkoek", pasta dishes, pizza, cake, as well as the unhealthy snacks mentioned for the 1–<3-year-olds. The strongest pattern (*Mostly unhealthy 1*) in the 6–<10-year-old group comprised unhealthy snacks such as crisps, salty biscuits, SSBs, and sweets (fruit and unrefined breakfast cereal also loaded on this pattern, but with lower PLs). Pattern 3 (*Mostly unhealthy 2*) in this age group comprised unhealthy meal items such as refined breakfast cereals and foods/items such as pies, "vetkoek", pasta dishes, and pizza, as well as the mentioned unhealthy snacks. Pattern 4 (*White bread and processed meat*) in this age group comprised white bread, a saturated/unsaturated fat (margarine) spread, and processed meat.

Unhealth dietary patterns consumed by South African children have been reported by a few groups. Hooper et al. [53] indicated that the diets of 8–13-year-olds in urban areas in the Western Cape included unhealthy items such as fried potatoes, sausages, tinned fruit salad, custard, and jelly. Faber et al. [51] described a *More westernized pattern* in children under 2 years of age in KwaZulu Natal and the North West province, which was found to be positively associated with unhealthy nutrients such as cholesterol and saturated fat, emphasizing the importance of interventions to address unhealthy food choices. White bread flour in South Africa is fortified with eight micronutrients [54], but the fortification mix does not include calcium; vitamins C, D, and E; and other biologically active compounds found in unrefined cereals, fruit, and vegetables. White bread with a margarine spread and polony seems to be a recurring pattern in young children in the Western Cape, either served as a meal at home or included in the school lunch box [55]. Together with refined breakfast cereals, white bread may reflect the presence of poor food choice patterns, as they were combined with unhealthy snacks and foods in the patterns we identified.

A higher socio-economic status (higher wealth index, father and/or mother with grade 12/post-grade 12 qualification, and mother employed) were significant promoters of the unhealthy patterns identified in the present study. Temple and Steyn [56] and Heady et al. [57] showed that healthy foods such as milk, animal proteins, vegetables, and fruits, are more expensive than unhealthy foods. However, in their comparison of relative caloric prices (RCPs) of healthy and unhealthy across income levels and continents, Heady et al. [57] also categorized soft drinks, fruit juice, and salty snacks as expensive, and processed meats as very expensive. The Health Promotion Levy on sugary beverages was legislated 2017, with the aim of reducing SSBs consumption [58]. If households were to continue purchasing these items, as is evident from the PDIS results where these drinks were the fifth most consumed item (31% in 1–<3-year-olds; 42% in 3–<6-year-olds; 50% in 6–<10-year-olds) [36], it would come with the extra cost. The continued use of SSBs in South African communities is further illustrated in the results of the household inventory conducted by O-Halloran et al. [55] in low-income households in the Cape Town Metropole, where fizzy drinks were present in 66.6% of surveyed households. Of note is that fruit, which is classified as expensive [56,57], also loaded on Pattern 3 in the 1–<3-year-olds and Pattern 1 in the 6–<10-year-olds, albeit with lower PLs than the unhealthy items. Socio-demographic predictors of a *Sweet tooth dietary pattern* in Ghanaian adolescents also included household wealth, living with parents, and going to school with pocket money [59]. Money taken to school is usually spent at school tuckshops, where mostly unhealthy items, including crisps, sweets, chocolates, and fizzy SSB are sold to children in higher and lower socio-economic areas [55,60].

Each of the three age groups had a dietary pattern that was composed mainly of one or a combination of starches. This could reflect the South African Food-Based Dietary Guideline of making starchy foods part of most meals [61]. Maize porridge is typically given to young children in South Africa, especially in under 2-year-olds, either as a soft porridge at breakfast and/or in a stiffer consistency at lunch and/or supper with/without a sauce/soup containing some form of meat and/or vegetables [36,51]. It was thus not surprising that the strongest pattern in the 1–<3-year-olds in the youngest age group in the current study comprised maize porridge as the key starch and a soup/sauce (*Maize pap and soup/sauce pattern*). The nutrient density of this pattern may be acceptable if combined with a sauce/soup containing a quality protein, because maize meal is also fortified with eight micronutrients [54]. However, it is a concern that dairy does not load on this pattern in this age group. Pattern 3 (*Starches and poultry*) in the 4–<6-year-olds and Pattern 5 (*Starches*) in the 6–<10-year-olds comprised rice, pasta, and starchy vegetables, with poultry also loading on Pattern 3 in the 3–<6-year-old group. Starch-based diets could be low in nutrient density if they do not include a fortified cereal and are not combined with quality protein, fruit, and vegetables. Being at risk of or experiencing hunger promoted the starchy patterns in the youngest and oldest age group, while there were no promotors of the pattern in the 4–<6-year-old group. Starchy dietary patterns have been shown to be linked to poverty [62,63].

Patterns that included mostly healthy food items were Pattern 5 (*Vegetables and Fish*) in the 1–<3-year-old group and Patterns 4 (*Breakfast items*) and 5 (*Legumes and Vegetables*) in the 4–<6-year-old group. Pattern 5 in the youngest age group included unsaturated fats/oils, vegetables (except starchy vegetables), and fish, with the fats/oils most probably used in the preparation of the fish and vegetables. Healthy food items that loaded on Pattern 4 (*Breakfast items*) in the 3–<6-year-old group were dairy, fruit, and cheese. However, refined breakfast cereal, which is typically not classified as healthy, also loaded on this pattern. Based on the profiles of expensive versus cheap foods outlined by Temple et al. and Heady et al. [56,57], these two patterns can be categorized as expensive. The promotors of greater adherence to these patterns were higher WI (both patterns), higher qualified mother (both patterns), and father and an employed mother, confirming that income plays a role in the establishment and maintenance of these mostly healthy patterns. Pattern 5 in the 4–<6-year-olds included legumes and vegetables (excluding starchy vegetables), which are classified as healthy food choices [64,65]. However, salty spreads and condiments also loaded equally on this pattern. Faber et al. [51] identified a *Rice and legume pattern* in in 12–17-month-olds from low socio-economic populations in South Africa in Kwa-Zulu Natal and the Northwest province, which was positively associated with fibre, plant protein, and polyunsaturated fat [51]. Of note is that, for some children a pattern consisting of either legumes and vegetables, or rice and legumes, this may not be combined with high-quality protein which would not sustain optimal growth and development. Confirming this notion, Pisa et al. [66] found a positive association between BMI-for-age z-scores and a dietary pattern characterized by animal products and a second pattern comprising starch and folate. Our results show that hunger in the household promoted greater adherence to the *Legume and vegetable pattern* in this age group, suggesting that there is a possibility that quality proteins may be lacking.

It is important to note that not a single pattern that included mostly healthy items emerged in the 6–<10-year-old group. This illustrates a potential decrease in dietary quality with increasing age among primary school children. The results on adequacy of micronutrient intake in the PDIS sample confirms that 6–<10-year-olds had a lower intake of calcium, phosphorus, zinc, and vitamin C than the younger age groups [67]. Dietary pattern identification using data from a quantified food frequency questionnaire (recall period past 7 days) and principal component (PC) analysis in 9–11-year-old children in urban areas from 12 countries across the world (n = 7199) resulted in two strong components, namely a *Healthy pattern* and an *Unhealthy pattern*. The *Healthy pattern* included dark green vegetables, orange vegetables, other vegetables, berries, and fruits while the *Unhealthy pattern* included fast foods, fried food, French fries ("slap chips"), and SSBs. The researchers concluded that the same "healthier" and "unhealthier" foods tend to be consumed in similar combinations among 9–11-year-old children in different countries, despite variation in food culture, geographical location, ethnic background, and economic development [68].

A few studies in children and adolescents have shown associations between dietary patterns and undernutrition outcomes. Three patterns that included a quality animal protein were identified in children aged < 5 years living in rural Burkina Faso, but the *Leaves-based diet* did not result in improvements in wasting and stunting (dietary method used: semi-quantitative food frequency questionnaire; recall period—past 7 days) [69]. Analysis of dietary data for children younger than 5 years of age from the Demographic Surveillance System conducted in Kwale County, Kenya, showed that the *Traditional pattern* (minimal animal protein) showed a higher risk for stunting compared with the *Protein-rich pattern* (dietary method used: semi-quantitative food frequency questionnaire; recall period—past month) [70]. A greater adherence to a *Dairy, vegetable, and fruit pattern* was found to be associated with increased HAZ and reduced risk of stunting in younger than 5-year-olds in Ethiopia. However, no significant associations between the *Egg, meat, poultry, and legume pattern* with HAZ and stunting were found (dietary method: single 24 h recall of frequency of intake of seven or nine food groups depending on age) [23]. In 6–19-year-old Nigerian children and adolescents, a *Traditional dietary pattern* (containing mainly cereals/starchy food and legumes, and thus no quality animal protein) increased, while a diversified dietary pattern (containing all food groups) reduced the odds for thinness (dietary method: food frequency questionnaire; recall period—past month) [70].

In the present study, at least one of the five dietary patterns identified for each of the three age groups included a quality animal protein. Egg loaded positively on pattern 4, and fish on pattern 5 in 1–<3-year-olds; poultry on pattern 3, and dairy and cheese on pattern 4 in 3–<6-year-olds; and dairy on pattern 1, and red meat on pattern 2 in 6–<10-year-olds. Based on findings by others, the expectation was that one or more of these patterns would protect against stunting. However, we did not find any associations between stunting and any of the dietary patterns in the three age groups.

Associations between energy-dense dietary patterns and overweight/obesity have also been reported, albeit mostly in adolescents. Keding et al. [52] stated that an average of 10% of urban children in Kenya and Tanzania were overweight or obese. According to these researchers, this is mainly due to a *Purchase dietary pattern*, which is dominated by bought and processed foods. In the Nigerian study mentioned above, the *Traditional dietary pattern* did not only promote thinness, but also increased the odds of being overweight or obese, reflecting a double burden of malnutrition linked to this dietary pattern [71]. As for stunting, no associations between any of the dietary patterns and overweight/obesity were found in the current study.

Finally, in the present study, the BMI of the mothers showed interesting associations with the dietary patterns of their children. In the 1–<3-year-olds, an obese mother predicted greater adherence to the *Healthy pattern* (unsaturated fats/oils, vegetables, and fish) and lesser adherence to the *White bread and topping pattern* (processed meat and salty spreads), and in the 3–<6-year-olds greater adherence to the mostly healthy *Breakfast item pattern*. In the 6–<10-year-olds, an obese mother predicted increased adherence to the strongest pattern, which was *Mostly unhealthy pattern 1*. Various elements of the energy-dense, low-nutrient Western dietary pattern have been linked to obesity in adults [72–74]. From their systematic review of risk factors for overweight and obesity within the home environment of preschool children in sub-Saharan Africa, Kwansa et al. concluded that the home food environment, through the types of foods offered, and greater maternal BMI, were key aspects contributing to overweight and obesity among pre-schoolers [74]. Our results may indicate that these effects were not yet in play in the 1–<6-year-olds in the Western Cape but affected the dietary patterns of the 6–<10-year-olds.

Limitations

Although the dietary data in the present study were collected using an adjusted 24 h recall to remove intra-individual variability, the recall period does not necessarily reflect the dietary patterns of children in early life. This may explain the lack of association between dietary patterns and anthropometric outcomes in the present study. We suggest that an investigation of the dietary intake of a cohort of children from one to 18 years of age may provide better insights into long-term dietary patterns and associations with anthropometric indicators and other health outcomes. However, the feasibility of this type of study design in an LMIC is questionable due to limited funding and resources, as well as the logistics of tracing often migrating children over such a period of time. Periodic cross-sectional surveys add value in terms of insights into the dietary patterns of children that can be used in intervention planning and assessment. Despite the limitation linked to the recall period, the dietary patterns that emerged are in line with expectations when considering the nutrition transition [14,15], most commonly consumed foods [31,36,55], and the dietary patterns reported by Faber et al. [51] for South African children. PC analysis for dietary pattern identification does not come without limitations, including subjective decisions on how to interpret and name patterns, the number of components to retain, and the threshold for factor loadings to be used in naming patterns [65]. The retained patterns typically also explain less than 50% of the variance explained. The five patterns in each age group in the present study explained almost a third of the variance, which compares well with the variance explained by the two patterns per age group by Faber et al. [51] (just more than a third). Principal factor analysis aims to maximize the fraction of variance explained by a weight linear combination of original variables, which does not necessarily increase the ability to discriminate between subjects with disease (malnutrition) or not. The present study's data were self-reported and a cross-sectional design was used; hence, no causal links can be implied.

5. Conclusions

Bearing in mind the limitations of this study, we conclude that the dietary patterns in 1–<10-year-old children in the Western Cape contain strong elements of energy-dense, nutrient-poor Western dietary patterns, as at least one such a pattern was included in the top three strongest patterns in all three age groups that were investigated. Few of the dietary patterns included vegetables other than starchy vegetables, or fruit, dairy, quality proteins, and unrefined carbohydrates. Key predictors of greater adherence to the mostly unhealthy patterns included indicators of a higher socio-economic status in all three age groups, as well as having an obese mother in the 6–12-year-old group. As dietary habits and patterns formed in childhood continue into adulthood [10], the findings of this research point to an urgent need for review of the effectiveness of current policy and interventions aimed at ensuring child food security and well-being, as well as a review of policy and legislation aimed at supporting a healthy food environment, to identify drivers of the nutrition transition that may need further actioning to improve the dietary patterns of children in the country.

Key predictors of greater adherence to the mostly healthy patterns were a higher wealth index and having an obese mother in the two younger groups, with no predictors in the 6–<10-year-old group. There were no associations between any of the dietary patterns and stunting or overweight/obesity in the children. We recommend that interventions to improve the dietary intake of children should be directed at both poorer and higher income communities.

We foresee that the methodology for extraction of dietary patterns from dietary datasets that we established and described in detail in this study will be of value to others in the same field of research. The dietary patterns and socio-demographic predictors for 1–<10-year-olds we reported in this paper may inform the need for, and design of, further research for the monitoring of dietary patterns of South African children.

Author Contributions: Conceptualization, M.S.; methodology, M.S., J.H.N., G.E. and N.P.S.; formal analysis, J.H.N. and M.S.; writing—M.S., N.P.S., J.H.N. and G.E.; original draft preparation, M.S.; writing—review and editing, M.S., N.P.S., J.H.N. and G.E.; supervision, N.P.S.; project administration, N.P.S.; funding acquisition, N.P.S. All authors have read and agreed to the published version of the manuscript.

Funding: This research was mainly funded by the International Life Sciences Institute of South Africa and their industry partners and the South African National Research Foundation.

Institutional Review Board Statement: This study was conducted according to the guidelines of the Declaration of Helsinki and approved by the Faculty of Health Sciences Human Research Ethics Committee at UCT on 18 July (HREC REF: 326/2018).

Informed Consent Statement: Informed consent was obtained from all subjects involved in the study.

Data Availability Statement: The data presented in this study are available on request from the corresponding author pending ethical approval from the Faculty of Health Sciences Human Research Ethics Committee, University of Cape Town.

Acknowledgments: We acknowledge the major contributions of the field workers in GTG and WC and the dieticians who supervised L Drummond and Sonya Malczyk. We also acknowledge Busi Booi-Shologu for considerable assistance with administrative work and overall support.

Conflicts of Interest: The authors declare no conflict of interest.

References

1. Ten Hove, H.; Guo, X.; Bakker, S.; Herens, M. *Addressing Overweight and Obesity in LMICs in Rural Development and Food Systems: A Comprehensive Literature Review*; Wageningen Centre for Development Innovation: Wageningen, The Netherlands, 2023. [CrossRef]
2. Shisana, O.; Labadarios, D.; Rehle, T.; Simbayi, L.; Zuma, K.; Dhansay, A.; Reddy, P.; Parker, W.; Hoosain, E.; Naidoo, P.; et al. *The South African National Health and Nutrition Examination Survey, 2012: SANHANES-1: The Health and Nutritional Status of the Nation*; HSRC Press: Cape Town, South Africa, 2014; Available online: http://repository.hsrc.ac.za/handle/20.500.11910/2864 (accessed on 20 January 2022).
3. Vorster, H.H.; Kruger, A. Poverty, malnutrition, underdevelopment and cardiovascular disease: A South African perspective. *Cardiovasc. J. Afr.* **2007**, *18*, 321–324.
4. Liberali, E.; Kupek, E.; Altenburg de Assis, M.A. Dietary patterns and childhood obesity risk: A systematic review. *Child. Obes.* **2020**, *16*, 70–85. [CrossRef]
5. Umer, A.; Kelley, G.A.; Cottrell, L.E.; Giacobbi, P.; Innes, K.E.; Lilly, C. Childhood obesity and adult cardiovascular disease risk factors: A systematic review with meta-analysis. *BMC Public Health* **2017**, *17*, 683. [CrossRef] [PubMed]
6. Katzmarzyk, P.; Barlow, S.; Bouchard, C.; Catalano, P.M.; Hsia, D.S.; Inge, T.H.; Lovelady, C.; Raynor, H.; Redman, L.M.; Staino, A.E.; et al. An evolving scientific basis for the prevention and treatment of pediatric obesity. *Int. J. Obes.* **2014**, *38*, 887–905. [CrossRef] [PubMed]
7. Senekal, M.; Nel, J.H.; Malczyk, S.; Drummond, L.; Harbron, J.; Steyn, N.P. Provincial Dietary Intake Study (PDIS): Prevalence and Sociodemographic Determinants of the Double Burden of Malnutrition in A Representative Sample of 1 to Under 10-Year-Old Children from Two Urbanized and Economically Active Provinces in South Africa. *Int. J. Environ. Res. Public Health* **2019**, *16*, 3334. [CrossRef]
8. Steyn, N.P.; Nel, J.H.; Drummond, L.; Malczyk, S.; Senekal, M. Has Food Security and Nutritional Status Improved in Children 1–<10 Years in two Provinces of South Africa between 1999 (National Food Consumption Survey) and 2018 (Provincial Dietary Intake Study (PDIS)). *Int. J. Environ. Res. Public Health* **2022**, *19*, 1038. [CrossRef]
9. Lobstein, T.; Brinsden, H.; Neveux, M. *World Obesity Atlas*; World Obesity Federation: London, UK, 2022.
10. WHO; FAO. *Diet, Nutrition and the Prevention of Chronic Diseases*; WHO: Geneva, Switzerland, 2003.
11. WHO. *Good Maternal Nutrition the Best Start in Life*; WHO: Geneva, Switzerland, 2016. Available online: https://iris.who.int/handle/10665/329459 (accessed on 1 May 2023).
12. Mikkelsen, B.; Williams, J.; Rakovac, I.; Wickramasing, K.; Hennis, A.; Hai-Rim Shin, H.R.; Breda, J.; Huber, M.; Borges, C.; Berdzuli, N.; et al. Life course approach to prevention and control of non-communicable diseases. *BMJ* **2019**, *364*, l257. [CrossRef]
13. GBD Diet Collaborators. Health effects of dietary risks in 195 countries, 1990–2017: A systematic analysis for the Global Burden of Disease Study 2017. *Lancet* **2019**, *393*, 1958–1972. [CrossRef]
14. Popkin, B.M.; Ng, S.W. The nutrition transition to a stage of high obesity and noncommunicable disease prevalence dominated by ultra-processed foods is not inevitable. *Obes. Rev.* **2022**, *23*, e13366. [CrossRef]
15. Popkin, B. Nutrition transition and the global diabetes epidemic. *Curr. Diabetes Rep.* **2015**, *15*, 1–8. [CrossRef] [PubMed]
16. Shang, X.; Li, Y.; Liu, A.; Zhang, Q.; Hu, X.; Du, S.; Ma, J.; Xu, G.; Ling, Y.; Guo, H.; et al. Dietary pattern and its association with the prevalence of obesity and related cardiometabolic risk factors among Chinese children. *PLoS ONE* **2012**, *7*, e43183. [CrossRef]
17. Monteiro, C.A.; Cannon, G.; Moubarac, J.C.; Levy, R.B.; Louzada, M.L.C.; Jaime, P.C. The UN decade of nutrition, the NOVA food classification and the trouble with ultra-processing. *Public Health Nutr.* **2018**, *21*, 5. [CrossRef] [PubMed]
18. Hoffmann, K.; Schulze, M.; Boeing, H.; Altenburg, H.P. Dietary patterns: Report of an International Workshop. *Public Health Nutr.* **2018**, *5*, 89–90. [CrossRef]
19. Schwerin, H.S.; Stanton, J.L.; Smith, J.L.; Riley, A.M., Jr.; Brett, B.E. Food, eating habits, and health: A further examination of the relationship between food eating patterns and nutritional health. *Am. J. Clin. Nutr.* **1982**, *35*, 1319–1325. [CrossRef] [PubMed]
20. Appel, L.J.; Moore, T.J.; Obarzanek, E.; Vollmer, W.M.; Svetkey, L.P.; Sacks, F.M.; Bray, G.A.; Vogt, T.M.; Cutler, J.A.; Windhauser, M.M.; et al. A clinical trial of the effects of dietary patterns on blood pressure. DASH collaborative research group. *N. Engl. J. Med.* **1997**, *336*, 1117–1124. [CrossRef]
21. Olinto, M.T.; Willett, W.C.; Gigante, D.P.; Victora, C.G. Sociodemographic and lifestyle characteristics in relation to dietary patterns among young Brazilian adults. *Public Health Nutr.* **2010**, *25*, 1–10. [CrossRef] [PubMed]
22. Marques-Vidal, P.; Waeber, G.; Vollenweider, P.; Guessous, I. Socio-demographic and lifestyle determinants of dietary patterns in French-speaking Switzerland, 2009–2012. *BMC Public Health* **2018**, *18*, 131. [CrossRef]
23. Melaku, Y.A.; Gill, T.K.; Taylor, A.W.; Adams, R.; Shi, Z.; Worku, A. Associations of childhood, maternal and household dietary patterns with childhood stunting in, E.thiopia: Proposing an alternative and plausible dietary analysis method to dietary diversity scores. *Nutr. J.* **2018**, *17*, 1–15. [CrossRef]
24. Johnson, L.; Toumpakari, Z.; Papadaki, A. Social gradients and physical activity trends in an obesogenic dietary pattern: Cross-sectional analysis of the UK National Diet and Nutrition Survey 2008–2014. *Nutrients* **2018**, *10*, 388. [CrossRef]
25. Nel, J.; Steyn, N.P. The Nutrition Transition and the Double Burden of Malnutrition in Sub-Saharan African Countries: How Do These Countries Compare with the Recommended Lancet Commission Global Diet? *Int. J. Environ. Res. Public Health* **2022**, *19*, 16791. [CrossRef]
26. FAO. Food-Based Dietary Guidelines. 2018. Available online: www.fao.org/nutrition/education/food-dietaryguidelines/home/en/ (accessed on 30 January 2018).

27. Bourne, L. South African paediatric food-based dietary guidelines. *Matern. Child Nutr.* **2007**, *3*, 3227–3229. [CrossRef] [PubMed]
28. Vorster, H.H.; Badham, J.B.; Venter, C.S. An introduction to the revised food-based dietary guidelines for South Africa. *S. Afr. J. Clin. Nutr.* **2013**, *3*, S5–S12.
29. Statistics South Africa. Mid-Year Population Estimates 2018. Available online: http://www.statssa.gov.za/?p=11341 (accessed on 6 March 2019).
30. Statistics South Africa Census 2011 Metadata. Available online: http://www.statssa.gov.za/census/census_2011/census_products/Census_2011_Metadata.pdf (accessed on 30 January 2022).
31. Labadarios, D.; Steyn, N.P.; Maunder, E.; MacIntryre, U.; Gericke, G.; Swart, R.; Huskisson, J.; Dannhauser, A.; Vorster, H.H.; Nesamvuni, A.E. The National Food Consumption Survey (NFCS): South Africa, 1999. *Public Health Nutr.* **2005**, *8*, 533–543. [CrossRef]
32. Filmer, D.; Pritchett, L. Estimating wealth effects without expenditure data—Or tears: With an application to educational enrollments in the states of India. In *The World Bank Development Research Group*; The World Bank: Washington, DC, USA, 1998.
33. South African Medical Research Council. *South African Demographic and Health Survey: 2016*; South Africa Medical Research Council: Pretoria, South Africa, 2017.
34. Wehler, C.; Scott, R.; Anderson, J. The community childhood hunger identification project: A model of domestic hunger-demonstration. *J. Nutr. Educ.* **1992**, *24*, 295–355. [CrossRef]
35. Moshfegh, A.J.; Rhodes, D.G.; Baer, D.J.; Muray, T.; Clemens, J.C.; Rumpler, W.V.; Paul, D.R.; Sebastian, R.S.; Kuczynski, K.J.; Ingwersen, L.A.; et al. The US Department of Agriculture Automated Multiple-Pass Method reduces bias in the collection of energy intakes. *Am. J. Clin. Nutr.* **2008**, *88*, 324–332. [CrossRef]
36. Steyn, N.P.; Nel, J.H.; Malczyk, S.; Drummond, L.; Senekal, M. Provincial Dietary Intake Study (PDIS): Energy and macronutrient intakes of children in a representative/random sample of 1–<10-year-old children in two economically active and urbanized provinces in South Africa. *Int. J. Environ. Res. Public Health* **2020**, *17*, 1717. [CrossRef]
37. Burrows, T.L.; Martin, R.J.; Collins, C.E. A systematic review of the validity of dietary assessment methods in children when compared with the method of doubly labelled water. *J. Am. Diet Assoc.* **2010**, *110*, 1501–1510. [CrossRef]
38. Steyn, N.P.; Senekal, M. *The Dietary Assessment and Education Kit (DAEK) The Chronic Diseases of Lifestyle Unit of the South African Medical Research Council*; MRC: Cape Town, South Africa, 2004.
39. Steyn, N.P.; Senekal, M.; Norris, S.A.; Whati, L.; Mackeown, J.M.; Nel, J.H. How well do adolescents determine portion sizes of foods and beverages? *Asia Pac. J. Clin. Nutr.* **2006**, *15*, 35–42. [PubMed]
40. Neville, M.C.; Allen, J.C.; Archer, P.C.; Casey, C.E.; Seacat, J.; Keller, R.P.; Lutes, V.; Rasbach, J.; Neifert, M. Studies in human lactation: Milk volume and nutrient composition during weaning and lactogenesis. *Am. J. Clin. Nutr.* **1991**, *54*, 81–92. [CrossRef] [PubMed]
41. Tooze, J.A.; Kipnis, V.; Buckman, D.W.; Carroll, R.J.; Freedman, L.S.; Guenther, P.M.; Krebs-Smith, S.M.; Subar, A.F.; Dodd, K.W. A mixed-effects model approach for estimating the distribution of usual intake of nutrients: The NCI method. *Stat. Med.* **2010**, *29*, 2857–2868. [CrossRef]
42. Herrick, K.A.; Rossen, L.M.; Parsons, R.; Dodd, K.W. Estimating usual dietary intake from National Health and Nutrition Examination 5. Survey data using the National Cancer Institute method. National Center for Health Statistics. *Vital Health Stat.* **2018**, *2*, 1–63.
43. Van Graan, A.E.; Chetty, J.M.; Links, M.R. *Food Composition Tables for South Africa*, 5th ed.; South African Medical Research Council: Cape Town, South Africa, 2017.
44. O'Rourke, N.; Hatcher, L. *A Step by Step Approach to Using SAS System for Factor Analysis and Structural Equation Modelling*, 2nd ed.; SAS Press: Cary, NC, USA, 2013.
45. Steyn, N.P.; Maunder, E.M.; Labadarios, D.; Nel, J.H. Foods and beverages that make a significant contribution to macro- and micronutrient intakes of children in South Africa- do they meet the food-based dietary guidelines? *S. Afr. J. Clin. Nutr.* **2006**, *19*, 66–76. [CrossRef]
46. Maunder, E.M.W.; Nel, J.H.; Steyn, N.P.; Kruger, H.S.; Labadarios, D. Added Sugar, Macro- and Micronutrient Intakes and Anthropometry of Children in a Developing World Context. *PLoS ONE* **2015**, *10*, e0142059. [CrossRef] [PubMed]
47. Okeyo, A.; Seekoe, E.; de Villiers, A.; Faber, M.; Nel, J.H.; Steyn, N.P. Dietary Practices and Adolescent Obesity in Secondary School Learners at Disadvantaged Schools in South Africa: Urban–Rural and Gender Differences. *Int. J. Environ. Res. Public Health* **2020**, *17*, 5864. [CrossRef] [PubMed]
48. Kaiser, H.F. A second generation little jiffy. *Psychometrika* **1970**, *35*, 401–415. [CrossRef]
49. Blom, G. *Statistical Estimates and Transformed Beta-Variables*; Wiley: New York, NY, USA, 1958.
50. Ricci, C.; Baumgartner, J.; Wentzel-viljoen, E.; Smuts, C.M. Food and nutrient pattern assessment using principal component analysis applied to food questionnaires. Pitfalls, tips and tricks. *Int. J. Food Sci.Nutr.* **2019**, *70*, 738–748. [CrossRef]
51. Faber, M.; Rothman, M.; Laubscher, R.; Smuts, C.M. Dietary patterns of 6–24-month-old children are associated with nutrient content and quality of the diet. *Matern. Child Nutr.* **2019**, *16*, e12901. [CrossRef] [PubMed]
52. Keding, G. Nutrition Transition in Rural Tanzania and Kenya. Hidden Hunger. Malnutrition and the First 1000 Days of Life: Causes, Consequences and Solutions. In *World Review of Nutrition and Dietetics*; Biesalski, H.K., Black, R.E., Eds.; Karger: Basel, Switzerland, 2016; Volume 115, pp. 68–81. [CrossRef]

53. Hooper, R.; Calvert, J.; Thompson, R.L.; Deetlefs, M.E.; Burney, P. Urban/rural differences in diet and atopy in South Africa. *Allergy* **2008**, *63*, 425–431. [CrossRef] [PubMed]
54. Department of Health. *Government Notice: Foodstuffs, Cosmetics and Disinfectants Act, No. R 2003. (Act No. 54 of 1972). Regulations Relating to the Fortification of Certain Foodstuffs*; Department of Health: Pretoria, South Africa, 2003.
55. O'Halloran, S.A.; Eksteen, G.; Polayya, N.; Ropertz, M.; Senekal, M. The food environment of primary school learners in a low-to-middle-income area in Cape Town, South Africa. *Nutrients* **2021**, *13*, 2043. [CrossRef] [PubMed]
56. Temple, N.J.; Steyn, N.P.; De Villiers, A. Price and availability of healthy food: A study in rural South Africa. *Nutr. J.* **2011**, *27*, 55–58. [CrossRef]
57. Headey, D.D.; Alderman, H.A. The relative caloric prices of healthy and unhealthy foods differ systematically across income levels and continents. *J. Nutr.* **2019**, *149*, 2020–2033. [CrossRef]
58. National Treasury. Policy Paper and Proposals on the Taxation of Sugar Sweetened Beverages-8 July 2016. Available online: https://www.gov.za/documents/taxation-sugar-sweetened-beverages-policy-paper-8-jul-2016-0000 (accessed on 3 May 2023).
59. Abizari, A.R.; Ali, Z. Dietary patterns and associated factors of schooling Ghanaian adolescents. *J. Health Popul. Nutr.* **2019**, *38*, 5. [CrossRef] [PubMed]
60. Okeyo, A.P.; Seekoe, E.; de Villiers, A.; Faber, M.; Nel, J.H.; Steyn, N.P. The food and nutrition environment at secondary schools in the Eastern Cape, South Africa as reported by learners. *Int. J. Environ. Res. Public Health* **2020**, *17*, 4038. [CrossRef] [PubMed]
61. Vorster, H.H. "Make starchy foods part of most meals": A food-based dietary guideline for South Africa. *S. Afr. J. Clin. Nutr.* **2013**, *3*, S28–S35.
62. Heady, D.; Hirvonen, K.; Hoddinott Stifel, D. Rural food markets and child nutrition. *Am. J. Agri. Econ.* **2019**, *101*, 1311–1327. [CrossRef] [PubMed]
63. Fitchen, K.M. Hunger, malnutrition, and poverty in the contemporary United States: Some observations on their social and cultural context. *Food Foodways* **1998**, *2*, 309–333. [CrossRef]
64. Venter, C.S.; Ochse, R.; Swart, R. "Eat dry beans, split peas, lentils and soya regularly": A food-based dietary guideline for South Africa. *S. Afr. J. Clin. Nutr.* **2013**, *3*, S36–S45.
65. Naude, C.E. "Eat plenty of vegetables and fruit every day": A food-based dietary guideline for South Africa. *S. Afr. J. Clin. Nutr.* **2013**, *3*, S46–S56.
66. Pisa, P.T.; Pedro, T.M.; Kahn, K.; Tollman, S.M.; Pettifor, J.M.; Norris, S.A. Nutrient patterns and their association with socio-demographic, lifestyle factors and obesity risk in rural South African adolescents. *Nutrients* **2015**, *7*, 3464–3482. [CrossRef]
67. Senekal, M.; Nel, J.H.; Malczyk, S.; Drummond, L.; Steyn, N.P. Provincial Dietary Intake Study (PDIS): Micronutrient intakes of children in a representative/random sample of 1- to <10 year old children in two economically active and urbanised provinces in South Africa. *Int. J. Environ. Res. Public Health* **2020**, *17*, 5924. [CrossRef]
68. Mikkilä, V.; Vepsäläinen, H.; Saloheimo, T.; Gonzalez, S.A.; Meisel, J.D.; Hu, G.; Champagne, C.M.; Chaput, J.P.; Church, T.S.; Katzmarzyk, P.T.; et al. International comparison of dietary patterns in 9-11-year-old children. *Int. J. Obes. Suppl.* **2015**, *5* (Suppl. 2), S17–S21. [CrossRef]
69. Mank, I.; Vandormael, A.; Traore, I.; Quedraogo, W.A.; Sauerborn, R.; Danquah, I. Dietary habits associated with growth development of children aged < 5 years in the Nouna Health and Demographic Surveillance System, Burkina Faso. *Nutr. J.* **2020**, *19*, 1–14.
70. Tanaka, J.; Yoshizawa, K.; Hirayama, K.; Karama, M.; Wanjihia, V.; Changoma, M.S.; Kaneko, S. Relationship between dietary patterns and stunting in preschool children: A cohort analysis from Kwale, Kenya. *Public Health* **2019**, *173*, 58–68. [CrossRef]
71. Adeomi, A.A.; Fatusi, A.; Klipstein-Grobusch, K. Food security, dietary diversity, dietary patterns and the double burden of malnutrition among school-aged children and adolescents in two Nigerian States. *Nutrients* **2022**, *14*, 789. [CrossRef]
72. Holmes, M.D.; Dalal, S.; Sewram, V.; Diamond, M.B.; Adebamowo, S.N.; Ajayi, I.O.; Adebamowo, C.; Chiwanga, F.S.; Njelekela, M.; Laurence, C.; et al. Consumption of processed food dietary patterns in four African populations. *Public Health Nutr.* **2018**, *21*, 1529–1537. [CrossRef]
73. Malik, V.S.; Hu, F.B. The role of sugar-sweetened beverages in the global epidemics of obesity and chronic diseases. *Nat. Rev Endocrinol.* **2022**, *18*, 205–218. [CrossRef]
74. Kwansa, A.L.; Akparibo, R.; Cecil, J.E.; Infield, S.G.; Caton, S.J. Risk factors for overweight and obesity within the home environment of preschool children in Sub-Saharan Africa: A systematic review. *Nutrients* **2022**, *14*, 1706. [CrossRef]

nutrients

MDPI

Article

Sex Differences in Dietary Patterns of Adults and Their Associations with the Double Burden of Malnutrition: A Population-Based National Survey in the Philippines

Aileen Rodil de Juras [1,2], Wan-Chen Hsu [1], Yu-Yao Cheng [3], Li-Jung Elizabeth Ku [1], Tsung Yu [1], Cheau-Jane Peng [4] and Susan C. Hu [1,*]

1 Department of Public Health, College of Medicine, National Cheng Kung University, Tainan City 701, Taiwan
2 Institute of Human Nutrition and Food, College of Human Ecology, University of the Philippines Los Baños, Los Baños 4030, Philippines
3 Department of Health and Nutrition, Chia Nan University of Pharmacy and Science, Tainan City 717, Taiwan
4 Department of Senior Welfare and Services, Southern Taiwan University of Science and Technology, Tainan City 701, Taiwan
* Correspondence: shuhu@mail.ncku.edu.tw; Tel.: +886-6-2353535 (ext. 5599)

Abstract: A dietary pattern transition is a risk factor for the double burden of malnutrition (DBM), but related information is limited. This study aimed to identify sex differences in dietary patterns of adults in a low–middle income country and to examine their association with DBM. A total of 8957 adults (4465 men and 4492 non-pregnant and non-lactating women) who participated in the 2013 Philippine National Nutrition Survey were included in the analysis. Logistic regression models were formulated to investigate the relationship between dietary patterns and DBM. The factor analysis derived seven dietary patterns for males and six patterns for females. Results showed that approximately 30% of Filipino adults suffered from DBM. The rice pattern was associated with lower odds of DBM for males only. The meat and sugar pattern in males and the protein-rich foods, cereal, and sugar pattern in females decreased DBM likelihood. An inverse relationship was observed for the vegetables and corn patterns, wherein females had an increased risk for DBM. Our findings suggest that rice-based and meat-containing patterns could play protective roles in DBM development among adults in the Philippines. Understanding sex-specific dietary patterns can be utilized to guide public health nutrition interventions in the prevention of malnutrition in all its forms.

Keywords: dietary patterns; double burden of malnutrition; adults; Philippines; low–middle income country

Citation: de Juras, A.R.; Hsu, W.-C.; Cheng, Y.-Y.; Ku, L.-J.E.; Yu, T.; Peng, C.-J.; Hu, S.C. Sex Differences in Dietary Patterns of Adults and Their Associations with the Double Burden of Malnutrition: A Population-Based National Survey in the Philippines. *Nutrients* **2022**, *14*, 3495. https://doi.org/10.3390/nu14173495

Academic Editors: Andriana Kaliora, Chara Tzavara and Charalampia Amerikanou

Received: 26 July 2022
Accepted: 23 August 2022
Published: 25 August 2022

1. Introduction

Dietary patterns are shifting considerably in low- and middle-income countries, as exemplified by the displacement of staple-food-based diets with increased meat, fat, salt, and added sugar intakes. Consequently, the transition in food patterns is a key driver for the double burden of malnutrition (DBM), defined as the co-existence of undernutrition with overnutrition and diet-related non-communicable diseases across the life course [1,2]. Several studies have been conducted using the dietary pattern approach in order to understand the complex etiology of DBM among adults [3–9]. Evidence suggests that less diverse diets and a traditional dietary pattern were risk factors for individual-level DBM [7–9].

The Philippines is continuously facing DBM. In particular, Filipino adults suffer from malnutrition in all its forms [10–12]. Transformations in food consumption have also been evident in the country [13]. What is known to date on the nexus of the dietary pattern–double burden of malnutrition is largely on a national scale. Hence, this study aimed to identify the distinct dietary patterns of male and female community-dwelling adults in a low–middle income setting and to examine the relation of dietary patterns to DBM using the Philippines as an example.

2. Materials and Methods

2.1. Data Source and Subjects

We analyzed the data from the 8th Philippine National Nutrition Survey (PNNS), a cross-sectional study that is accessible through http://enutrition.fnri.dost.gov.ph/site/home.php (accessed on 3 September 2020) [14]. Briefly, the survey was carried out from 2013 to 2014 by the Department of Science and Technology–Food and Nutrition Research Institute to determine the nutrition and health status of Filipinos. The PNNS has a stratified multistage sampling design representative at the national, regional, and provincial levels. The objectives, design, and procedures of PNNS have been detailed elsewhere [15,16].

The study participants were restricted to male and female adults (≥20 years old) with complete subject identification in the six survey components (i.e., dietary, anthropometry, biochemical, clinical, socioeconomic individual, and socioeconomic household). Pregnant women, lactating mothers, and those with missing data on cardiometabolic risk factors (CMRF), hemoglobin, serum retinol, and urinary iodine excretion (UIE) were excluded. Participants with high energy intake (greater than 5 standard deviations of mean energy intake) were also excluded [17]. No participants had low energy intake or lower than 5 standard deviations of mean energy intake. As a result, a total of 8957 adults were included in the analysis. The flowchart of the selection process for the study samples is illustrated in Figure 1.

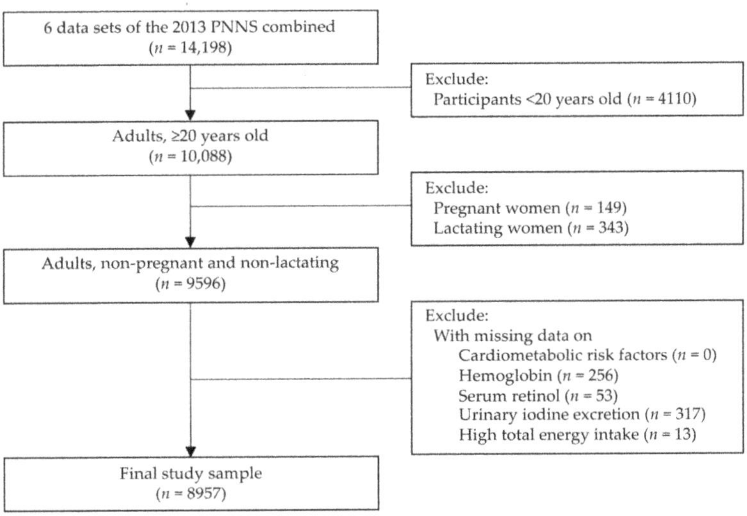

Figure 1. Selection of study participants. (Cardiometabolic risk factors were defined as an individual with any of the following factors: (1) overweight/obesity or abdominal obesity, (2) hypertension, (3) hyperglycemia, or (4) dyslipidemia [low high-density lipoprotein (HDL) cholesterol or hypertriacylglycerolemia]. There were no study participants with missing values on cardiometabolic risk factors).

2.2. Dietary Intake Assessment and Dietary Pattern Analysis

Dietary intake was assessed with 24-h food recall in the 8th PNNS. Registered nutritionist-dietitians administered the food recalls on two non-consecutive days. Common household measurements or food sample weighing was utilized to estimate the amount of food and beverages consumed. Calibrated kitchen utensils, rulers, and a photo compilation of foods were used as aids. Subsequently, the weights were converted to purchased values, and energy intakes were computed utilizing the Philippine Food Composition Table. The food items were then aggregated into food groups [18].

Dietary patterns were derived separately for males and females through factor analysis (principal axis factoring method with varimax rotation in R software) based on the mean intake of 18 food groups (Supplementary Table S1). To avoid too many zero values in the data and irrelevant results, only the food groups that were consumed by more than 10% of the study population were included in the analysis [19–22]. The number of factors retained was determined considering the scree plot results, components with eigenvalue >1.0, and factor interpretability (Supplementary Figure S1). A factor loading of $\geq$0.25 was the cut-off value for identifying food groups that strongly contribute to a particular dietary pattern [22–24]. The naming of dietary patterns was decided according to published studies and data interpretation. Additionally, when a food group was loaded in more than one dietary pattern, the group with the higher or positive coefficient was accounted for in the labeling [25]. Factor scores were then calculated and divided into tertile intervals. The bottom tertile (T1) corresponds to low adherence in a dietary pattern, the middle tertile (T2) corresponds to medium adherence, and the upper tertile (T3) corresponds to high adherence. The Kaiser-Meyer-Olkin measure of sampling adequacy and Bartlett's test of sphericity were done before factor analysis to evaluate data suitability.

2.3. Undernutrition, Cardiometabolic Risk Factors, and Double Burden of Malnutrition

A study participant was considered to be experiencing undernutrition if at least one of the following conditions was present: (1) underweight, (2) anemia, (3) vitamin A deficiency, or (4) iodine insufficiency. An underweight categorization was assessed utilizing body mass index (BMI) and calculated as the weight in kilograms divided by the height in meters squared. The weight and height of the participants were obtained by employing mechanical platform beam balance scales (Detecto™) and microtoise (Seca™), respectively. Furthermore, the BMI classification applied was done according to the World Health Organization (WHO) [26]. Biochemical indicators for three micronutrient deficiencies, i.e., anemia, vitamin A deficiency, and iodine insufficiency, were collected during the survey. Anemia was examined from hemoglobin utilizing a spectrophotometer [27]. Hemoglobin values <13 g/dL for males and <12 g/dL for females indicated anemia [28]. On the other hand, vitamin A deficiency was tested from serum retinol by High-Performance Liquid Chromatography [29] and distinguished as serum retinol <10 µg/dL [30]. Iodine insufficiency was determined from UIE levels through the acid digestion/colorimetric method [31]. The cut-off used was UIE <50 µg/dL [32].

The criteria used for having a CMRF were adopted from Zeba and colleagues [9]. It was defined as having any of the following factors: (1) overweight/obesity or abdominal obesity, (2) hypertension, (3) hyperglycemia, or (4) dyslipidemia [low high-density lipoprotein (HDL) cholesterol or hypertriacylglycerolemia]. Overweight/obesity and abdominal obesity were categorized based on the WHO guidelines [26,33]. Overweight and obesity were evaluated by computing the BMI. For waist circumference, calibrated tape measures were utilized [18]. Hypertension was denoted by a blood pressure measurement of $\geq$140/$\geq$90 mmHg [34]. Blood pressure readings were performed with a calibrated non-mercurial sphygmomanometer (A&D Um-101™) and stethoscope [18]. Hyperglycemia was characterized as a fasting blood glucose $\geq$110 mg/dL [35], and dyslipidemia was characterized as having an HDL cholesterol <40 mg/dL for males or <50 mg/dL for females, or triglyceride $\geq$150 mg/dL [36,37]. Plasma blood glucose was analyzed for hyperglycemia, while serum blood lipids were assessed for dyslipidemia via the enzymatic colorimetric method [18]. We described the total double burden of malnutrition (total DBM) at the individual level as the concomitance of various forms of undernutrition (underweight, anemia, and vitamin A deficiency or iodine insufficiency) and at least one CMRF [12].

2.4. Other Co-Variates

The other co-variates in this study were the sociodemographic and lifestyle characteristics obtained through one-on-one interviews. Sociodemographic information encompasses sex (male or female), age (20–39, 40–59, and $\geq$60 years), educational attainment (elementary and lower, high school, college and higher), marital status (single, married or with partner, and others or widowed/separated/annulled/divorced), employment status (employed or unemployed), and wealth quintile (poorest, poor, middle, rich, richest). Household size was created from the socioeconomic datasets and categorized as 1–3, 4–6, and ≥7. The lifestyle factors of smoking (current smoker or not), alcohol consumption (current drinker or not), and physical activity (low or high) were likewise controlled in the analysis and classified utilizing WHO standards [38,39].

2.5. Statistical Analysis

All data analyses were conducted in R software version 4.0.3 (R Foundation for Statistical Computing, Vienna, Austria). The percentages of sociodemographic characteristics, lifestyle factors, and total DBM were generated according to sex and compared using the Chi-square test. A binary logistic regression analysis was employed to evaluate the relationship between the tertiles of dietary pattern scores and total DBM for males and females separately, since there was significant interaction with sex and some outcome variables. The formulated models were adjusted for all co-variates and the energy intake. The reference group for each dietary pattern was the bottom tertile (T1). Multicollinearity was assessed in all models. The multi-level sampling design of the survey was considered in the regression analysis, i.e., sampling weights were employed to generate results representative of the adult population in the Philippines. The significance level was set at $p < 0.05$.

3. Results

3.1. Participants' Characteristics

A total of 8957 adults were included in this study with a balance between male and female participants (Table 1). The study sample mostly belonged to the 20–39 years old age group (46.5%), finished high school education (37.9%), were married (66.6%), and were employed (59.5%). There were slightly more females in the older age group (16.6%) and more females who attained college education or higher (32.4%) relative to males. Alternatively, more males were single or unmarried (26.4%) and employed (76.3%). The median household size was four and no sex differences were noted. There were more males in the poorest and poor quintiles than females. In terms of lifestyle factors, 26.9% were current smokers, 51.3% were current alcohol drinkers, and 44.2% had low physical activity. A noticeably greater percentage of males were smokers and alcohol drinkers, whereas more females had low physical activity.

Table 1. Characteristics of the participants according to sex.

Variables [1]	Total (n = 8957)	Male (n = 4465)	Female (n = 4492)	p-Value
Age group				<0.001
20–39 years	46.5	48.8	44.2	
40–59 years	38.3	37.3	39.3	
≥60 years	15.2	13.8	16.6	
Educational attainment				0.005
≤Elementary	31.9	33.9	29.9	
High school	37.9	38.2	37.6	
≥College	30.2	27.8	32.4	
Marital status				<0.001
Single	23.4	26.4	20.4	
Married	66.6	67.9	65.2	
Others	10.1	5.7	14.4	
Employment status				<0.001
Employed	59.5	76.3	43.0	
Unemployed	40.5	23.7	57.0	
Household size				0.840
1–3	33.3	33.3	33.3	
4–6	45.1	44.9	45.3	
≥7	21.6	21.8	21.4	

Table 1. *Cont.*

Variables [1]	Total (n = 8957)	Male (n = 4465)	Female (n = 4492)	p-Value
Wealth quintile				<0.001
Poorest	17.6	19.4	15.8	
Poor	19.3	20.0	18.6	
Middle	20.6	20.7	20.4	
Rich	20.6	19.9	21.2	
Richest	22.0	20.0	24.0	
Current smoker				<0.001
Yes	26.9	46.0	8.1	
No	73.1	54.0	91.9	
Current alcohol drinker				<0.001
Yes	51.3	72.4	30.6	
No	48.7	27.6	69.4	
Physical activity				<0.001
Low	44.2	36.5	51.7	
High	55.8	63.5	48.3	

Values are weighted percentages (%). [1] Variables with missing observations: educational attainment (n = 44), smoking and drinking status (n = 521), physical activity classification (n = 614).

About 36% of the participants were suffering from undernutrition and 84.5% had CMRF with a significant sex difference (Table 2). Iodine insufficiency (23.8%) and low HDL cholesterol (70.1%) had the highest prevalence among the indicators of undernutrition and CMRF, respectively. Correspondingly, the individual-level DBM affected approximately one-third of the adult population (29.5%).

Table 2. Distributions of undernutrition, cardiometabolic risk factors, and double burden of malnutrition.

Variables	Total (n = 8957)	Male (n = 4465)	Female (n = 4492)	p-Value
Undernutrition				
Underweight	11.2	10.4	12.0	<0.001
Anemia	6.5	5.4	7.5	0.014
Vitamin A deficiency	0.1	0.0	0.1	0.430
Iodine insufficiency	23.8	21.0	26.5	0.001
Cardiometabolic risk factors				
Overweight/Obesity	29.4	25.9	32.9	0.001
Abdominal obesity	13.4	3.4	23.2	<0.001
Hypertension	22.5	24.3	20.6	0.008
Hyperglycemia	10.2	11.1	9.3	0.047
Low HDL cholesterol	70.1	61.4	78.8	<0.001
Hypertriacylglycerolemia	39.5	46.5	32.6	<0.001
≥1 Undernutrition [1]	35.5	32.1	38.9	0.001
≥1 Cardiometabolic risk factor [2,3]	84.5	81.9	87.1	0.002
Total DBM [4]	29.5	25.6	33.3	<0.001

Values are weighted percentages (%). [1] Having any of the following conditions: (1) underweight, (2) anemia, (3) vitamin A deficiency, (4) iodine insufficiency. [2] Having any of the following factors: (1) overweight/obesity or abdominal obesity, (2) hypertension, (3) hyperglycemia, (4) dyslipidemia (low HDL cholesterol or hypertriacylglycerolemia). [3] Cardiometabolic risk factor with missing observations: both body mass index and abdominal obesity (n = 220), hypertension (n = 36), hyperglycemia (n = 352), both low high-density lipoprotein (HDL) cholesterol and hypertriacylglycerolemia (n = 110). [4] Total DBM, total double burden of malnutrition or the co-existence of underweight or anemia or vitamin A deficiency or iodine insufficiency and at least one cardiometabolic risk factor.

3.2. Dietary Patterns

Tables 3 and 4 present the dietary patterns that were extracted through factor analysis for males and females. Seven dietary patterns explaining 25.5% of the total variance in the consumption of food groups were derived for males. For females, 6 dietary patterns were generated, representing 20.5% of the variance in food intake. The three dietary patterns composed of: (1) the rice pattern (with high positive loading in the rice and rice products food group), (2) the fruits and miscellaneous food pattern (consisting of fruits and other miscellaneous food groups), and (3) the fish pattern (the fish and fish products food group had high factor loading), which were common for both males and females. The results of the factor analysis also demonstrated a number of sex differences. For example, the meat and sugar pattern; the vegetables pattern; the cereal, egg, and oils pattern; and the beverage pattern emerged among males but not females. On the contrary, the dietary patterns labeled as protein-rich food, cereal and sugar, vegetables and corn, and fats and oils were seen only among females.

Table 3. Factor loadings for the seven dietary patterns identified among males.

Food Groups	Dietary Patterns [1]						
	Rice	Meat and Sugar	Fruits and Miscellaneous Food	Fish	Vegetables	Cereal, Egg, and Oils	Beverage
Rice and rice products	**0.964**	0.071	0.007	0.077	0.076	−0.002	−0.038
Corn and corn products	−0.411	−0.022	−0.029	0.043	0.236	−0.118	−0.007
Other cereal products	−0.026	0.252	0.032	−0.056	−0.066	**0.381**	−0.031
Starchy roots and tubers	−0.067	0.017	0.015	0.008	0.120	−0.008	−0.008
Sugar and syrups	0.027	**0.460**	0.005	−0.018	−0.025	0.172	−0.007
Dried beans, nuts, and seeds	0.034	0.016	−0.014	−0.074	0.021	0.065	0.021
Green leafy and yellow vegetables	−0.035	−0.091	0.024	−0.012	**0.547**	−0.079	−0.012
Other vegetables	0.099	−0.079	0.024	−0.134	**0.282**	0.074	−0.001
Fruits	0.005	0.028	**0.548**	0.043	0.061	0.052	−0.015
Fish and fish products	0.113	−0.001	0.001	**0.860**	−0.072	−0.012	0.012
Meat and meat products	0.057	**0.351**	0.009	−0.159	−0.101	0.117	0.262
Poultry	0.058	0.174	−0.011	−0.115	−0.036	0.141	0.079
Eggs	0.092	0.002	−0.002	−0.079	−0.065	**0.307**	0.016
Milk and milk products	−0.026	0.083	0.097	−0.023	0.021	0.228	0.037
Fats and oils	0.019	0.041	0.013	0.050	0.022	**0.377**	0.018
Beverages	−0.025	0.103	−0.009	0.008	−0.008	0.041	**0.531**
Condiments and spices	−0.012	0.178	0.026	0.111	−0.016	−0.004	0.063
Other miscellaneous	0.016	0.005	**0.639**	−0.013	0.012	0.051	0.005
Proportion variance, %	6.4	2.8	4.0	4.6	2.7	3.0	2.0
Cumulative variance, %	6.4	9.2	13.2	17.8	20.5	23.5	25.5

Bold values represent food groups kept in their related dietary pattern. [1] Dietary patterns are labeled based on the factor loadings with the value of 0.25 or greater.

Table 4. Factor loadings for the six dietary patterns identified among females.

Food Groups	Dietary Patterns [1]					
	Rice	Protein-Rich Foods, Cereal, and Sugar	Fruits and Miscellaneous Food	Fish	Vegetables and Corn	Fats and Oils
Rice and rice products	**0.889**	−0.308	0.019	0.131	0.029	0.114
Corn and corn products	−0.299	−0.067	−0.043	0.028	**0.327**	−0.055
Other cereal products	−0.040	**0.456**	−0.028	−0.001	−0.078	0.122
Starchy roots and tubers	−0.031	0.028	0.065	0.002	0.149	−0.012
Sugar and syrups	0.057	**0.374**	0.056	−0.007	0.040	0.125
Dried beans, nuts, and seeds	0.032	0.037	−0.042	−0.057	0.070	0.205
Green leafy and yellow vegetables	0.018	−0.150	−0.010	0.010	**0.405**	−0.032
Other vegetables	0.074	−0.030	0.053	−0.118	**0.331**	0.112
Fruits	0.000	0.112	**0.536**	0.012	0.080	−0.011
Fish and fish products	0.097	−0.093	0.028	**0.669**	−0.091	−0.070
Meat and meat products	0.051	**0.385**	0.011	−0.156	−0.057	0.110
Poultry	0.047	**0.303**	0.081	−0.050	−0.042	0.069
Eggs	0.045	0.046	0.058	−0.055	−0.041	0.189
Milk and milk products	−0.040	**0.345**	0.107	0.036	0.007	0.063
Fats and oils	−0.018	0.135	0.035	0.090	−0.024	**0.362**
Beverages	−0.028	0.206	0.033	−0.074	−0.031	0.022
Condiments and spices	−0.010	**0.344**	−0.045	0.100	−0.017	−0.087
Other miscellaneous	0.030	0.036	**0.457**	0.002	0.051	0.057
Proportion variance, %	5.1	5.2	3.0	3.0	2.5	1.7
Cumulative variance, %	5.1	10.3	13.3	16.3	18.8	20.5

Bold values represent food groups kept in their related dietary pattern. [1] Dietary patterns are labeled based on the factor loadings with the value of 0.25 or greater.

3.3. Association of Dietary Patterns and Double Burden of Malnutrition

The relationship between total DBM and tertiles of dietary pattern scores was examined using a logistic regression analysis. The rice pattern and the meat and sugar pattern were associated with DBM in males (Table 5). Those in the middle tertile (T2) of the rice pattern were less likely to have DBM—after adjusting for sociodemographic characteristics, lifestyle factors, and energy intake, unlike males in the bottom tertile (T1). Similarly, male adults with medium and high adherence (T2 and T3) to the meat and sugar pattern had a lower risk for total DBM. The remaining dietary patterns showed no significant associations with DBM among males.

Table 5. Logistic regression models for double burden of malnutrition across tertiles of dietary pattern scores among males [1].

Dietary Patterns	Total DBM [2] (*n* = 1250) OR (95% CI)
Rice pattern (ref. = tertile 1)	
Tertile 2	**0.82 (0.68, 0.99)**
Tertile 3	1.06 (0.80, 1.41)
Meat and sugar pattern (ref. = tertile 1)	
Tertile 2	**0.78 (0.64, 0.96)**
Tertile 3	**0.75 (0.58, 0.97)**
Fruits and miscellaneous food pattern (ref. = tertile 1)	
Tertile 2	0.99 (0.82, 1.19)
Tertile 3	0.96 (0.79, 1.16)
Fish pattern (ref. = tertile 1)	
Tertile 2	1.02 (0.86, 1.22)
Tertile 3	1.00 (0.82, 1.23)
Vegetables pattern (ref. = tertile 1)	
Tertile 2	0.97 (0.80, 1.18)
Tertile 3	1.12 (0.91, 1.38)
Cereal, egg, and oils pattern (ref. = tertile 1)	
Tertile 2	0.89 (0.73, 1.09)
Tertile 3	0.93 (0.75, 1.16)
Beverage pattern (ref. = tertile 1)	
Tertile 2	1.05 (0.87, 1.25)
Tertile 3	0.93 (0.75, 1.14)

[1] Values in bold are significantly different at a level of $p < 0.05$. Models were adjusted for sociodemographic characteristics, lifestyle factors, and energy intake. [2] Total DBM, total double burden of malnutrition or the co-existence of underweight or anemia or vitamin A deficiency or iodine insufficiency and at least one cardiometabolic risk factor.

Regarding females, two dietary patterns were found to be associated with DBM (Table 6). Those in the upper tertile (T3) of the protein-rich foods, cereal, and sugar patterns had a lower likelihood of having DBM in the regression models that controlled for the co-variates. An inverse relationship was noted in the vegetables and corn pattern. Female adults with medium and high adherence (T2 and T3) to the latter pattern had higher odds of developing DBM.

Table 6. Logistic regression models for double burden of malnutrition across tertiles of dietary pattern scores among females [1].

Dietary Patterns	Total DBM [2] (*n* = 1654) OR (95% CI)
Rice pattern (ref. = tertile 1)	
Tertile 2	1.16 (0.97, 1.39)
Tertile 3	1.02 (0.80, 1.30)
Protein-rich foods, cereal, and sugar pattern (ref. = tertile 1)	
Tertile 2	0.94 (0.77, 1.15)
Tertile 3	**0.78 (0.61, 0.99)**
Fruits and miscellaneous food pattern (ref. = tertile 1)	
Tertile 2	0.99 (0.82, 1.19)
Tertile 3	0.99 (0.82, 1.18)
Fish pattern (ref. = tertile 1)	
Tertile 2	0.95 (0.79, 1.13)
Tertile 3	0.86 (0.71, 1.04)
Vegetables and corn pattern (ref. = tertile 1)	
Tertile 2	**1.28 (1.07, 1.53)**
Tertile 3	**1.36 (1.12, 1.64)**
Fats and oils pattern (ref. = tertile 1)	
Tertile 2	0.84 (0.69, 1.02)
Tertile 3	0.83 (0.66, 1.03)

[1] Values in bold are significantly different at a level of $p < 0.05$. Models were adjusted for sociodemographic characteristics, lifestyle factors, and energy intake. [2] Total DBM, total double burden of malnutrition or the co-existence of underweight or anemia or vitamin A deficiency or iodine insufficiency and at least one cardiometabolic risk factor.

4. Discussion

In this study, seven dietary patterns emerged through factor analysis for males and six for females. The rice and fish patterns were also ascertained in previous research [40–43]. In the same manner, the fruits and miscellaneous food pattern, the vegetables pattern, and the vegetables and corn pattern were consistent with past literature [44,45]. The key food groups in the meat and sugar pattern and cereal, egg, and oils pattern of males and the protein-rich foods, cereal, and sugar pattern of females resembled the dietary patterns pertained as unhealthy [46–48], Western [49], and high fat and sugar [50]. The beverage pattern and the fats and oils pattern were described in earlier studies as well [51–54].

Individual-level DBM affected about three in every ten adults (29.5%) and was higher than the estimates of Zeba and colleagues [9]. Data analysis revealed that dietary patterns had mixed effects on total DBM. Filipino male adults consuming a diet high in rice had a decreased susceptibility for total DBM. The meat and sugar pattern identified among males and the protein-rich foods, cereal, and sugar pattern derived among females were associated with a decreased risk for DBM. Interestingly, the vegetables and corn dietary patterns increased the risk for DBM in females.

Studies on the relationship between rice intake and DBM are scarce and frequently draw on metabolic syndromes or its components as outcomes. In a recent meta-analysis, rice intake was positively correlated with metabolic syndrome [55]. A pooled analysis of three US cohorts observed no associations between white and brown rice consumption and cardiovascular health [56]. Eshak and colleagues [57] likewise explored the relation between white rice and major cardiovascular diseases among men and reported an inverse correlation. The latter study was coherent with our findings. Probable reasons for lower cardiometabolic risk are the varying rice starch compositions [58], processing and cooking methods [58], and complementary dishes eaten with rice [59]. For micronutrient deficiencies, the Philippines has been implementing the fortification of rice with iron since 2000 [60]. Production of healthier rice varieties, such as high-iron and high-zinc types, are also being carried out as part of the biofortification efforts [61]. These recommended public health strategies have been implemented to address micronutrient malnutrition in the country [62]. However, it is important to note that high consumption of rice and rice products alone is not recommended without ensuring diet diversity, and that total carbohydrate intake is within the acceptable macronutrient distribution range.

Male adults adhering to the meat and sugar pattern and female adults favoring the protein-rich foods, cereal, and sugar pattern had a lower likelihood of having DBM. This can be supported on a number of accounts. First, these dietary patterns are comprised of animal proteins, mainly from meat, poultry, milk, and dairy products that have been found to be negatively associated with blood pressure, insulin resistance, and obesity in previous literature [63,64]. Second, animal-based protein foods, specifically meat, are high in heme-iron, zinc, and vitamin B12 [65]. Third, high amounts of sugar and syrups are present in these dietary patterns. Available evidence has illustrated that dietary sugars do not cause obesity and diet-related disease, but rather sugar consumption in excess of energy requirements [66]. Khan and colleagues further substantiated that the kind of sugar, sucrose in particular, was associated with a reduction in cardiovascular disease mortality [67].

It is widely known that vegetable-containing dietary patterns are favorable for lessening the risk of non-communicable diseases due to the dietary fiber, antioxidants, and phytochemicals it contains. These bioactive compounds regulate insulin secretion, lipid profile fluctuations, oxidative stress, and inflammatory and immune status [68–75]. Moreover, some vegetables are rich sources of essential vitamins and minerals, though less bioavailable than animal sources [76]. The positive association between the vegetables and corn pattern and DBM can be substantiated by the low vegetable consumption among Filipino adults, i.e., daily per capita vegetable intake (68.5–68.9 g/day) [77,78], which may have counteracted the hypothesized benefits. Collectively, a balanced dietary pattern with the appropriate combination of food groups should be put forward together with the current nutritional guidelines.

Our findings also had limitations. Firstly, causality and lifetime dietary intake cannot be drawn, given the nature of the study. Secondly, reporting bias and measurement errors are inherent in assessing dietary intakes. Thirdly, factor analysis involves several subjective decisions to be made. Fourthly, the dietary patterns explained the low variability in total food intake (25.5% for males and 20.5% for females). Fifthly, the nutrient intakes of the male and female adults by tertile of the dietary pattern scores were not calculated. Finally, there were unmeasured confounders, thus warranting careful interpretation of the results when generalizing to the general population of adults. Despite these limitations, our study is one of the few nationally-representative epidemiological investigations focusing on the impact of dietary patterns on malnutrition in all its forms among community-dwelling adults.

5. Conclusions

In conclusion, this population-based study identified sex-specific dietary patterns that were significantly associated with DBM development among adults in the Philippines. Our findings suggest that rice-based and meat-containing food patterns may potentially exert protection against the risk of developing nutritional deficiencies and cardiometabolic diseases among Filipino adults simultaneously. These unique dietary patterns can be utilized to guide public health nutrition interventions directed toward DBM prevention. Further research is necessary to validate our findings.

Supplementary Materials: The following supporting information can be downloaded at: https://www.mdpi.com/article/10.3390/nu14173495/s1, Table S1: Food groups used in the dietary pattern analysis; Figure S1: Scree plots showing the eigenvalues of components extracted using factor analysis by sex.

Author Contributions: Conceptualization, A.R.d.J., W.-C.H. and S.C.H.; writing—original draft preparation, A.R.d.J.; data analysis, W.-C.H.; supervision, S.C.H.; writing—review and editing, A.R.d.J., W.-C.H., Y.-Y.C., L.-J.E.K., T.Y., C.-J.P. and S.C.H. All authors have read and agreed to the published version of the manuscript.

Funding: This research received no external funding.

Institutional Review Board Statement: The study was conducted according to the guidelines of the Declaration of Helsinki and certified for exemption from the Human Research Ethics Committee of National Cheng Kung University (HREC No. 110-280, date of approval: 27 May 2021).

Informed Consent Statement: The 2013 National Nutrition Survey obtained ethical clearance from the Institutional Ethics Review Committee of the Department of Science and Technology-Food and Nutrition Research Institute, Manila, Philippines before the survey was conducted. Written informed consent was obtained from all the subjects involved in the study.

Data Availability Statement: Publicly available data sets were analyzed in this study. This data can be found here: http://enutrition.fnri.dost.gov.ph/site/home.php (accessed on 3 September 2020).

Acknowledgments: We are grateful to the Food and Nutrition Research Institute of the Department of Science and Technology, Philippines for providing access to the 2013 National Nutrition Survey data.

Conflicts of Interest: The authors declare no conflict of interest.

References

1. Winichagoon, P.; Margetts, B.M. The Double Burden of Malnutrition in Low-and Middle-Income Countries. Available online: https://publications.iarc.fr/Book-And-Report-Series/Iarc-Working-Group-Reports/Energy-Balance-And-Obesity-2017 (accessed on 26 September 2020).
2. Popkin, B.M.; Corvalan, C.; Grummer-Strawn, L.M. Dynamics of the double burden of malnutrition and the changing nutrition reality. *Lancet* **2020**, *395*, 65–74. [CrossRef]
3. Bertin, M.; Touvier, M.; Dubuisson, C.; Dufiur, A.; Havard, S.; Lafay, L.; Volatier, J.L.; Lioret, S. Dietary patterns of French adults: Associations with demographic, socio-economic and behavioural factors. *J. Hum. Nutr. Diet.* **2016**, *29*, 241–254. [CrossRef] [PubMed]
4. Knudsen, V.K.; Matthiessen, J.; Biltoft-Jensen, A.; Sørensen, M.R.; Groth, M.V.; Christensen, T.; Fagt, S. Identifying dietary patterns and associated health-related lifestyle factors in the adult Danish population. *Eur. J. Clin. Nutr.* **2014**, *68*, 736–740. [CrossRef] [PubMed]

5. Muga, M.A.; Owili, P.O.; Hsu, C.Y.; Rau, H.H.; Chao, J.C.J. Dietary patterns, gender, and weight status among middle-aged and older adults in Taiwan: A cross-sectional study. *BMC Geriatr.* **2017**, *17*, 1–10. [CrossRef]
6. Khor, G.L.; Sharif, Z.M. Dual forms of malnutrition in the same households in Malaysia—A case study among Malay rural households. *Asia Pac. J. Clin. Nutr.* **2003**, *12*, 427–437.
7. Lee, S.J.; Ryu, H.K. Relationship between dietary intakes and the double burden of malnutrition in adults of Malang, Indonesia: An exploratory study. *Nutr. Res. Pract.* **2018**, *12*, 426–435. [CrossRef]
8. Melby, C.L.; Orozco, F.; Averett, J.; Muñoz, F.; Romero, M.J.; Barahona, A. Agricultural food production diversity and dietary diversity among female small holder farmers in a region of the Ecuadorian Andes experiencing nutrition transition. *Nutrients* **2020**, *12*, 2454. [CrossRef]
9. Zeba, A.N.; Delisle, H.F.; Renier, G. Dietary patterns and physical inactivity, two contributing factors for the double burden of malnutrition among adults in Burkina Faso, West Africa. *J. Nutr. Sci.* **2014**, *3*, e50. [CrossRef]
10. Department of Science and Technology-Food and Nutrition Research Institute. Philippine Facts and Figures: 2018 Expanded National Nutrition Survey. Available online: http://enutrition.fnri.dost.gov.ph/site/uploads/2018%20Expanded%20National%20Nutrition%20Survey-TAGUIG.pdf (accessed on 3 August 2020).
11. De Juras, A.R.; Hsu, W.C.; Hu, S.C. Prevalence and determinants of the co-occurrence of overweight or obesity and micronutrient deficiencies among adults in the Philippines: Results from a National Representative Survey. *Nutrients* **2021**, *13*, 2339. [CrossRef]
12. De Juras, A.R.; Hsu, W.C.; Hu, S.C. The double burden of malnutrition at the individual level among adults: A nationwide survey in the Philippines. *Front. Nutr.* **2021**, *8*, 760437. [CrossRef]
13. Pedro, M.R.; Barba, C.; Benavides-de Leon, R. Nutrition Transition in the Philippines. Available online: https://www.semanticscholar.org/paper/Nutrition-Transition-in-the-Philippines-Pedro-Barba/d3aa78884070a2a53fa99cd801d847992fcfb90b (accessed on 27 November 2020).
14. Department of Science and Technology-Food and Nutrition Research Institute. Public Use File. Available online: http://enutrition.fnri.dost.gov.ph/site/home.php (accessed on 3 September 2020).
15. Barcenas, M.L. The Development of the 2003 Master Sample (MS) for Philippine Household Surveys. In Proceedings of the 9th National Nutrition on Statistics, Manila, Philippines, 4–5 October 2004.
16. Patalen, C.F.; Ikeda, N.; Angeles-Agdeppa, I.; Vargas, M.B.; Nishi, N.; Duante, C.A.; Capanzana, M.V. Data Resource Profile: The Philippine National Nutrition Survey (NNS). *Int. J. Epidemiol.* **2020**, *49*, 742–743f. [CrossRef] [PubMed]
17. Flores, M.; Macias, N.; Rivera, M.; Lozada, A.; Barquera, S.; Rivera-Dommarco, J.; Tucker, K.L. Dietary patterns in Mexican adults are associated with risk of being overweight or obese. *J. Nutr.* **2010**, *140*, 1869–1873. [CrossRef] [PubMed]
18. Department of Science and Technology-Food Nutrition Research Institute. eNutrition, Facts and Figure 2013 National Nutrition Survey. Available online: https://www.fnri.dost.gov.ph/index.php/19-nutrition-statistic/175-national-nutrition-survey#facts-and-figures (accessed on 3 September 2020).
19. Gazan, R.; Béchaux, C.; Crépet, A.; Sirot, V.; Drouillet-Pinard, P.; Dubuisson, C.; Harvard, S. Dietary patterns in the French adult population: A study from the second French national cross-sectional dietary survey (INCA2) (2006–2007). *Br. J. Nutr.* **2016**, *116*, 300–315. [CrossRef] [PubMed]
20. De Gavelle, E.; Huneau, J.F.; Mariotti, F. Patterns of protein food intake are associated with nutrient adequacy in the general French adult population. *Nutrients* **2018**, *10*, 226. [CrossRef]
21. Zárate-Ortiz, A.G.; Melse-Boonstra, A.; Rodríguez-Ramírez, S.; Hernández-Cordero, S.; Feskens, E.J. Dietary patterns and the double burden of malnutrition in Mexican adolescents: Results from ENSANUT-2006. *Nutrients* **2019**, *11*, 2753. [CrossRef]
22. De Juras, A.R.; Hsu, W.C.; Hu, S.C. Dietary patterns and their association with sociodemographic and lifestyle factors in Filipino adults. *Nutrients* **2022**, *14*, 886. [CrossRef]
23. Angeles-Agdeppa, I.; Sun, Y.; Tanda, K.V. Dietary pattern and nutrient intakes in association with non-communicable disease risk factors among Filipino adults: A cross-sectional study. *Nutr. J.* **2020**, *19*, 1–13. [CrossRef]
24. Bell, L.K.; Edwards, S.; Grieger, J.A. The relationship between dietary patterns and metabolic health in a representative sample of adult Australians. *Nutrients* **2015**, *7*, 6491–6505. [CrossRef]
25. Cai, J.X.; Nuli, R.; Zhang, Y.; Zhang, Y.Y.; Abudusemaiti, M.; Kadeer, A.; Tian, X.; Xiao, H. Association of dietary patterns with type 2 diabetes mellitus among middle-aged adults in Uygur population of Xinjiang region. *J. Nutr. Sci. Vitaminol.* **2019**, *65*, 362–374. [CrossRef]
26. World Health Organization. WHO Technical Report Series on Obesity: Preventing and Managing the Global Epidemic. Available online: https://apps.who.int/iris/handle/10665/42330 (accessed on 21 September 2020).
27. International Committee for Standardization in Haematology. International Committee for Standardization in Haematology: Protocol for type testing equipment and apparatus used for haematological analysis. *J. Clin. Pathol.* **1978**, *31*, 275–279. [CrossRef]
28. World Health Organization; United Nations Children's Fund; United Nations University. Iron Deficiency Anaemia: Assessment, Prevention and Control, a Guide for Programme Managers. Available online: https://www.ihs.org.in/SHG/WHO-Anemia%20detection%20guidelines.pdf (accessed on 26 September 2020).
29. Furr, H.C.; Tanmihardjo, S.A.; Olson, J.A. *Training Manual for Assessing Vitamin a Status by Use of the Modified Relative Dose Response and Relative Dose Response Assays*; United States Agency for International Development: Washington, DC, USA, 1992.
30. Sommer, A. Vitamin A Deficiency and Its Consequences: A Field Guide to Detection and Control. Available online: https://apps.who.int/iris/handle/10665/40535 (accessed on 26 September 2020).

nutrients

MDPI

Article

Association of Body Mass Index and Plant-Based Diet with Cognitive Impairment among Older Chinese Adults: A Prospective, Nationwide Cohort Study

Fang Liang [1,†], Jialin Fu [1,†], Gabrielle Turner-McGrievy [2], Yechuang Wang [1], Nan Qiu [1], Kai Ding [1], Jing Zeng [1], Justin B. Moore [3] and Rui Li [1,4,*]

[1] School of Public Health, Wuhan University, Wuhan 430071, China; fliang@whu.edu.cn (F.L.); fjl0708@whu.edu.cn (J.F.); ywang20@whu.edu.cn (Y.W.); 2013302170051@whu.edu.cn (N.Q.); 2021203050024@whu.edu.cn (K.D.); 2021283050065@whu.edu.cn (J.Z.)
[2] Department of Health Promotion, Education, and Behavior, Arnold School of Public Health, University of South Carolina, Columbia, SC 29208, USA; brie@sc.edu
[3] Department of Implementation Science, Division of Public Health Sciences, Wake Forest School of Medicine, Winston-Salem, NC 27101, USA; jusmoore@wakehealth.edu
[4] School of Nursing, Wuhan University, Wuhan 430071, China
* Correspondence: rli@whu.edu.cn; Tel.: +86-27-68759901; Fax: +86-27-68758648
† These authors have contributed equally to this work.

Citation: Liang, F.; Fu, J.; Turner-McGrievy, G.; Wang, Y.; Qiu, N.; Ding, K.; Zeng, J.; Moore, J.B.; Li, R. Association of Body Mass Index and Plant-Based Diet with Cognitive Impairment among Older Chinese Adults: A Prospective, Nationwide Cohort Study. *Nutrients* **2022**, *14*, 3132. https://doi.org/10.3390/nu14153132

Academic Editor: Martina Barchitta

Received: 21 June 2022
Accepted: 27 July 2022
Published: 29 July 2022

Abstract: To examine the association of body mass index (BMI) and a plant-based diet (PBD) with cognitive impairment in older adults, this cohort study used data from the Chinese Longitudinal Healthy Longevity Survey (CLHLS), a national, community-based, longitudinal, prospective study in China. Cognitive function was evaluated via the Mini-Mental State Examination (MMSE). Diet was assessed using a simplified food frequency questionnaire (FFQ), and PBD patterns were estimated using the overall plant-based diet index (PDI), the healthful plant-based diet index (hPDI), and the unhealthful plant-based diet index (uPDI). BMI was measured objectively during the physical examination. Cox proportional hazard models and restricted cubic spline analyses were used. A total of 4792 participants with normal cognition at baseline were included, and 1077 participants were identified as having developed cognitive impairment during the 24,156 person-years of follow-up. A reverse J-shaped association was observed between BMI and cognitive impairment ($p = 0.005$ for nonlinearity). Participants who were overweight (HR = 0.79; 95% CI 0.66–0.95) and obese (HR = 0.72; 95% CI 0.54–0.96) had a decreased risk of cognitive impairment, while those who were underweight (HR = 1.42; 95% CI 1.21–1.66) had an increased risk. Lower PDI, lower hPDI, and higher uPDI were associated with an increased risk of cognitive impairment (HR = 1.32; 95% CI 1.16–1.50 for PDI; HR = 1.46; 95% CI 1.29–1.66 for hPDI; HR = 1.21; 95% CI 1.06–1.38 for uPDI). The protective effect of being overweight on cognitive impairment was more pronounced among participants with a higher PDI (HR = 0.74; 95% CI 0.57–0.95) than those with a lower PDI (HR = 0.87; 95% CI 0.67–1.12), among participants with a higher hPDI (HR = 0.73; 95% CI 0.57–0.94) than those with a lower hPDI (HR = 0.93; 95% CI 0.72–1.10), and among participants with a lower uPDI (HR = 0.61; 95% CI 0.46–0.80) than those with a higher uPDI (HR = 1.01; 95% CI 0.80–1.27). Our results support the positive associations of overweight status, obesity, an overall PBD, and a healthful PBD with cognitive function in older adults. A lower adherence to an overall PBD, a healthful PBD, and a higher adherence to an unhealthful PBD may attenuate the protective effect of being overweight on cognitive function.

Keywords: cognitive impairment; body mass index; plant-based dietary pattern; older Chinese adults; cohort

1. Introduction

With the global population ageing, the number of older adults with dementia is set to rise substantially across the world. Nearly 46 million individuals were affected by dementia worldwide in 2015, and that number is predicted to reach 152 million in 2050 [1]. Dementia is a common and serious neurodegenerative disorder of older adults which impairs quality of later life and imposes a heavy burden on the affected individuals, their families, and the economy [2]. As there are currently no effective treatments for dementia, prevention is of major importance in fighting this disease [3]. Cognitive impairment is a prodromal phase of dementia that provides an opportunity to take steps to prevent dementia [4]. Therefore, the recognition of possibly modifiable risk factors for cognitive impairment is of great importance for dementia prevention.

Increasing attention has been paid to associations between weight status, measured by body mass index (BMI), and cognitive function in older populations. Although the mechanism has not been completely explained, it has been widely proposed that unfavorable weight status may affect metabolic functions, promote inflammation, and disrupt the balance of gut microbiota, which could increase the risk of poor cognitive function [5]. However, previous epidemiological studies have shown conflicting results between BMI and cognitive function, with some research suggesting that higher BMI contributes to poor cognitive function [6–12], and other studies observing an apparent beneficial effect of higher BMI on cognitive function [13–21]. Also, few large, prospective cohort studies have been conducted in the older Chinese population. Our previous results suggested that a larger BMI and a BMI-defined overweight status were related to slower cognitive decline [13]. A cohort study reported that BMI-defined overweight status was associated with a lower risk of cognitive impairment [22], and another recent cohort study suggested that a BMI-defined underweight status was related to a higher risk of cognitive impairment [23].

Plant-based foods are a rich source of antioxidants and anti-inflammatory nutrients that could reduce inflammation and oxidative stress in the central nervous system [24–26]. Several studies have linked plant-based diets (PBDs), which are characterized by a higher consumption of plant-based foods and a lower or no intake of animal-based foods, with better neurological health [26,27]. However, previous studies on PBDs are somewhat limited because due to the lack of differentiation between the quality of plant-based foods. Recent research further defined three plant-based diet indices (PDIs), including the overall plant-based diet index (PDI), the healthful plant-based diet index (hPDI), and the unhealthful plant-based diet index (uPDI), so as to consider the dietary quality of a PBD. For instance, the PDI assesses alignment with diets higher in plant-based foods and lower in animal-based foods, the hPDI emphasizes a high consumption of healthful plant-based foods and a low consumption of unhealthful plant-based foods, and the uPDI is the opposite of the hPDI in that it emphasizes a high consumption of unhealthful plant-based foods within the context of an overall PBD [28–30]. Previous research has shown that healthful plant-based foods (e.g., fresh vegetables and fresh fruits) were related to better neurological health, while unhealthful plant-based foods (e.g., preserved vegetables and added sugars) were related to poor neurological health [27,31]. To date, relatively little research has investigated the relationship between plant-based dietary patterns (overall PBD, healthful PBD, and unhealthful PBD) and cognitive function [32,33].

Currently, the evidence for a potential moderating role of a PBD in the relationship between BMI and cognitive function is scarce. To fill this knowledge gap, we utilized a nationally representative sample of older Chinese adults to prospectively evaluate the association of BMI with cognitive impairment, explore the associations of three plant-based dietary patterns with cognitive impairment, and examine the potential moderating role of a PBD in the association between BMI and cognitive impairment.

2. Methods

2.1. Study Population

As detailed elsewhere [34,35], the Chinese Longitudinal Healthy Longevity Survey (CLHLS) is an ongoing, prospective cohort study among Chinese adults aged 65 years and older that was established in 1998 using multistage cluster sampling, and recruiting participants from 23 out of the 31 provinces in China, thus covering about 85% of the total population in China. Follow-up surveys were conducted every 3 or 4 years. All participants signed written informed consent for the baseline and follow-up surveys. The CLHLS study was approved by the Biomedical Ethics Committee of Peking University, Beijing, China (IRB00001052–13074).

Since the height and weight information were first objectively measured in the sixth wave (2011), our research considered the sixth wave (2011) as the baseline. The seventh wave (2014) and the eighth wave (2018) were considered as the follow-up. Figure 1 shows the detailed flowchart of participant selection for the current study. A total of 9765 participants attended the 2011 cycle survey of the CLHLS. Of these, 360 were excluded for the following reasons: they had missing height or weight measurements (n = 247), they did not complete the cognitive measurements (n = 54), they did not complete the dietary assessments (n = 2), or they were younger than age 65 at baseline (n = 57). An additional 2245 participants were excluded due to cognitive impairment at baseline, and an additional 360 participants had a confirmed diagnosis of dementia and/or Alzheimer's disease at baseline. In addition, 2008 participants without at least one follow-up assessment of cognition were excluded. The remaining 4792 individuals were included in the analyses.

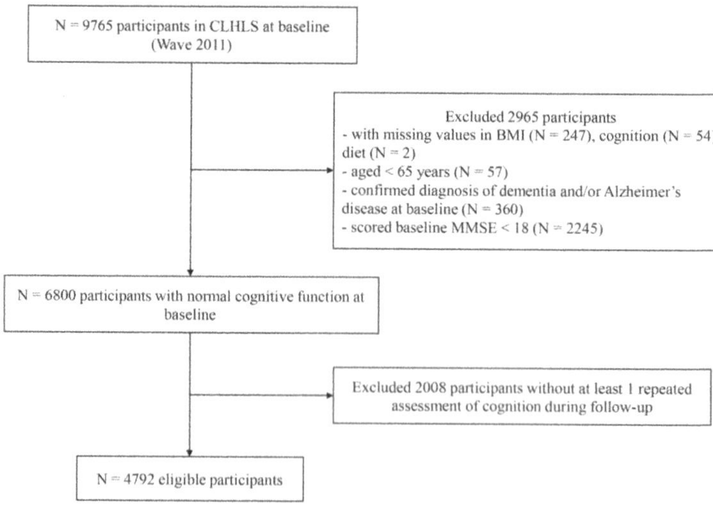

Figure 1. Flow chart of participants.

2.2. Measurement and Calculation of Body Mass Index

Body weight (in kilograms) and height (in centimeters) were measured by trained assessors following standardized procedures. BMI, defined as the weight (kg) in kilograms divided by the height (m) squared, was categorized as: underweight (BMI < 18.5 kg/m^2), normal (18.5 $\leq$ BMI < 24 kg/m^2), overweight (24 $\leq$ BMI < 28 kg/m^2) and obese (BMI $\geq$ 28 kg/m^2) [21].

2.3. Assessment of Cognitive Function

The CLHLS used the Chinese version of the Mini-Mental State Examination (MMSE) to evaluate cognitive function. The MMSE contains a total of 30 items that assess orientation, registration, attention and calculation, recall, and language, with a score range from zero to 30 [36,37]. Use of the

MMSE in the CLHLS is well-documented as both reliable and valid [22,35,38–41]. Since MMSE scores might be influenced by education level [40], participants were defined as cognitively impaired following education-based MMSE cutoff points. Specifically, we used the MMSE scores of 18, 20, and 24 as the cut-off points for subjects with no formal education, only a primary school education (1–6 years), and a middle-school or higher education (>6 years), respectively [32,40].

2.4. Measurement and Calculation of Plant-Based Diet Indices

Each participant's dietary information was collected using a simplified food frequency questionnaire (FFQ). The questionnaire has been broadly used, with its reliability and validity both well-supported [28,40,42–44]. The simplified FFQ in the CLHLS included 16 food groups which are commonly consumed in China. In the present study, we divided the 16 food groups into 3 categories according to their potentially divergent health effects, including healthful plant-based foods (whole grains, fresh fruits, fresh vegetables, legumes, garlic, vegetable oils, nuts, and tea), unhealthful plant-based foods (refined grains, preserved vegetables, and sugar (white granulated sugar or candies)), and animal-based foods (animal fat, eggs, fish and aquatic products, meat, and milk and dairy products) [29,45,46]. For legumes; garlic; nuts; tea; salted, preserved vegetables; sugar (white granulated sugar or candies); eggs; fish; meat; and milk, the questionnaire had 5 options, including "almost every day", "≥1 time per week", "≥1 time per month", "occasionally", or "rarely or never". For whole grains, refined grains, vegetable oil, and animal fats, the questionnaire had two options, including "yes" and "no". For fruits and fresh vegetables, the questionnaire had four options, including "almost every day", "quite often", "occasionally", or "rarely or never".

Using this dietary data, we computed the PDI, the hPDI, and the uPDI to evaluate the overall PBD pattern, the healthful PBD pattern, and the unhealthful PBD pattern, respectively [28,29,40]. Intake frequencies of the 16 food groups were assigned a score between 1 and 5. For the PDI, plant-based food groups were given positive scores (1 for the least frequent consumption and 5 for the most frequent consumption), whereas animal-based food groups were given reverse scores (5 for the least frequent consumption and 1 for the most frequent consumption). For the hPDI, healthful plant-based foods were given positive scores, but unhealthful plant-based foods and animal-based foods were reverse scored. For the uPDI, healthful plant-based foods and animal-based foods were reverse scored, but unhealthful plant-based foods were given positive scores. We summed the 16 food-group scores for everyone to derive the PDI, hPDI, and uPDI, with a theoretical range of 16 to 80. More detailed information on calculating the PDI, hPDI, and uPDI are provided in Table A1. In the present study, the PDI, hPDI, and uPDI were classified into 2 halves based on the median level, including a lower half (lower PDI, lower hPDI, and lower uPDI) and a higher half (higher PDI, higher hPDI, and higher uPDI), respectively.

2.5. Assessment of Covariates

Covariates shown by prior research that could alter the associations of the BMI and a PBD with cognitive function were adjusted in our analyses. Potential confounders included age (years), sex (male or female), type of residence (city, town, or country), education (illiterate or literate), main occupation before 60, smoking status (current, former, or never), drinking status (current, former, or never), financial status (financial dependence or independence), regular exercise (yes or no), and health conditions. Health conditions were evaluated by taking into consideration six diseases: hypertension, diabetes, heart disease, stroke, cancer, and respiratory disease. Each disease was scored 1 (present) or 0 (not present).

2.6. Statistical Analysis

Descriptive statistics were used to summarize the baseline characteristics. Cox proportional hazard models were conducted to evaluate the association of baseline BMI with cognitive impairment using categories of BMI with the normal group as the reference. We also used Cox proportional hazard models to examine the associations of PDIs, hPDIs, and uPDIs with cognitive impairment. The follow-up period for each individual was computed from baseline to the date of the first occurrence of cognitive impairment, to the date of death, lost-to-follow-up, or to the end of follow-up, whichever occurred first. The proportional hazards assumption was verified by using a global test for zero slope of the scaled Schoenfeld residuals over time. In addition, we performed a restricted cubic spline with 4 knots placed at the 5th, 35th, 65th, and 95th percentiles, and we used the median value of the baseline BMI as a reference point to test the potential non-linear association of the baseline BMI with cognitive impairment. We performed stratified analyses by PDI, hPDI, and uPDI score to assess whether the associations of BMI and cognitive impairment varied with PDI, hPDI, and uPDI scores. The regression models included sex, age, residence, education, occupation, smoking status, drinking status, regular exercise, financial independence, and health conditions.

Data were analyzed using STATA 16 (StataCorp, College Station, TX, USA) and R software, version 3.4.2 (R Foundation, Vienna, Austria). Tests were two-sided with the statistical significance set as $p < 0.05$.

3. Results

Of the 4792 participants included, 2425 (50.61%) were men, and there was a mean age of 80.70 ± 9.58 years old at baseline. In total, 2493 (52.02%) participants were living in rural locations, 2339 (48.81%) were illiterate, 2972 (62.02%) were never smokers, and 3133 (65.38%) participants were never drinkers. The mean baseline BMI was 22.02 ± 4.46 kg/m^2, and the percentages of participants classified as underweight, normal, overweight, and obese were 18.53%, 55.46%, 19.39%, and 6.62%, respectively. The mean PDI, hPDI, and uPDI were 48.71 ± 6.05, 54.09 ± 5.38, and 42.78 ± 6.65 at baseline, respectively. The distribution of baseline covariates by baseline BMI level is shown in Table 1.

During the 24156 person-years of follow-up, 1077 participants developed cognitive impairment. As shown in Table 2, after multivariable adjustment, as compared with the normal weight group, the HRs of cognitive impairment were 1.42 (95% CI = 1.21–1.66, $p < 0.001$) in the underweight group, 0.79 (95% CI = 0.66–0.95, $p = 0.010$) in the overweight group, and 0.72 (95% CI = 0.54–0.96, $p = 0.026$) in the obese group. Baseline BMI was non-linearly correlated to the risk of cognitive impairment, with a reverse J-shaped relationship (p for non-linear trend = 0.005). (See Figure 2.)

After multivariable adjustment, a lower PDI, a lower hPDI, and a higher uPDI were related to an increased risk of cognitive impairment. The HRs of cognitive impairment were 1.32 (95% CI = 1.16–1.50, $p < 0.001$) in the lower PDI group compared with the higher PDI group; the HRs of cognitive impairment were 1.46 (95% CI = 1.29–1.66, $p < 0.001$) in the lower hPDI group as compared with the higher hPDI group, and the HRs of cognitive impairment were 1.21 (95% CI = 1.06–1.38, $p = 0.004$) in the higher uPDI group as compared with the lower uPDI group (Table 3).

We observed a significant interaction between baseline BMIs and PDIs, with the corresponding associations of an overweight status being much more pronounced among participants with a higher PDI than those with a lower PDI, among participants with a higher hPDI than those with a lower hPDI, and among participants with a lower uPDI than those with a higher hPDI (Figure 3). Specifically, the protective effect of being overweight on cognitive impairment was attenuated with a 13% (95% CI = 0.67–1.12, $p = 0.267$) decreased risk, which was not significant among those with a lower PDI, in contrast with a 26% (95% CI = 0.57–0.95, $p = 0.017$) decreased risk, which was significant among those with a higher PDI. Similarly, the protective effect of an overweight status on cognitive impairment was attenuated with a 7% (95% CI = 0.72–1.10, $p = 0.568$) non-significant decrease in risk among those with a lower hPDI, in contrast with a 27% (95% CI = 0.57–0.94, $p = 0.013$) significantly decreased risk for those with a higher hPDI. In addition, the protective effect of an overweight status on cognitive impairment was attenuated with a 1% (95% CI = 0.89–1.61, $p = 0.234$)

non-significant increase in risk among those with a higher uPDI, in contrast with a 39% (95% CI = 0.46–0.80, $p < 0.001$) significant decrease in risk among those with a lower uPDI (Table 4).

Table 1. Characteristics of the study population at baseline.

Characteristics	Total	Underweight	Normal	Overweight	Obese	*p* Value
N	4792	888	2658	929	317	
BMI (kg/m^2) *	22.02 ± 4.46	16.98 ± 1.28	21.20 ± 1.52	25.58 ± 1.12	32.53 ± 6.94	<0.001
PDI score *	48.71 ± 6.05	47.02 ± 6.34	48.69 ± 6.02	49.90 ± 5.43	50.12 ± 5.98	<0.001
hPDI score *	54.09 ± 5.38	52.56 ± 5.49	54.06 ± 5.40	55.32 ± 4.83	55.07 ± 5.31	<0.001
uPDI score *	42.78 ± 6.65	44.43 ± 6.53	42.87 ± 6.55	41.50 ± 6.58	41.18 ± 6.95	<0.001
Age, years *	80.70 ± 9.58	84.38 ± 9.87	80.71 ± 9.49	78.11 ± 8.76	77.91 ± 8.37	<0.001
Sex, male **	2425 (50.61)	390 (43.92)	1447 (54.44)	466 (50.16)	122 (38.49)	<0.001
Residence **						<0.001
City	782 (16.32)	78 (8.78)	400 (15.05)	224 (24.11)	80 (25.24)	
Town	1517 (31.66)	261 (29.39)	861 (32.39)	287 (30.89)	108 (34.07)	
Rural	2493 (52.02)	549 (61.82)	1397 (52.56)	418 (44.99)	129 (40.69)	
Illiterate **	2339 (48.81)	508 (57.21)	1280 (48.16)	406 (43.70)	145 (45.74)	<0.001
Financial independence **	1157 (24.14)	114 (12.84)	595 (22.39)	324 (34.88)	124 (39.12)	<0.001
With regular exercise **	1997 (41.67)	301 (33.90)	1090 (41.01)	439 (47.26)	147 (46.37)	<0.001
Smoking status **						<0.001
Never smoker	2972 (62.02)	560 (63.06)	1571 (59.10)	604 (65.02)	237 (74.76)	
Former smoker	772 (16.11)	122 (13.74)	460 (17.31)	153 (16.47)	37 (11.67)	
Current smoker	1048 (21.87)	206 (23.20)	627 (23.59)	172 (18.51)	43 (13.56)	
Alcohol consumption **						<0.001
Never drinker	3133 (65.38)	612 (68.92)	1676 (63.05)	616 (66.31)	229 (72.24)	
Former drinker	681 (14.21)	97 (10.92)	403 (15.16)	134 (14.42)	47 (14.83)	
Current drinker	978 (20.41)	179 (20.16)	579 (21.78)	179 (19.27)	41 (12.93)	
Occupation **						0.156
Professional and technical personnel	201 (4.19)	18 (2.03)	111 (4.18)	55 (5.92)	17 (5.36)	
Governmental, institutional, or managerial personnel	165 (3.44)	13 (1.46)	78 (2.93)	52 (5.60)	22 (6.94)	
Commercial, service, or industrial worker	578 (12.06)	56 (6.31)	300 (11.29)	161 (17.33)	61 (19.24)	
Self-employed	81 (1.69)	12 (1.35)	42 (1.58)	22 (2.37)	5 (1.58)	
Agricultural, forestry, animal husbandry, or fishery worker	2972 (62.02)	638 (71.85)	1692 (63.66)	482 (51.88)	160 (50.47)	
Houseworker	213 (4.44)	50 (5.63)	101 (3.80)	47 (5.06)	15 (4.73)	
Military personnel	32 (0.67)	3 (0.34)	22 (0.83)	5 (0.54)	2 (0.63)	
Never worked	16 (0.33)	2 (0.23)	9 (0.34)	3 (0.32)	2 (0.63)	
Others	534 (11.14)	96 (10.81)	303 (11.40)	102 (10.98)	33 (10.41)	
Disease score ***	1 (1)	1 (1)	1 (1)	1 (1)	1 (1)	<0.001
Hypertension **	1480 (30.88)	177 (19.93)	749 (28.18)	394 (42.41)	160 (50.47)	<0.001
Diabetes **	230 (4.80)	16 (1.80)	95 (3.57)	83 (8.93)	36 (11.36)	<0.001
Heart diseases **	332 (6.93)	42 (4.73)	174 (6.55)	90 (9.69)	26 (8.20)	<0.001
Stroke **	342 (7.14)	44 (4.95)	44 (1.66)	109 (11.73)	189 (59.62)	0.001
Cancer **	27 (0.56)	4 (0.45)	13 (0.49)	10 (1.08)	0 (0.00)	0.093
Respiratory disease **	534 (11.14)	124 (13.96)	281 (10.57)	90 (9.69)	39 (12.30)	0.008

Abbreviations: BMI: body mass index; PDI: plant-based diet index; hPDI: healthful plant-based diet index; uPDI: unhealthful plant-based diet index. *: mean (standard deviation) was reported; **: Number (%) was reported; ***: median (interquartile range) was reported.

Table 2. Association of baseline BMI with incidence of cognitive impairment risk.

	Events	Participants	Person-Years	HR (95% CI) [a]	*p* Value
Underweight	263	888	4072	1.42 (1.21–1.66)	<0.001
Normal	579	2658	13,498	1.00	
Overweight	172	929	4891	0.79 (0.66–0.95)	0.010
Obese	63	317	1695	0.72 (0.54–0.96)	0.026

HR: hazard ratio; CI: confidence interval. [a]: Adjusted for sex, age, residence, education, occupation, smoking status, alcohol consumption, regular exercise, financial independence, and health conditions.

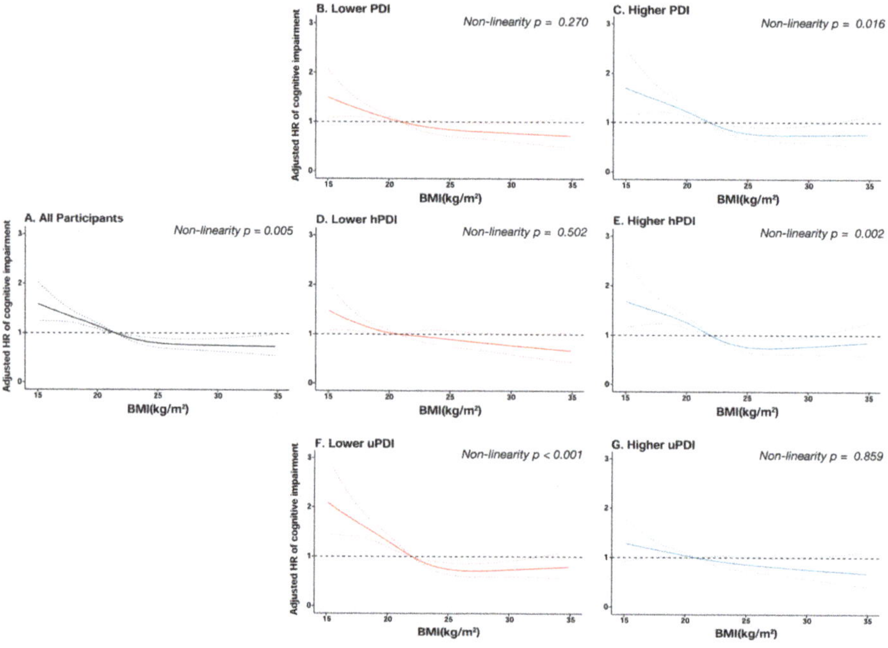

Figure 2. Cubic splines for the associations of baseline BMI with cognitive impairment, stratified by plant-based diet indices. (**A**): all participants; (**B**): lower plant-based diet index (PDI); (**C**): higher plant-based diet index (PDI); (**D**): lower healthful plant-based diet index (hPDI); (**E**): higher healthful plant-based diet index (hPDI); (**F**): lower unhealthful plant-based diet index (uPDI); (**G**): higher unhealthful plant-based diet index (uPDI).

Table 3. Associations of baseline plant-based diet indices with cognitive impairment risk.

	Events	Participants	Person-Years	HR (95% CI) [a]	*p* Value
Stratified by PDI					
Lower PDI	594	2274	11,330	1.32 (1.16–1.50)	<0.001
Higher PDI	483	2518	12,826	1.00	
Stratified by hPDI					
Lower hPDI	561	2081	10,295	1.46 (1.29–1.66)	<0.001
Higher hPDI	516	2711	13,861	1.00	
Stratified by uPDI					
Lower uPDI	480	2462	12,490	1.00	
Higher uPDI	597	2330	11,666	1.21 (1.06–1.38)	0.004

HR: hazard ratio; CI: confidence interval; PDI: plant-based dietary index; hPDI: healthful plant-based dietary index; uPDI: unhealthful plant-based dietary index. [a]: Adjusted for sex, age, residence, education, occupation, smoking status, alcohol consumption, regular exercise, financial independence, and health conditions.

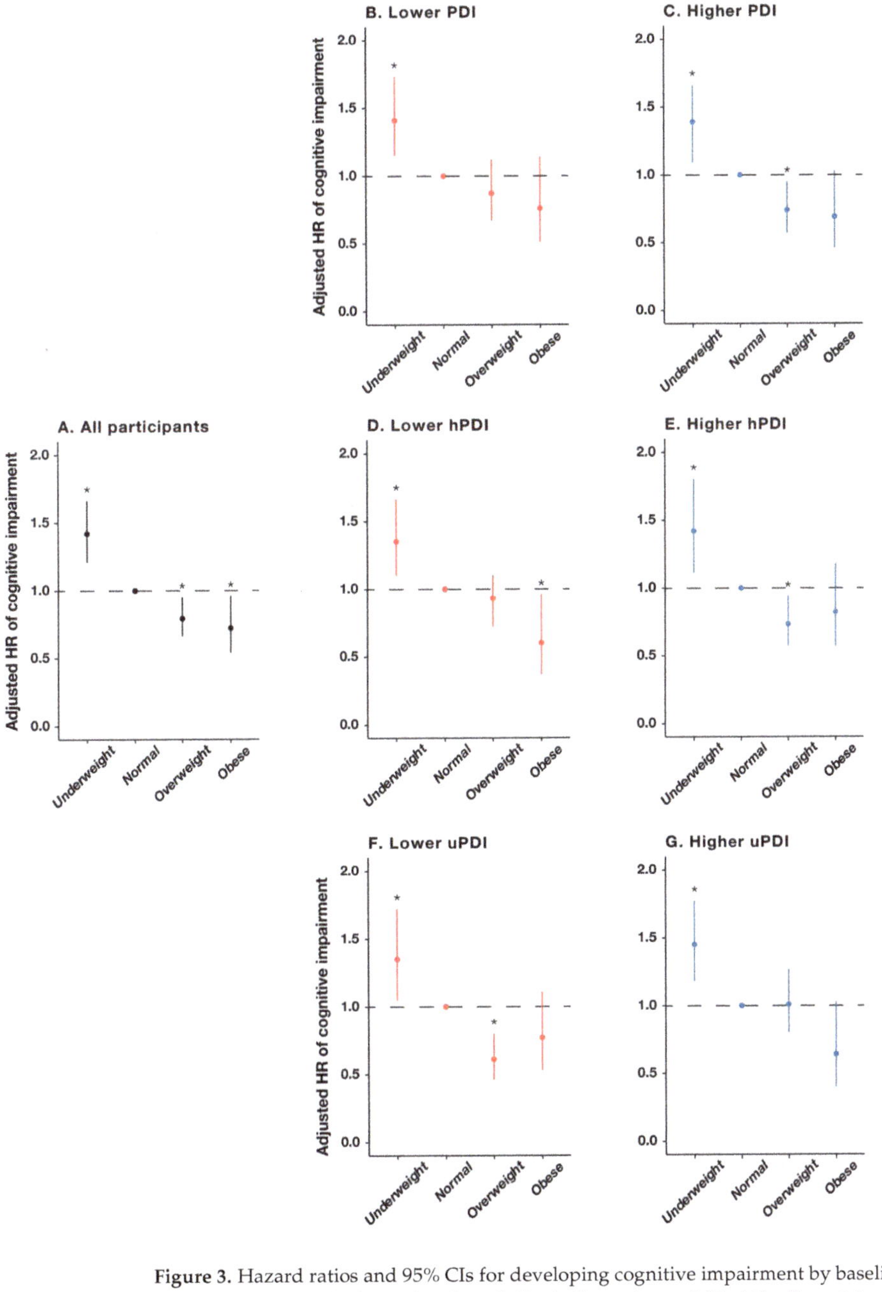

Figure 3. Hazard ratios and 95% CIs for developing cognitive impairment by baseline body-mass-index groups, stratified by plant-based diet indices. *: $p < 0.05$. (**A**): all participants; (**B**): lower plant-based diet index (PDI); (**C**): higher plant-based diet index (PDI); (**D**): lower healthful plant-based diet index (hPDI); (**E**): higher healthful plant-based diet index (hPDI); (**F**): lower unhealthful plant-based diet index (uPDI); (**G**): higher unhealthful plant-based diet index (uPDI).

term status. Second, diet was assessed using a simple FFQ without information on portion sizes; hence, we cannot calculate and adjust for total energy intake. In addition, dietary assessment via FFQ may have been subject to recall bias. Third, detailed information for several food items (e.g., potatoes, honey, and berries) was not available in the FFQ in the CLHLS. Further research with more-detailed dietary assessments is required to validate the observed findings. Fourth, the contribution of dietary supplements was not considered in the present study, which could have caused a bias in our results. Fifth, residual, unknown confounding factors cannot be entirely ruled out. All included participants were from China, which limits the extrapolation of our conclusions to other nationalities and ethnic groups. Sixth, given the observational study design, no causal association can be proved.

5. Conclusions

Based on a national, community-based, longitudinal prospective study in China, we found that BMI-defined statuses of overweight and obese were related to a decreased risk of cognitive impairment, while an underweight status was related to increased risk. Lower PDI, lower hPDI, and higher uPDI were associated with an increased risk of cognitive impairment. Furthermore, we first demonstrated that a lower adherence to an overall and a healthful PBD and a higher adherence to an unhealthful PBD may attenuate the protective effect of being overweight on cognitive impairment. Our findings are informative in facilitating the development of tailored body-weight-management and dietary recommendations for preventing cognitive impairment in an elderly population.

Author Contributions: F.L. and R.L. conceived and designed the study; Y.W., N.Q., J.Z. and K.D. collected and processed the data; F.L. and J.F. interpreted and analyzed the data; F.L. and J.F. drafted the manuscript; G.T.-M., J.B.M. and R.L. revised the manuscript. All authors have read and agreed to the published version of the manuscript.

Funding: This research received no external funding.

Institutional Review Board Statement: The CLHLS study was approved by the Research Ethics Committee of Peking University (IRB00001052-13074), and all participants or their proxy respondents provided written informed consent.

Informed Consent Statement: Informed consent was obtained from all subjects involved in the study.

Data Availability Statement: The data of this study are available to researchers upon reasonable request to corresponding author.

Acknowledgments: The authors would like to thank the CLHLS research team and all of the participants for their contributions.

Conflicts of Interest: The authors report no conflict of interest.

Appendix A

Table A1. Plant-based diet index scoring.

Food Category		Food Groups	Frequency	PDI	hPDI	uPDI
Plant-based food						
	Healthful	Whole grain	Yes	5	5	1
			No	1	1	5
		Vegetable oils	Yes	5	5	1
			No	1	1	5
		Fresh fruits	Almost everyday	5	5	1
			Quite often	4	4	2
			Occasionally	2	2	4
			Rarely or never	1	1	5

Table A1. *Cont.*

Food Category		Food Groups	Frequency	PDI	hPDI	uPDI
Plant-based food	Healthful	Fresh vegetables	Almost everyday	5	5	1
			Quite often	4	4	2
			Occasionally	2	2	4
			Rarely or never	1	1	5
		Legumes	Almost everyday	5	5	1
			≥1 time/week	4	4	2
			≥1 time/month	3	3	3
			Occasionally	2	2	4
			Rarely or never	1	1	5
		Garlic	Almost everyday	5	5	1
			≥1 time/week	4	4	2
			≥1 time/month	3	3	3
			Occasionally	2	2	4
			Rarely or never	1	1	5
		Nuts	Almost everyday	5	5	1
			≥1 time/week	4	4	2
			≥1 time/month	3	3	3
			Occasionally	2	2	4
			Rarely or never	1	1	5
		Tea	Almost everyday	5	5	1
			≥1 time/week	4	4	2
			≥1 time/month	3	3	3
			Occasionally	2	2	4
			Rarely or never	1	1	5
	Unhealthful	Refined grains	Yes	5	1	5
			No	1	5	1
		Sugar (white granulated sugar or candies)	Almost everyday	5	1	5
			≥1 time/week	4	2	4
			≥1 time/month	3	3	3
			Occasionally	2	4	2
			Rarely or never	1	5	1
		Preserved vegetables	Almost everyday	5	1	5
			≥1 time/week	4	2	4
			≥1 time/month	3	3	3
			Occasionally	2	4	2
			Rarely or never	1	5	1
Animal-based food		Animal fat	Yes	1	1	1
			No	5	5	5
		Meat	Almost everyday	1	1	1
			≥1 time/week	2	2	2
			≥1 time/month	3	3	3
			Occasionally	4	4	4
			Rarely or never	5	5	5
		Fish	Almost everyday	1	1	1
			≥1 time/week	2	2	2
			≥1 time/month	3	3	3
			Occasionally	4	4	4
			Rarely or never	5	5	5
		Eggs	Almost everyday	1	1	1
			≥1 time/month	3	3	3
			Occasionally	4	4	4
			Rarely or never	5	5	5
		Dairy products	Almost everyday	1	1	1
			≥1 time/week	2	2	2
			≥1 time/month	3	3	3
			Occasionally	4	4	4
			Rarely or never	5	5	5

Abbreviations: PDI: plant-based diet index; hPDI: healthful plant-based diet index; uPDI: unhealthful plant-based diet index.

References

1. Livingston, G.; Sommerlad, A.; Orgeta, V.; Costafreda, S.G.; Huntley, J.; Ames, D.; Ballard, C.; Banerjee, S.; Burns, A.; Cohen-Mansfield, J.; et al. Dementia prevention, intervention, and care. *Lancet* **2017**, *390*, 2673–2734. [CrossRef]
2. Alzheimer's, A. 2018 Alzheimer's disease facts and figures. *Alzheimers Dement.* **2018**, *14*, 367–425. [CrossRef]
3. Livingston, G.; Huntley, J.; Sommerlad, A.; Ames, D.; Ballard, C.; Banerjee, S.; Brayne, C.; Burns, A.; Cohen-Mansfield, J.; Cooper, C.; et al. Dementia prevention, intervention, and care: 2020 report of the Lancet Commission. *Lancet* **2020**, *396*, 413–446. [CrossRef]
4. Rabin, L.A.; Smart, C.M.; Amariglio, R.E. Subjective Cognitive Decline in Preclinical Alzheimer's Disease. *Annu. Rev. Clin. Psychol.* **2017**, *13*, 369–396. [CrossRef]
5. Grillner, S. Human Brain Project. *Nat. Neurosci.* **2016**, *19*, 1118. [CrossRef]
6. Rubin, L.H.; Gustafson, D.; Hawkins, K.L.; Zhang, L.; Jacobson, L.P.; Becker, J.T.; Munro, C.A.; Lake, J.E.; Martin, E.; Levine, A.; et al. Midlife adiposity predicts cognitive decline in the prospective Multicenter AIDS Cohort Study. *Neurology* **2019**, *93*, E261–E271. [CrossRef]
7. Liu, Z.Z.; Yang, H.Q.; Chen, S.Y.; Cai, J.; Huang, Z.J. The association between body mass index, waist circumference, waist-hip ratio and cognitive disorder in older adults. *J. Public Health* **2019**, *41*, 305–312. [CrossRef]
8. Gunstad, J.; Lhotsky, A.; Wendell, C.R.; Ferrucci, L.; Zonderman, A.B. Longitudinal Examination of Obesity and Cognitive Function: Results from the Baltimore Longitudinal Study of Aging. *Neuroepidemiology* **2010**, *34*, 222–229. [CrossRef]
9. Ntlholang, O.; McCarroll, K.; Laird, E.; Molloy, A.M.; Ward, M.; McNult, H.; Hory, L.; Hughes, C.F.; Strain, J.J.; Casey, M.; et al. The relationship between adiposity and cognitive function in a large community-dwelling population: Data from the Trinity Ulster Department of Agriculture (TUDA) ageing cohort study. *Br. J. Nutr.* **2018**, *120*, 517–527. [CrossRef]
10. West, N.A.; Lirette, S.T.; Cannon, V.A.; Turner, S.T.; Mosley, T.H.; Windham, B.G. Adiposity, Change in Adiposity, and Cognitive Decline in Mid- and Late Life. *J. Am. Geriatr. Soc.* **2017**, *65*, 1282–1288. [CrossRef]
11. Arvanitakis, Z.; Capuano, A.W.; Bennett, D.A.; Barnes, L.L. Body Mass Index and Decline in Cognitive Function in Older Black and White Persons. *J. Gerontol. Ser. A Biol. Sci. Med. Sci.* **2018**, *73*, 198–203. [CrossRef]
12. Nianogo, R.A.; Rosenwohl-Mack, A.; Yaffe, K.; Carrasco, A.; Hoffmann, C.M.; Barnes, D.E. Risk Factors Associated With Alzheimer Disease and Related Dementias by Sex and Race and Ethnicity in the US. *JAMA Neurol.* **2022**, *79*, 584–591. [CrossRef]
13. Liang, F.; Fu, J.L.; Moore, J.B.; Zhang, X.E.; Xu, Y.J.; Qiu, N.; Wang, Y.C.; Li, R. Body Mass Index, Waist Circumference, and Cognitive Decline Among Chinese Older Adults: A Nationwide Retrospective Cohort Study. *Front. Aging Neurosci.* **2022**, *14*, 9. [CrossRef]
14. Kim, S.; Kim, Y.; Park, S.M. Body Mass Index and Decline of Cognitive Function. *PLoS ONE* **2016**, *11*, 14. [CrossRef]
15. Aiken-Morgan, A.T.; Capuano, A.W.; Arvanitakis, Z.; Barnes, L.L. Changes in Body Mass Index Are Related to Faster Cognitive Decline Among African American Older Adults. *J. Am. Geriatr. Soc.* **2020**, *68*, 2662–2667. [CrossRef]
16. Michaud, T.L.; Siahpush, M.; Farazi, P.A.; Kim, J.; Yu, F.; Su, D.J.; Murman, D.L. The Association Between Body Mass Index, and Cognitive, Functional, and Behavioral Declines for Incident Dementia. *J. Alzheimers Dis.* **2018**, *66*, 1507–1517. [CrossRef]
17. Rodriguez-Fernandez, J.M.; Danies, E.; Martinez-Ortega, J.; Chen, W.C. Cognitive Decline, Body Mass Index, and Waist Circumference in Community-Dwelling Elderly Participants: Results From a Nationally Representative Sample. *J. Geriatr. Psychiatry Neurol.* **2017**, *30*, 67–76. [CrossRef]
18. Deckers, K.; van Boxtel, M.P.J.; Verhey, F.R.J.; Kohler, S. Obesity and cognitive decline in adults: Effect of methodological choices and confounding by age in a longitudinal study. *J. Nutr. Health Aging* **2017**, *21*, 546–553. [CrossRef]
19. Kim, G.; Choi, S.; Lyu, J. Body mass index and trajectories of cognitive decline among older Korean adults. *Aging Ment. Health* **2020**, *24*, 758–764. [CrossRef]
20. Tolppanen, A.M.; Ngandu, T.; Kareholt, I.; Laatikainen, T.; Rusanen, M.; Soininen, H.; Kivipelto, M. Midlife and Late-Life Body Mass Index and Late-Life Dementia: Results from a Prospective Population-Based Cohort. *J. Alzheimers Dis.* **2014**, *38*, 201–209. [CrossRef]
21. Hou, Q.T.; Guan, Y.; Yu, W.H.; Liu, X.T.; Wu, L.H.; Xiao, M.Z.; Lu, Y. Associations between obesity and cognitive impairment in the Chinese elderly: An observational study. *Clin. Interv. Aging* **2019**, *14*, 367–373. [CrossRef]
22. Wu, S.S.; Lv, X.Z.; Shen, J.; Chen, H.; Ma, Y.; Jin, X.R.; Yang, J.X.; Cao, Y.Y.; Zong, G.; Wang, H.L.; et al. Association between body mass index, its change and cognitive impairment among Chinese older adults: A community-based, 9-year prospective cohort study. *Eur. J. Epidemiol.* **2021**, *36*, 1043–1054. [CrossRef]
23. Ren, Z.; Li, Y.Y.; Li, X.R.; Shi, H.; Zhao, H.F.; He, M.F.; Zha, S.; Qiao, S.Y.; Pu, Y.J.; Liu, H.J.; et al. Associations of body mass index, waist circumference and waist-to-height ratio with cognitive impairment among Chinese older adults: Based on the CLHLS. *J. Affect. Disord.* **2021**, *295*, 463–470. [CrossRef]
24. Zhang, H.; Tsao, R. Dietary polyphenols, oxidative stress and antioxidant and anti-inflammatory effects. *Curr. Opin. Food Sci.* **2016**, *8*, 33–42. [CrossRef]
25. Ricker, M.A.; Haas, W.C. Anti-Inflammatory Diet in Clinical Practice: A Review. *Nutr. Clin. Pract.* **2017**, *32*, 318–325. [CrossRef]
26. Medawar, E.; Huhn, S.; Villringer, A.; Witte, A.V. The effects of plant-based diets on the body and the brain: A systematic review. *Transl. Psychiatr.* **2019**, *9*, 17. [CrossRef]
27. Rajaram, S.; Jones, J.; Lee, G.J. Plant-Based Dietary Patterns, Plant Foods, and Age-Related Cognitive Decline. *Adv. Nutr.* **2019**, *10*, S422–S436. [CrossRef]

28. Chen, H.; Shen, J.; Xuan, J.; Zhu, A.; Ji, J.S.; Liu, X.; Cao, Y.; Zong, G.; Zeng, Y.; Wang, X.J.N.A. Plant-based dietary patterns in relation to mortality among older adults in China. *Nat. Aging* **2022**, *2*, 1–7. [CrossRef]
29. Satija, A.; Bhupathiraju, S.N.; Rimm, E.B.; Spiegelman, D.; Chiuve, S.E.; Borgi, L.; Willett, W.C.; Manson, J.E.; Sun, Q.; Hu, F.B. Plant-Based Dietary Patterns and Incidence of Type 2 Diabetes in US Men and Women: Results from Three Prospective Cohort Studies. *PLos Med.* **2016**, *13*, 18. [CrossRef]
30. Baden, M.Y.; Liu, G.; Satija, A.; Li, Y.P.; Sun, Q.; Fung, T.T.; Rimm, E.B.; Willett, W.C.; Hu, F.B.; Bhupathiraju, S.N. Changes in Plant-Based Diet Quality and Total and Cause-Specific Mortality. *Circulation* **2019**, *140*, 979–991. [CrossRef]
31. Scarmeas, N.; Anastasiou, C.A.; Yannakoulia, M. Nutrition and prevention of cognitive impairment. *Lancet Neurol.* **2018**, *17*, 1006–1015. [CrossRef]
32. Wu, J.; Song, X.Y.; Chen, G.C.; Neelakantan, N.; van Dam, R.M.; Feng, L.; Yuan, J.M.; Pan, A.; Koh, W.P. Dietary pattern in midlife and cognitive impairment in late life: A prospective study in Chinese adults. *Am. J. Clin. Nutr.* **2019**, *110*, 912–920. [CrossRef] [PubMed]
33. Liu, X.; Dhana, K.; Barnes, L.; Tangney, C.; Aggarwal, P.; Agarwal, N.; Holland, T.; Rajan, K.J.C. A Healthy Plant-based Diet Was Associated With Slower Cognitive Decline In African Americans: A Biracial Community-based Cohort Of Older Adults. *Circulation* **2022**, *145*, A066. [CrossRef]
34. Yi, Z. Introduction to the chinese longitudinal healthy longevity survey (CLHLS). In *Healthy Longevity in China*; Springer: Dordrecht, The Netherlands, 2008; pp. 23–38.
35. Zeng, Y.; Feng, Q.; Hesketh, T.; Christensen, K.; Vaupel, J.W.J.T.L. Survival, disabilities in activities of daily living, and physical and cognitive functioning among the oldest-old in China: A cohort study. *Lancet* **2017**, *389*, 1619–1629. [CrossRef]
36. Folstein, M.F.; Folstein, S.E.; McHugh, P.R. "Mini-mental state": A practical method for grading the cognitive state of patients for the clinician. *J. Psychiatr. Res.* **1975**, *12*, 189–198. [CrossRef]
37. Tombaugh, T.N.; McIntyre, N.J. The mini-mental-state-examination—A comprehensive review. *J. Am. Geriatr. Soc.* **1992**, *40*, 922–935. [CrossRef]
38. Sun, R.J.; Gu, D.N. Air Pollution, Economic Development of Communities, and Health Status Among the Elderly in Urban China. *Am. J. Epidemiol.* **2008**, *168*, 1311–1318. [CrossRef]
39. Wang, J.; Li, T.; Lv, Y.; Kraus, V.B.; Zhang, Y.; Mao, C.; Yin, Z.; Shi, W.; Zhou, J.; Zheng, T.; et al. Fine Particulate Matter and Poor Cognitive Function among Chinese Older Adults: Evidence from a Community-Based, 12-Year Prospective Cohort Study. *Environ. Health Perspect.* **2020**, *128*, 67013. [CrossRef]
40. Zhu, A.; Chen, H.; Shen, J.; Wang, X.; Li, Z.; Zhao, A.; Shi, X.; Yan, L.; Zeng, Y.; Yuan, C.; et al. Interaction between plant-based dietary pattern and air pollution on cognitive function: A prospective cohort analysis of Chinese older adults. *Lancet Reg. Health West. Pac.* **2022**, *20*, 100372. [CrossRef]
41. Lv, Y.B.; Gao, X.; Yin, Z.X.; Chen, H.S.; Luo, J.S.; Brasher, M.S.; Kraus, V.B.; Li, T.T.; Zeng, Y.; Shi, X.M. Revisiting the association of blood pressure with mortality in oldest old people in China: Community based, longitudinal prospective study. *BMJ Br. Med. J.* **2018**, *361*, 11. [CrossRef]
42. Jin, X.R.; He, W.Y.; Zhang, Y.; Gong, E.; Niu, Z.M.; Ji, J.; Li, Y.X.; Zeng, Y.; Yan, L.L.J. Association of APOE epsilon 4 genotype and lifestyle with cognitive function among Chinese adults aged 80 years and older: A cross-sectional study. *PLoS Med.* **2021**, *18*, 18. [CrossRef]
43. Zhang, Y.; Jin, X.R.; Lutz, M.W.; Ju, S.Y.; Liu, K.Y.; Guo, G.; Zeng, Y.; Yao, Y. Interaction between APOE epsilon 4 and dietary protein intake on cognitive decline: A longitudinal cohort study. *Clin. Nutr.* **2021**, *40*, 2716–2725. [CrossRef]
44. Wang, Z.B.; Pang, Y.J.; Liu, J.; Wang, J.; Xie, Z.; Huang, T. Association of healthy lifestyle with cognitive function among Chinese older adults. *Eur. J. Clin. Nutr.* **2021**, *75*, 325–334. [CrossRef]
45. Satija, A.; Malik, V.; Rimm, E.B.; Sacks, F.; Willett, W.; Hu, F.B. Changes in intake of plant-based diets and weight change: Results from 3 prospective cohort studies. *Am. J. Clin. Nutr.* **2019**, *110*, 574–582. [CrossRef]
46. Satija, A.; Bhupathiraju, S.N.; Spiegelman, D.; Chiuve, S.E.; Manson, J.E.; Willett, W.; Rexrode, K.M.; Rimm, E.B.; Hu, F.B. Healthful and Unhealthful Plant-Based Diets and the Risk of Coronary Heart Disease in US Adults. *J. Am. Coll. Cardiol.* **2017**, *70*, 411–422. [CrossRef]
47. Power, B.D.; Alfonso, H.; Flicker, L.; Hankey, G.J.; Yeap, B.B.; Almeida, O.P. Changes in body mass in later life and incident dementia. *Int. Psychogeriatr.* **2013**, *25*, 467–478. [CrossRef]
48. Coin, A.; Veronese, N.; De Rui, M.; Mosele, M.; Bolzetta, F.; Girardi, A.; Manzato, E.; Sergi, G. Nutritional predictors of cognitive impairment severity in demented elderly patients: The key role of BMI. *J. Nutr. Health Aging* **2012**, *16*, 553–556. [CrossRef]
49. Yamamoto, H.; Kato, Y. Relationship between plasma insulin-like growth factor-i (igf-i) levels and body-mass index (bmi) in adults. *Endocr. J.* **1993**, *40*, 41–45. [CrossRef]
50. Harvey, J.; Solovyova, N.; Irving, A. Leptin and its role in hippocampal synaptic plasticity. *Prog. Lipid Res.* **2006**, *45*, 369–378. [CrossRef]
51. Singh, M.; Dykens, J.A.; Simpkins, J.W. Novel mechanisms for estrogen-induced neuroprotection. *Exp. Biol. Med.* **2006**, *231*, 514–521. [CrossRef]
52. Oomura, Y.; Hori, N.; Shiraishi, T.; Fukunaga, K.; Takeda, H.; Tsuji, M.; Matsurihiya, T.; Ishibashi, M.; Aou, S.; Li, X.L.; et al. Leptin facilitates learning and memory performance and enhances hippocampal CA1 long-term potentiation and CaMK II phosphorylation in rats. *Peptides* **2006**, *27*, 2738–2749. [CrossRef]

53. Okereke, O.; Kang, J.H.; Ma, J.; Hankinson, S.E.; Pollak, M.N.; Grodstein, F. Plasma IGF-I levels and cognitive performance in older women. *Neurobiol. Aging* **2007**, *28*, 135–142. [CrossRef]
54. Snijder, M.B.; Dekker, J.M.; Visser, M.; Bouter, L.M.; Stehouwer, C.D.A.; Yudkin, J.S.; Heine, R.; Nijpels, G.; Seidell, J.C. Trunk fat and log fat have independent and opposite associations with fasting and postload glucose levels—The Hoorn Study. *Diabetes Care* **2004**, *27*, 372–377. [CrossRef]
55. Butterfield, D.A.; Halliwell, B. Oxidative stress, dysfunctional glucose metabolism and Alzheimer disease. *Nat. Rev. Neurosci.* **2019**, *20*, 148–160. [CrossRef]
56. Chen, H.L.; Mosley, T.H.; Alonso, A.; Huang, X.M. Plasma Urate and Parkinson's Disease in the Atherosclerosis Risk in Communities (ARIC) Study. *Am. J. Epidemiol.* **2009**, *169*, 1064–1069. [CrossRef]
57. Van den Brink, A.C.; Brouwer-Brolsma, E.M.; Berendsen, A.A.M.; van de Rest, O. The Mediterranean, Dietary Approaches to Stop Hypertension (DASH), and Mediterranean-DASH Intervention for Neurodegenerative Delay (MIND) Diets Are Associated with Less Cognitive Decline and a Lower Risk of Alzheimer's Disease—A Review. *Adv. Nutr.* **2019**, *10*, 1040–1065. [CrossRef]
58. Colho, H.J.; Trichopoulou, A.; Panza, F. Cross-sectional and longitudinal associations between adherence to Mediterranean diet with physical performance and cognitive function in older adults: A systematic review and meta-analysis. *Ageing Res. Rev.* **2021**, *70*, 20. [CrossRef]
59. Shishtar, E.; Rogers, G.T.; Blumberg, J.B.; Au, R.D.; Jacques, P.F. Long-term dietary flavonoid intake and change in cognitive function in the Framingham Offspring cohort. *Public Health Nutr.* **2020**, *23*, 1576–1588. [CrossRef]
60. Suh, S.W.; Kim, H.S.; Han, J.H.; Bae, J.B.; Oh, D.J.; Han, J.W.; Kim, K.W. Efficacy of Vitamins on Cognitive Function of Non-Demented People: A Systematic Review and Meta-Analysis. *Nutrients* **2020**, *12*, 17. [CrossRef]
61. Calder, P.C. Marine omega-3 fatty acids and inflammatory processes: Effects, mechanisms and clinical relevance. Biochim. *Biophys. Acta Mol. Cell Biol. Lipids* **2015**, *1851*, 469–484. [CrossRef]
62. Duvall, M.G.; Levy, B.D. DHA- and EPA-derived resolvins, protectins, and maresins in airway inflammation. *Eur. J. Pharmacol.* **2016**, *785*, 144–155. [CrossRef] [PubMed]
63. Aleksandrova, K.; Koelman, L.; Rodrigues, C.E. Dietary patterns and biomarkers of oxidative stress and inflammation: A systematic review of observational and intervention studies. *Redox Biol.* **2021**, *42*, 16. [CrossRef] [PubMed]
64. Marsland, A.L.; Gianaros, P.J.; Kuan, D.C.H.; Sheu, L.K.; Krajina, K.; Manuck, S.B. Brain morphology links systemic inflammation to cognitive function in midlife adults. *Brain Behav. Immun.* **2015**, *48*, 195–204. [CrossRef] [PubMed]
65. Liu, Z.W.; Zhou, T.Y.; Ziegler, A.C.; Dimitrion, P.; Zuo, L. Oxidative Stress in Neurodegenerative Diseases: From Molecular Mechanisms to Clinical Applications. *Oxidative Med. Cell. Longev.* **2017**, *2017*, 11. [CrossRef] [PubMed]
66. Mohan, D.; Yap, K.H.; Reidpath, D.; Soh, Y.C.; McGrattan, A.; Stephan, B.C.M.; Robinson, L.; Chaiyakunapruk, N.; Siervo, M.; De, P.E.C.T. Link Between Dietary Sodium Intake, Cognitive Function, and Dementia Risk in Middle-Aged and Older Adults: A Systematic Review. *J. Alzheimers Dis.* **2020**, *76*, 1347–1373. [CrossRef]
67. Muth, A.K.; Park, S.Q. The impact of dietary macronutrient intake on cognitive function and the brain. *Clin. Nutr.* **2021**, *40*, 3999–4010. [CrossRef]
68. Pearson, T.A.; Mensah, G.A.; Alexander, R.W.; Anderson, J.L.; Cannon, R.O.; Criqui, M.; Fadl, Y.Y.; Fortmann, S.P.; Hong, Y.; Myers, G.L.; et al. Markers of inflammation and cardiovascular disease application to clinical and public health practice—A statement for healthcare professionals from the centers for disease control and prevention and the American Heart Association. *Circulation* **2003**, *107*, 499–511. [CrossRef]
69. Biessels, G.J.; Whitmer, R.A. Cognitive dysfunction in diabetes: How to implement emerging guidelines. *Diabetologia* **2020**, *63*, 3–9. [CrossRef]
70. Villringer, A.; Laufs, U. Heart failure, cognition, and brain damage. *Eur. Heart J.* **2021**, *42*, 1579–1581. [CrossRef]

nutrients MDPI

Systematic Review

Dietary Effects on Pain Symptoms in Patients with Fibromyalgia Syndrome: Systematic Review and Future Directions

Emma K. Maddox [1], Shawn C. Massoni [2], Cara M. Hoffart [3,4] and Yumie Takata [5,*]

[1] BioHealth Science Program, College of Science, Oregon State University, Corvallis, OR 97331, USA
[2] Department of Microbiology, College of Science, Oregon State University, Corvallis, OR 97331, USA
[3] Department of Pediatrics, Division of Rheumatology, Children's Mercy Kansas City, Kansas City, MO 64108, USA
[4] School of Medicine, University of Missouri-Kansas City, Kansas City, MO 64108, USA
[5] School of Biological and Population Health Sciences, College of Public Health and Human Sciences, Oregon State University, Corvallis, OR 97331, USA
* Correspondence: yumie.takata@oregonstate.edu; Tel.: +1-541-737-1606

Abstract: Fibromyalgia syndrome (FMS) is recognized for its difficulty to diagnose and its subjective symptomatology. There is neither a known cure nor a recommended therapeutic diet to aid in the multidisciplinary treatment. We conducted a systematic review to investigate if diets can improve pain symptoms of fibromyalgia. Through the PubMed search in March 2022, 126 abstracts were identified. We included both intervention and observational studies of diets and pain symptoms among patients with FMS. After screening titles, abstracts, and full-texts, 12 studies, including 11 intervention and one observational study, were selected. These studies included 546 participants and investigated plant-based diets (n = 3), anti-inflammatory diets (n = 1), gluten-free diets (n = 2), and elimination/restrictive diets (n = 6). These studies assessed pain symptoms through visual analogue scale for pain, fibromyalgia impact questionnaire/revised fibromyalgia impact questionnaire, tender point count, pain pressure threshold, and/or total myalgic score. Nine studies, including all three plant-based diet studies, reported statistically significant beneficial effects of their respective diets on pain symptom measurements. Given the small sample size and short intervention duration of the included studies, limited evidence currently exists to recommend any specific diet to patients with FMS. Further research is warranted to clarify specific diets to recommend and explore their potential mechanisms.

Keywords: fibromyalgia syndrome; fibromyalgia patients; diet; pain symptoms

Citation: Maddox, E.K.; Massoni, S.C.; Hoffart, C.M.; Takata, Y. Dietary Effects on Pain Symptoms in Patients with Fibromyalgia Syndrome: Systematic Review and Future Directions. *Nutrients* **2023**, *15*, 716. https://doi.org/10.3390/nu15030716

Academic Editor: Andriana Kaliora

Received: 27 December 2022
Revised: 23 January 2023
Accepted: 27 January 2023
Published: 31 January 2023

1. Introduction

Fibromyalgia is a complex chronic syndrome that is largely characterized by subjective symptoms. There is no known cause or cure although symptoms may be managed via multimodal treatment strategies [1]. The most frequent symptoms of fibromyalgia syndrome (FMS) include widespread pain and tenderness, fatigue, stiffness, headaches, issues with sleep and cognitive functions. Among the less common symptoms are digestive problems, tingling or numbness of extremities, and face and jaw pain [1]. This chronic syndrome affects around 2–3% of the world's population [2]. Fibromyalgia predominately affects women to men at a ratio of 3:1 [2]. Some individuals experience a rapid onset of FMS, which often occurs after an illness or traumatic incident, whereas others may not have a triggering event [1].

The diagnostic process identifying this disease can be difficult, due to widespread, subjective symptoms and the lack of a known clear etiology [2]. Theories explaining underlying causes are evolving and recently fibromyalgia has been described as a central sensitization disorder, in that those with FMS have a heightened sensitivity to pain due to improper pain signal processing by the central nervous system [1]. Fibromyalgia may be caused by environmental and genetic factors operating in tandem [1]. It is hypothesized that the central nervous system and the peripheral nervous system are involved in the pathological mechanisms of FMS [2]. Given the likely complex biopsychosocial pathology of this condition, a comprehensive approach, including physical and psychological interventions, are the ideal treatment of FMS. Fibromyalgia patients may suffer from whole body pain, fatigue, stiffness, hypersensitivity to external stimuli, and autonomic disturbances [2]. Additionally, fibromyalgia can have a substantial impact on mental health; patients can develop anxiety, depression, and post-traumatic stress disorder [2]. Potential cognitive difficulties include memory deficits, concentration difficulties, and sleep disturbances [2]. Consequently, there are multiple treatment modalities required due to the complexity of the syndrome [1].

Currently, exercise is the most recommended treatment method as it has been shown to reduce pain symptoms and fatigue; however, there is currently no known diet or vitamin supplementation recommended for the treatment of FMS [2]. Dietary interventions are a tool used for the treatment of many diseases, due to healthy diets improving physical fitness, mental health, and cognitive abilities [3]. For example, plant-based diets have been used to treat hypertension as they significantly lower systolic and diastolic blood pressure [4]. However, the current status of the evidence on dietary influences on FMS is not known. Therefore, we investigated if diet affects pain symptoms for individuals with fibromyalgia through a systematic literature review of published studies and summarized future research directions.

2. Materials and Methods

A systematic literature review was conducted using the PubMed online database to identify published studies. The following search terms were applied for (all fields): diet AND fibromyalgia. This search was conducted with filters for studies conducted in humans and articles written in English and with no limits on publication years. The final search was completed on 7 March 2022, and a total of 126 abstracts were retrieved and reviewed independently by two authors (E.K.M. and Y.T.), who discussed discrepancies when they occurred and brought them to consensus. The full-text article of a study was not acquired if the abstract or title met at least one of the following exclusion criteria: (1) not including fibromyalgia participants; (2) not using visual analogue scale for pain (VAS), fibromyalgia impact questionnaire/revised fibromyalgia impact questionnaire (FIQ/FIQR), tender point count (TPC), pain pressure threshold (PPT), or total myalgic score (TMS) measurements; (3) not testing diet as intervention or exposure variable; and (4) not being an original research article such as a review or commentary. The studies that met at least one of the exclusion criteria were separated from the remaining abstracts. The full-text article was obtained for all the remaining abstracts that did not meet the exclusion criteria for the abstract or title to be further reviewed. The articles were excluded if they met at least one of the following exclusion criteria: (1) having an overlap of study participants between studies; (2) not including fibromyalgia participants; (3) not using VAS, TPC, PPT, FIQ/FIQR, or TMS measurements; (4) not testing diet as intervention or exposure variable; and (5) not being an original research article such as a review or commentary. The protocol of this systematic review was not registered.

Each study was cited, and the following information was compiled: first author; year of publication; country; study design; calendar year the study was conducted; number of participants; age; sex; race or ethnicity; length of intervention; presence of control group; and study results. Not all the information was included in the published article of all 12 studies. In these instances, the studies were examined for references cited, searching for related articles about the study through PubMed, and/or contacting the corresponding author to obtain missing information if possible. Two authors (E.K.M. and Y.T.) extracted the information mentioned above from each study independently. Inconsistencies were discussed and brought to consensus.

An outcome variable of this literature review was pain symptoms. Each study used the evaluation procedures including the VAS, TPC, PPT, TMS, and/or FIQ/FIQR. The FIQ allows patients to report the severity of their pain, fatigue, stiffness, anxiety, depression, morning tiredness, physical impairment, mood, and ability to go to and do work. The FIQR allows patients to report the severity of how FMS interferes with their function, overall impact, and symptoms, including pain and tenderness. The FIQ and FIQR values of each subsection were added and converted to fit into a 0–100 scoring range, with the higher values indicating more severe FMS symptoms. The VAS is used to measure the intensity of pain commonly on a 0–100 mm length line, where the 0 mm region is absence of pain, and the 100 mm region is the maximum pain imaginable [5]. The TPC was an essential measurement of the 1990 American College of Rheumatology (ACR) criteria but had since been phased out for the 2010 ACR criteria. A tender point is indicated by the patient feeling pain from a 4-kg palpitation on one of the 18 designated sites [6]. The PPT is commonly performed on the 18 tender sites and measures the amount of pressure over a given area, specifically the point where non-painful pressure changes into painful pressure. The TMS is the PPT score over the tender points. For all pain symptom measures, the average values and standard deviations for each group (e.g., intervention and control groups) at each time point (e.g., the baseline and the end) for intervention studies and, for a cross-sectional study, the average values of pain symptom measures by quartiles of dietary inflammatory index (DII) scores at one time point were extracted. When studies reported average values differently (e.g., average changes during the intervention, instead of the average values at the end), we estimated the average values using the available information.

The included studies were analyzed based on their participant inclusion and exclusion criteria; recruitment method; method of questionnaire/pain symptom measurement; type of diet that was investigated; how diet was administered and monitored; length of intervention; presence of control group; study results; study's strengths and limitations; and authors' conclusion. For the study results, we assessed the certainty of each study's results based on the reported statistical significance (p-values < 0.05). Risk of bias of each study was assessed through the National Heart, Lung, and Blood Institute (NHLBI) study quality assessment tools [7] and independently by two authors (E.K.M. and Y.T.). Discrepancies were discussed and brought to consensus.

3. Results

This review examined 126 abstracts that were published before March 2022 using the previously stated search terms. The abstracts and titles of these 126 articles were reviewed (Figure 1) based on the exclusion criteria; 20 articles were selected for full-text review. From the 20 full-text articles, 12 articles met the final inclusion criteria and were included in our literature review.

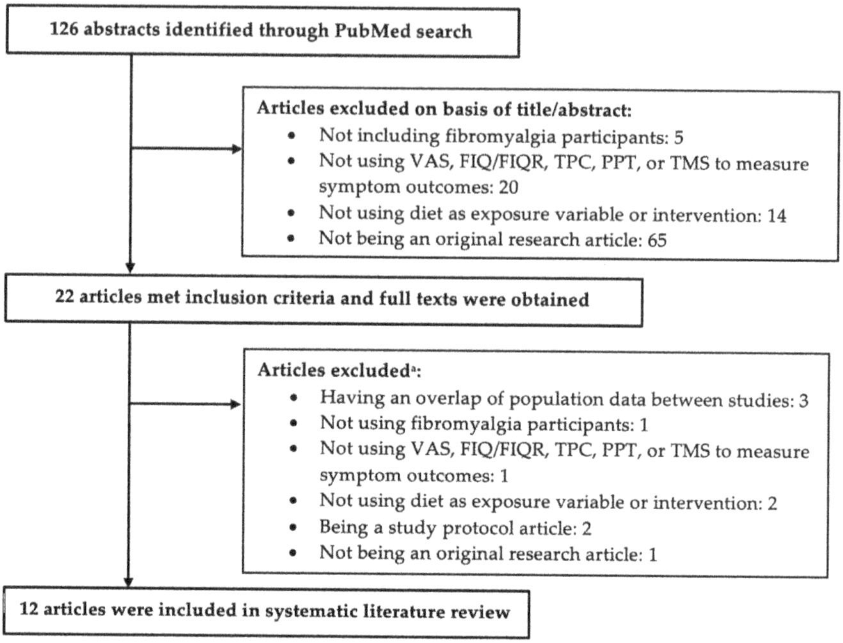

Figure 1. Flow chart of the systematic literature review. Abbreviations used: visual analogue scale for pain (VAS), fibromyalgia impact questionnaire/revised fibromyalgia impact questionnaire (FIQ/FIQR), tender point count (TPC), pain pressure threshold (PPT), or total myalgic score (TMS). [a] The following articles were excluded due to having an overlap of study participants between studies [8–10]; not including fibromyalgia participants [11]; not using VAS, TPC, PPT, FIQ/FIQR, or TMS measurements [12]; not testing diet as intervention or exposure variable [13,14]; and not being an original research article such as a review or commentary [15].

The 12 studies were published spanning from 2000 to 2020 and were primarily intervention studies except for one study [16], which was an observational, cross-sectional study (Table 1). These studies took place in seven different countries, four in Spain, three in the United States, and one in each of the following countries, Bangladesh, Finland, Portugal, Italy and Egypt. The sample sizes of all 12 articles were relatively small, with a minimum of 7 participants [17] and a maximum of 95 participants [16], totaling 546 participants. Five of the reviewed studies included solely female participants [16–20], and six studies included female participants as the majority with less than 25% male participants [21–26]. Hänninen et al. did not report the sex, race, or ethnicity of their participants. Only two studies [17,18] reported race or ethnicity and included all white participants. All 12 studies used ACR criteria for diagnostic inclusion criteria, specifically nine used ACR 1990 criteria, but three studies [17,19,24] used ACR 2010 criteria. Multiple studies had participants with comorbidities, as it is common for FMS patients to have additional illnesses. Marum et al. [19] reported that 88% of their participants had a gastrointestinal disorder and 60% had food intolerance in addition to their FMS diagnosis. Three other studies had at least one comorbidity as additional inclusion criteria [17,23,24]. In terms of study quality, three studies rated as good quality [24–26], five as fair quality [16,17,19,21,23] and four as poor quality [18,20,22,27], respectively.

Table 1. Characteristics of studies included in the systematic review.

First Author, Publication Year	Study Design	Year(s)	Country	Number of Participants	Sex [a]	Age [b]	Race or Ethnicity	Diagnosis [c]
Azad 2000 [21]	Intervention Control Trial	No data	Bangladesh	78	Female 78% Male 22%	30.9 (12–60)	No data	FMS
Correa-Rodríguez 2020 [16]	Observational, Cross-sectional	2018	Spain	95	Female 100%	55.76	No data	FMS
Donaldson 2001 [22]	Intervention Pre and Post Trial	No data	United States	20	Female 93% Male 7%	(45–54)	No data	FMS
Hänninen 2000 [27]	Intervention Pre and Post Control Trial	No data	Finland	33	No data	51	No data	FMS
Holton 2012 [23]	Intervention Pre and Post Trial	No data	United States	37	Female 92% Male 8%	51.6	No data	FMS and IBS
Lamb 2011 [18]	Intervention Cross-over Trial	2008	United States	8	Female 100%	55.6 (48–74)	White	FMS
Marum 2016 [19]	Intervention Pre and Post Trial	2015	Portugal	38	Female 100%	51	No data	FMS and GID
Pagliai 2020 [25]	Intervention Cross-over Trial	No data	Italy	20	Female 95% Male 5%	48.95	No data	FMS
Rodrigo 2013 [17]	Intervention Pre and Post Trial	2007–2012	Spain	7	Female 100%	49 (34–68)	White	FMS, IBS and CD
Senna 2012 [26]	Intervention Control Trial	2011	Egypt	83	Female 90% Male 10%	45.6	No data	FMS
Slim 2017 [24]	Intervention Parallel-group Trial	2012–2014	Spain	55	Female 97% Male 3%	HCD: 53 (32–65) GFD: 52 (36–66)	No data	FMS and GS
Vellisca 2014 [20]	Intervention Control Trial	No data	Spain	72	Female 100%	40.98 (24–65)	No data	FMS

[a] Based on 75 participants at the start of the trial for the Slim study. [b] Mean (range) is included; for the Donaldson study, values are from normal data range for women aged 45–54 noted in the publication [22]; for the Hänninen study, average age was estimated based on another publication of the same study [8]; and for the Slim study the median age (range) is included. [c] ACR 1990 Criteria for Fibromyalgia Syndrome and ROME III Criteria for Irritable Bowel Syndrome were used except for Marum, Pagliai and Slim studies. Abbreviations used: celiac disease (CD), fibromyalgia syndrome (FMS), gastrointestinal disorder (GID), gluten-free diet (GFD), gluten sensitivity (GS), hypocaloric diet (HCD), and irritable bowel syndrome (IBS).

The twelve studies were grouped into four categories based on the type of diet. The first category, plant-based diets, consisted of three studies: a vegetarian diet [21], a raw vegetarian diet [22] and a living food diet, which is defined as an uncooked vegan diet [27]. The second category, gluten-free diets, contained two studies [17,24]. The next category is an anti-inflammatory diet, which was solely comprised of one study [16]. The final category consisted of the remaining six studies that implemented elimination/restrictive diets. Marum et al. [19] introduced a diet that eliminated foods high in fermentable oligo-, di- or mono-saccharides and polyols (FODMAPs). Lamb et al. [18] implemented a diet that eliminated simple sugars, artificial colorings, flavorings, and sweeteners; caffeinated beverages; grains with gluten; eggs and dairy products; allergenic foods; and foods high in arachidonic acid. Both Vellisca and Holton studies [20,23] implemented excitotoxin elimination diets. Vellisca and Latorre [20] focused on eliminating monosodium glutamate (MSG) and aspartame. Holton et al. [23] had participants eliminate both MSG and aspartame but included a list of other excitotoxic food additives to avoid as well. Pagliai et al. [25] implemented the Khorasan Wheat Replacement diet where wheat products made with Khorasan wheat were provided to the intervention group and the same products made with regular wheat were provided to the control group. Another study implemented an energy-restrictive diet [26].

Among the three plant-based dietary intervention studies, in the Azad study [21], the vegetarian diet group's mean VAS score statistically significantly decreased from 5.7 at baseline to 5.0 at the end of the six-week intervention, which was a smaller change than the control group which decreased from 6.2 to 2.3. However, this control group was given amitriptyline to help with insomnia (Table 2). The mean TPC had a statistically significant decrease in the control group (from 16.1 to 6.4), but not in the vegetarian diet group (from 15.7 to 14.7) (Table 3). In the Donaldson study [22], the mean FIQ decreased from 51.4 at baseline to 27.6 at the end of seven months of living food diet intervention with a statistical significance. In the Hänninen study [27], the vegan diet intervention resulted in a statistically significant decrease in their mean VAS scores over three months (specific mean values not reported but presented in graphs).

Table 2. Effects of diets on pain symptoms measured by Revised Fibromyalgia Impact Questionnaire (FIQR) and Visual Analogue Scale for Pain (VAS).

First Author of Article	Dietary Variable or Intervention	Length of Intervention	FIQ/FIQR [c]			VAS [d]		
						Baseline		**End**
Azad	Vegetarian	6 weeks				5.7 ± 1.8		5.0 ± 1.8 *
	Control	6 weeks				6.2 ± 1.9		2.3 ± 1.3 *
			Cross-sectional			**Cross-sectional**		
Correa-Rodríguez [a]	Anti-Inflammatory diet: DII score Quartile 1	N/A	70.5 ± 13.3			7.20 ± 1.64		
	Quartile 2		79.9 ± 10.5			7.91 ± 1.57		
	Quartile 3		69.2 ± 20.1			7.48 ± 2.10		
	Quartile 4		71.9 ± 15.4			7.40 ± 1.14		
			Baseline	**2 months**	**End**	**Baseline**	**2 months**	**End**
Donaldson	Raw Vegetarian	7 months	51.4 ± 14.2	33.6 ± 15.6	27.6 ± 19.0 *			
						Baseline	**Middle**	**End**
Hänninen [b]	Living Food	3 months				6.0	3.0	3.2 *
	Control	3 months				5.8	4.8	6.5
			Baseline	**End**		**Baseline**		**End**
Holton	Dietary Additive Excitotoxin Elimination	4 weeks	58.6	36.4 *		13.1		7.7 *
			Baseline	**End**				
Lamb	Control	4 weeks	46.3 ± 3.4	43.6 ± 5.1				
	Modified Elimination	4 weeks		36.6 ± 8.2				
			Baseline	**End**		**Baseline**		**End**
Marum	Low FODMAP	4 weeks	61.6	47.9 *		6.6		4.9 *
			Baseline	**End**		**Baseline**		**End**
Rodrigo	Gluten-Free	1 year	74.3 ± 2.9	36.6 ± 4.0 *		8.0 ± 0.5		3.9 ± 1.0 *
			Baseline	**End**				
Senna	Energy-restricted	6 months	54.6 ± 13.1	47 ± 5.1 *				
	Control	6 months	53.2 ± 11.55	51.6 ± 9.4				
			Baseline	**End**				
Slim	Gluten-Free	24 weeks	69.5 ± 16.3	60.3 ± 19.6				
	Hypocaloric (Control)	24 weeks	70.4 ± 16.1	61.7 ± 22.2				

Table 2. *Cont.*

First Author of Article	Dietary Variable or Intervention	Length of Intervention	FIQ/FIQR [c]		VAS [d]			
			Baseline	End				
Pagliai	Khorasan Wheat Replacement	8 weeks	54.3	42.06 *				
	Control	8 weeks	54.06	53.87				
					Baseline	1 month	2 months	End
Vellisca	Control	3 months			5.63 ± 0.86	5.41 ± 0.73	5.05 ± 0.82	5.31 ± 0.88
	Dietary Additive Excitotoxin Elimination	3 months			5.58 ± 0.91	5.05 ± 0.82	4.88 ± 0.97	5.15 ± 0.95

[a] Both diet and pain symptoms were measured at the same time due to a cross-sectional study design. [b] Values were estimated based on the graph for Hänninen study. [c] FIQ 1991 version (FIQ) were used in Donaldson, Lamb and Rodrigo studies. [d] VAS scores were reported with a range 0–10 cm, except for a range 0–20 cm for Holton and a range 0–7 cm for Vellisca studies. * Statistically significant ($p \leq 0.05$).

Table 3. Effects of diets on pain symptoms measured by Tender Point Count (TPC) and Total Myalgic Score (TMS).

First Author of Article	Dietary Variable or Intervention	Length of Intervention	TPC		TMS	
			Baseline	End	Baseline	End
Azad	Vegetarian	6 weeks	15.7 ± 2.4	14.7 ± 3.6 *	-	-
	Control	6 weeks	16.1 ± 2.3	6.4 ± 3.0 *	-	-
Holton	Dietary Additive Excitotoxin Elimination	4 weeks	16.5	14.0 *	35.2	25.7 *
Rodrigo	Gluten-Free	1 year	16.3 ± 2.4	8.0 ± 1.6 *	-	-
Senna	Energy-restricted	6 months		4.9 ± 0.8	-	
	Control	6 months		5.7 ± 1		

Abbreviations used: Dietary Inflammatory Index (DII), Fibromyalgia Impact Questionnaire (FIQ), fermentable oligo-, di- or mono-saccharides and polyols (FODMAP), Revised Fibromyalgia Impact Questionnaire (FIQR), and Visual Analogue Scale for Pain (VAS) * Statistically significant ($p \leq 0.05$).

An anti-inflammatory diet was only investigated in one observational, cross-sectional study. Correa-Rodríguez et al. [16] completed a 24-h dietary recall and one-time measurement of FIQR, VAS, and PPT. The participants were categorized into quartiles based on their dietary inflammatory index (DII) scores where lower scores represented an anti-inflammatory diet and higher scores represented a pro-inflammatory diet. All locations of the PPT measurements were significantly associated with a lower DII quartile (Table 4). There was no significant association of lower FIQR and VAS scores with DII scores.

Table 4. Effects of diets on pain symptoms measured by Pain Pressure Threshold (PPT).

First Author of Article	Dietary Variable or Intervention	Length of Intervention	PPT																	
			Occiput		Trapezius		Zygapophyseal Joint		Supraspinatus		Second Rib		Epicondyle		Gluteus		Greater Trochanter		Knee	
			Cross-Sectional																	
Correa-Rodríguez [a]	Anti-Inflammatory: DII score Quartile 1	N/A	1.18 ± 0.79		1.37 ± 0.84		1.50 ± 1.12		1.61 ± 1.14		1.09 ± 0.49		1.24 ± 0.74		2.43 ± 1.63		2.39 ± 0.95		2.17 ± 1.00	
	Quartile 2		0.94 ± 0.75		0.97 ± 0.65		0.97 ± 0.72		1.31 ± 0.89		0.87 ± 0.52		0.95 ± 0.59		2.12 ± 1.70		2.36 ± 1.47		1.91 ± 1.43	
	Quartile 3		0.72 ± 0.44		0.83 ± 0.41		0.90 ± 0.53		1.08 ± 0.52		0.80 ± 0.37		0.85 ± 0.43		1.65 ± 1.32		1.78 ± 0.87		1.05 ± 1.05	
	Quartile 4		0.57 ± 0.37 *		0.73 ± 0.45 *		0.80 ± 0.53 *		0.99 ± 0.51		0.67 ± 0.25 *		0.78 ± 0.35		1.44 ± 0.78		1.65 ± 0.72 *		1.20 ± 0.65 *	
			BL	End	BL	End	BL	End	BL	End	BL	End	BL	End	BL	End	BL	End	BL	End
Lamb	Control	4 weeks	0.89	1.35	1.62	2.11	1.11	1.64	1.85	2.19	1.11	1.03	0.94	1.13	2.38	2.73	1.78	2.26	1.41	1.29
	Modified Elimination	4 weeks		1.48		2.33		2.05		2.50		0.93		1.54		2.81		2.41		1.87
			End		End		End		End		End		End		End		End		End	
Senna	Energy-restricted	6 months	4.5 ± 2.9		6 ± 2.3 *		4.7 ± 1.9		4.6 ± 2.1 *		5.8 ± 1.5		4.9 ± 1.8		4.3 ± 1.8 *		4.9 ± 2.1 *		4.2 ± 1.8 **	
	Control		5.1 ± 2.7		7.3 ± 2.4		4.5 ± 1.7		5.9 ± 2.7		6 ± 1.7		4.3 ± 2.2		5.7 ± 2.2		6.2 ± 2.4		6.1 ± 2	

[a] Both diet and pain symptoms were measured at the same time due to a cross-sectional study design. * Statistically significant ($p \leq 0.05$); ** Statistically significant $p \leq 0.01$. Abbreviations used: Baseline (BL), and Dietary Inflammatory Index (DII). Regarding the gluten-free diet, Rodrigo et al. [17] reported statistically significant decreases in FIQ, TPC, and VAS values after one year of the intervention among participants who were diagnosed with both FMS and Irritable Bowel Syndrome (IBS). In the Slim study [24] the gluten-free diet was compared to a control diet (hypocaloric diet) among patients with FMS and gluten sensitivity symptoms. At the end of the 24-week intervention period, the baseline mean FIQR score of 69.5 decreased to 60.3 in the gluten-free diet group. However, this change in FIQR score did not reveal a statistically significant difference from the hypocaloric control group.

As the last category, six studies used an elimination/restrictive dietary intervention. Marum et al. [19] used a low FODMAP diet and found that both the mean FIQR and VAS scores significantly decreased from 61.6 at baseline to 47.9 at the end of the four-week trial, and from 6.6 to 4.9, respectively. Lamb et al. [18] implemented a hypoallergenic, modified elimination diet with phytonutrient-rich medical food followed by a control period of a USDA food pyramid diet with a rice protein powder supplement for four weeks each in women with FMS. They reported a mean FIQ score of 46.3 at baseline, 43.6 at the end of the USDA food pyramid diet and 36.6 at the end of the intervention period. This decrease in the FIQ score was not statistically significant, although the mean FIQ sub-section pain score had a statistically significant decrease (5.5 at baseline, 5.94 at the end of USDA control diet and 4.92 at the end of the intervention period). No statistically significant differences were reported for various bodily areas of PPT measurements over time. Vellisca and Latorre [20] implemented a diet that eliminated MSG and aspartame for three months. Their VAS score scale was not conventional as it only ranged from 0–7. Although there were some improvements in pain in both the control and intervention groups over three months, neither group achieved statistical significance. Holton et al. [23] implemented a diet that eliminated additive excitotoxins from their patients with FMS and IBS diagnoses. Eight of the participants reduced their tender points to less than 11 after four weeks of the intervention. In the Holton study [23], their average FIQ, VAS (0–20 score scale), TPC, and TMS scores decreased by 22.2, 5.4, 2.5, and 9.5 after four weeks, respectively. Senna's [26] energy-restricted diet intervention for six months resulted in a statistically significant decrease in FIQR. Although no baseline measure was taken, TPC was lower in the intervention group than the control group at the end of the intervention. For PPT, five out of nine areas assessed had a statistically significantly lower PPT in the intervention than the control diet groups. Besides the FMS pain measures, the intervention group experienced more weight loss than the control group. In the Pagliai study [25], only FIQR was measured and FIQR scores at baseline did not differ between the Khorasan Wheat Replacement diet and control diet groups. After eight weeks of the intervention, the participants who received the Khorasan wheat products had a statistically significantly lower average FIQR score.

4. Discussion

We conducted this systematic literature review to investigate if diet has the potential to provide symptom relief for those with fibromyalgia based on published studies. Among the 12 studies reviewed, eight intervention trials reported statistically significant improvements in at least one of the pain measurements as a result of the intervention and one cross-sectional study observed an inverse association between DII score and pain measurements. By diet categories, all of the plant-based diet and anti-inflammatory diet studies found statistically significant improvement in pain measurements based on FIQ, VAS, TPC or PPT [16,21,22,27]. Inconsistent results were reported for gluten-free and elimination/restrictive diets. Only one out of the two gluten-free diet studies [17] reported statistically significant improvements in pain measurements including FIQ, VAS and TPC. Four out of the six elimination/restrictive diets reported statistically significant pain improvements in FIQR/FIQ, VAS, TPC, TMS or PPT [19,23,25,26].

Overall, it is encouraging that the majority of the studies included in this review reported a pain improvement in at least one of the measurements. Although we were able to cover a variety of diet types, specific diet types that are effective to alleviate pain symptoms among patients with FMS are not clear. In our systematic review, plant-based diets reported the most consistent results [21,22,27] and the only anti-inflammatory diet study also reported significant pain improvement [16]. Both diets were similar in terms of high consumption of vegetables, fruits, vegetable/olive oils and nuts, and low consumption of red meats. Furthermore, some elimination diets in our review share commonalities. Food additives were avoided in Holton, Lamb and Vellisca studies [18,20,23]. As food additives are contained in processed foods, the living food diet in the Donaldson study would also restrict the consumption of food additives [22]. Among these four studies, only two reported a statistically significant improvement in pain symptoms with the intervention diet [22,23]. Hence, the effectiveness of eliminating food additives on pain symptoms among patients with FMS is currently inconclusive and needs to be clarified in future studies. With regards to gluten-free diets, only one of the two trials reported a significant pain improvement [17]. The diet used in the Lamb study is an elimination/detoxification diet that also excluded gluten from the diet and no significant difference in the pain symptom changes between the intervention and control diets were reported [18]. Potential reasons for these inconsistent results are that gluten-free diets had no specific guidance on the amount of fruit, vegetables, nuts and red meats allowable, and they did not restrict consumption of processed foods high in food additives.

One potential mechanism that may explain our finding of more consistent results from plant-based and anti-inflammatory diets than gluten-free diets is weight loss. Among patients with FMS, weight loss is associated with improved pain symptoms [28]. Plant-based and vegetarian diets are generally associated with lower body weight [29]. For the Hänninen study [27], weight loss in the vegan diet group was reported [8], although body weights were not reported in the other two plant-based diet studies. In contrast, a gluten-free diet may not necessarily result in weight loss; instead, a healthful weight gain was reported among celiac patients following a gluten-free diet as it helped to alleviate malabsorption/digestive issues [30]. In the Slim study, the control diet, not the gluten-free diet, group experienced significant weight loss. Among elimination/restrictive diet studies, the Senna study among obese patients with FMS [26] reported both pain improvement and weight loss as a result of the energy-restricted diet intervention. In the Marum study [19], the low FODMAP diet group had significant weight loss [9]. Given that only four of the 12 studies reported the participants' weight or BMI changes during the diet intervention period, future studies are needed to report their weight changes to help elucidate this potential mechanism.

Besides body weight, other potential and hypothesized mechanisms are through decreased inflammation and decreased activation of neurotransmitters in central sensitization. In the Senna study, the intervention group had more pain improvements and weight loss as well as lower concentration of inflammatory markers (i.e., C-reactive protein and interleukin-6) than the control group. This study highlights the effects of body weight on pain symptoms, potentially mediated through inflammatory pathways. The finding from the Correa-Rodríguez study [16] also lends support for decreased inflammation as a potential mechanism, given that participants whose diets were characterized as higher anti-inflammatory dietary potentials had lower scores of pain measurements. The cross-sectional nature of this study limits us to consider the weight change over time, which was not reported, as an additional mechanism to explain their finding. Future studies including biomarkers to assess inflammation and central sensitization would also help to elucidate potential mechanisms. Additionally, diets other than those included in this review may also result in pain improvement. For instance, low-carbohydrate and ketogenic diets were recently hypothesized to alleviate chronic pain, also through inflammatory and nervous system pathways [31], and other diets also need to be investigated in future studies of pain symptoms among FMS patients.

In addition, future studies should take into consideration the variability of adherence to the respective diet being studied. The majority of studies in our review did not report the adherence rate to the diet regimen and a few studies noted challenges in adhering to the diet regimen experienced by study participants [21,22]. Moreover, all 12 studies included outpatient participants who may have experienced additional difficulties regarding adherence as they were having to make a major dietary change upon enrolling in a dietary intervention trial on top of managing one or more chronic conditions and completing tasks for their daily life. To increase their adherence, a meal plan service that is vetted by researchers could help alleviate the challenges that participants may experience. The majority of the studies reviewed did not report providing participants with adequate resources and introductory sessions with dietitians. As an exception, the Holton study provided detailed instructions for the dietary regimen (e.g., a list of excitotoxic food additives to avoid), individual professional dietary counseling sessions and food diaries to complete three days a week, all of which might have contributed to a significant pain improvement in a short period of four weeks [23]. Hence, future studies may also consider providing detailed instructions and support for participants to follow the dietary regimen as part of the intervention trial.

One strength of this systematic literature review is that the 12 studies took place over seven different countries and four continents with varying cultures, which raises the possibility that these results could potentially be generalized among multiple ethnicities. The variation of diets investigated in the studies reviewed is also a strength as it helps to compare different dietary aspects that may result in improved pain symptoms among FMS patients. Another strength is the inclusion of participants aged 12 to 74 years in the studies reviewed. This allows for the generalization of results over a wide age range of individuals with FMS. Eleven of the 12 studies were intervention trials, which allow for more control of potential confounding than observational studies.

There are several limitations in this systematic review. First, most studies did not provide information about the adherence to the diets, which might have affected the study results. Another limitation is that most of the participants were female in all 12 studies, with five studies having only female participants. In the six other studies men only made up 3–22% of the participants, which is not in line with the 3:1 female to male ratio of FMS prevalence. Future studies need to include more men to increase the generalizability to both sexes. An additional limitation is the variety of pain symptom assessments used across all 12 studies and within each of the diet categories. Therefore, it was difficult to make comparisons between studies that did not use the same assessments. One of the assessment types included in this review, TPC, was a major diagnostic criterion from the ACR 1990. The ACR was updated in 2010, some changes being the removal of TPC and the addition of a widespread pain index, as well as a symptom severity scale. For this review it was necessary to include studies that used TPC, as there were not sufficient studies on diet and pain symptoms among FMS patients after 2010. For future studies, the updated pain assessments should be included alongside FIQR, VAS, possibly TPC, TMS, and PPT as well. All 12 studies reviewed had sample sizes below 100, which reduced the power to detect a statistically significant effect of the diets on pain symptoms.

5. Conclusions

The results from this systematic literature review suggest that there is potential that diet can be helpful in improving pain symptoms in patients with FMS. From this review, plant-based diets seem to have more consistent and overall success in lessening pain symptoms than elimination/restrictive diets. Given that a limited number of studies have been conducted to date, findings from gluten-free and anti-inflammatory diet studies need to be followed up in future studies. Nevertheless, further studies should be conducted for all four diet categories included in this review and be completed with larger sample sizes and longer intervention periods. Furthermore, using dietary intervention implementation strategies to enhance participants' adherence to the diet regimen, and including body weight and biomarker measurements to explore potential biological mechanisms are other ways to advance research. In terms of clinical application, there is currently very weak to insufficient evidence for any of the four diet categories to change the status of 'no recommended diet' for FMS in the clinical practice.

Author Contributions: Conceptualization, E.K.M. and Y.T.; methodology, E.K.M. and Y.T.; investigation (screening and selection of articles and data extraction), E.K.M. and Y.T.; writing—original draft preparation, E.K.M. and Y.T.; writing—review and editing, S.C.M. and C.M.H.; supervision, Y.T.; and project administration, Y.T. All authors have read and agreed to the published version of the manuscript.

Funding: This study received no external funding.

Institutional Review Board Statement: Not applicable.

Informed Consent Statement: Not applicable.

Data Availability Statement: Data sharing not applicable.

Acknowledgments: E.K.M. completed this work as part of her Honors' thesis in the College of Science at Oregon State University.

Conflicts of Interest: All authors declare no conflict of interest.

References

1. Siracusa, R.; Paola, R.D.; Cuzzocrea, S.; Impellizzeri, D. Fibromyalgia: Pathogenesis, Mechanisms, Diagnosis and Treatment Options Update. *Int. J. Mol. Sci.* **2021**, *22*, 3891. [CrossRef] [PubMed]
2. Sarzi-Puttini, P.; Giorgi, V.; Marotto, D.; Atzeni, F. Fibromyalgia: An update on clinical characteristics, aetiopathogenesis and treatment. *Nat. Rev. Rheumatol.* **2020**, *16*, 645–660. [CrossRef] [PubMed]
3. Lewis, J.D.; Albenberg, L.; Lee, D.; Kratz, M.; Gottlieb, K.; Reinisch, W. The Importance and Challenges of Dietary Intervention Trials for Inflammatory Bowel Disease. *Inflamm. Bowel Dis.* **2017**, *23*, 181–191. [CrossRef] [PubMed]
4. Clem, J.; Barthel, B. A Look at Plant-Based Diets. *Missouri Med.* **2021**, *118*, 233–238. [PubMed]
5. Haefeli, M.; Elfering, A. Pain assessment. *Eur. Spine J.* **2006**, *15* (Suppl. 1), S17–S24. [CrossRef] [PubMed]
6. Wolfe, F.; Smythe, H.A.; Yunus, M.B.; Bennett, R.M.; Bombardier, C.; Goldenberg, D.L.; Tugwell, P.; Campbell, S.M.; Abeles, M.; Clark, P.; et al. The American College of Rheumatology 1990 Criteria for the Classification of Fibromyalgia. Report of the Multicenter Criteria Committee. *Arthritis Rheum.* **1990**, *33*, 160–172. [CrossRef] [PubMed]
7. National Heart Lung and Blood Institute. Study Quality Assessment Tools. Available online: https://www.nhlbi.nih.gov/health-topics/study-quality-assessment-tools (accessed on 5 July 2022).
8. Kaartinen, K.; Lammi, K.; Hypen, M.; Nenonen, M.; Hanninen, O.; Rauma, A.L. Vegan diet alleviates fibromyalgia symptoms. *Scand. J. Rheumatol.* **2000**, *29*, 308–313. [CrossRef]
9. Marum, A.P.; Moreira, C.; Tomas-Carus, P.; Saraiva, F.; Guerreiro, C.S. A low fermentable oligo-di-mono-saccharides and polyols (FODMAP) diet is a balanced therapy for fibromyalgia with nutritional and symptomatic benefits. *Nutr. Hosp.* **2017**, *34*, 667–674. [CrossRef]
10. Rodrigo, L.; Blanco, I.; Bobes, J.; de Serres, F.J. Effect of one year of a gluten-free diet on the clinical evolution of irritable bowel syndrome plus fibromyalgia in patients with associated lymphocytic enteritis: A case-control study. *Arthritis Res. Ther.* **2014**, *16*, 421. [CrossRef] [PubMed]
11. Schrepf, A.; Harte, S.E.; Miller, N.; Fowler, C.; Nay, C.; Williams, D.A.; Clauw, D.J.; Rothberg, A. Improvement in the Spatial Distribution of Pain, Somatic Symptoms, and Depression After a Weight Loss Intervention. *J. Pain* **2017**, *18*, 1542–1550. [CrossRef] [PubMed]
12. Isasi, C.; Colmenero, I.; Casco, F.; Tejerina, E.; Fernandez, N.; Serrano-Vela, J.I.; Castro, M.J.; Villa, L.F. Fibromyalgia and non-celiac gluten sensitivity: A description with remission of fibromyalgia. *Rheumatol. Int.* **2014**, *34*, 1607–1612. [CrossRef] [PubMed]
13. Rasmussen, L.B.; Mikkelsen, K.; Haugen, M.; Pripp, A.H.; Fields, J.Z.; Forre, O.T. Treatment of fibromyalgia at the Maharishi Ayurveda Health Centre in Norway II–a 24-month follow-up pilot study. *Clin. Rheumatol.* **2012**, *31*, 821–827. [CrossRef]
14. Rasmussen, L.B.; Mikkelsen, K.; Haugen, M.; Pripp, A.H.; Forre, O.T. Treatment of fibromyalgia at the Maharishi Ayurveda Health Centre in Norway. A six-month follow-up study. *Clin. Exp. Rheumatol.* **2009**, *27*, S46–S50.
15. Bennett, R.M. A raw vegetarian diet for patients with fibromyalgia. *Curr. Rheumatol. Rep.* **2002**, *4*, 284. [CrossRef] [PubMed]
16. Correa-Rodriguez, M.; Casas-Barragan, A.; Gonzalez-Jimenez, E.; Schmidt-RioValle, J.; Molina, F.; Aguilar-Ferrandiz, M.E. Dietary Inflammatory Index Scores Are Associated with Pressure Pain Hypersensitivity in Women with Fibromyalgia. *Pain Med.* **2020**, *21*, 586–594. [CrossRef] [PubMed]
17. Rodrigo, L.; Blanco, I.; Bobes, J.; de Serres, F.J. Clinical impact of a gluten-free diet on health-related quality of life in seven fibromyalgia syndrome patients with associated celiac disease. *BMC Gastroenterol.* **2013**, *13*, 157. [CrossRef]
18. Lamb, J.J.; Konda, V.R.; Quig, D.W.; Desai, A.; Minich, D.M.; Bouillon, L.; Chang, J.L.; Hsi, A.; Lerman, R.H.; Kornberg, J.; et al. A program consisting of a phytonutrient-rich medical food and an elimination diet ameliorated fibromyalgia symptoms and promoted toxic-element detoxification in a pilot trial. *Altern. Ther. Health Med.* **2011**, *17*, 36–44. [PubMed]
19. Marum, A.P.; Moreira, C.; Saraiva, F.; Tomas-Carus, P.; Sousa-Guerreiro, C. A low fermentable oligo-di-mono saccharides and polyols (FODMAP) diet reduced pain and improved daily life in fibromyalgia patients. *Scand. J. Pain* **2016**, *13*, 166–172. [CrossRef] [PubMed]
20. Vellisca, M.Y.; Latorre, J.I. Monosodium glutamate and aspartame in perceived pain in fibromyalgia. *Rheumatol. Int.* **2014**, *34*, 1011–1013. [CrossRef] [PubMed]
21. Azad, K.A.; Alam, M.N.; Haq, S.A.; Nahar, S.; Chowdhury, M.A.; Ali, S.M.; Ullah, A.K. Vegetarian diet in the treatment of fibromyalgia. *Bangladesh Med. Res. Counc. Bull.* **2000**, *26*, 41–47.

22. Donaldson, M.S.; Speight, N.; Loomis, S. Fibromyalgia syndrome improved using a mostly raw vegetarian diet: An observational study. *BMC Complement Altern. Med.* **2001**, *1*, 7. [CrossRef]
23. Holton, K.F.; Taren, D.L.; Thomson, C.A.; Bennett, R.M.; Jones, K.D. The effect of dietary glutamate on fibromyalgia and irritable bowel symptoms. *Clin. Exp. Rheumatol.* **2012**, *30*, 10–17. [PubMed]
24. Slim, M.; Calandre, E.P.; Garcia-Leiva, J.M.; Rico-Villademoros, F.; Molina-Barea, R.; Rodriguez-Lopez, C.M.; Morillas-Arques, P. The Effects of a Gluten-free Diet Versus a Hypocaloric Diet Among Patients with Fibromyalgia Experiencing Gluten Sensitivity-like Symptoms: A Pilot, Open-Label Randomized Clinical Trial. *J. Clin. Gastroenterol.* **2017**, *51*, 500–507. [CrossRef] [PubMed]
25. Pagliai, G.; Colombini, B.; Dinu, M.; Whittaker, A.; Masoni, A.; Danza, G.; Amedei, A.; Ballerini, G.; Benedettelli, S.; Sofi, F. Effectiveness of a Khorasan Wheat-Based Replacement on Pain Symptoms and Quality of Life in Patients with Fibromyalgia. *Pain Med.* **2020**, *21*, 2366–2372. [CrossRef] [PubMed]
26. Senna, M.K.; Sallam, R.A.; Ashour, H.S.; Elarman, M. Effect of weight reduction on the quality of life in obese patients with fibromyalgia syndrome: A randomized controlled trial. *Clin. Rheumatol.* **2012**, *31*, 1591–1597. [CrossRef] [PubMed]
27. Hanninen, O.; Kaartinen, K.; Rauma, A.L.; Nenonen, M.; Torronen, R.; Hakkinen, A.S.; Adlercreutz, H.; Laakso, J. Antioxidants in vegan diet and rheumatic disorders. *Toxicology* **2000**, *155*, 45–53. [CrossRef] [PubMed]
28. Kocyigit, B.F.; Okyay, R.A. The relationship between body mass index and pain, disease activity, depression and anxiety in women with fibromyalgia. *PeerJ* **2018**, *6*, e4917. [CrossRef] [PubMed]
29. Spencer, E.A.; Appleby, P.N.; Davey, G.K.; Key, T.J. Diet and body mass index in 38000 EPIC-Oxford meat-eaters, fish-eaters, vegetarians and vegans. *Int. J. Obes. Relat. Metab. Disord.* **2003**, *27*, 728–734. [CrossRef]
30. Barone, M.; Della Valle, N.; Rosania, R.; Facciorusso, A.; Trotta, A.; Cantatore, F.P.; Falco, S.; Pignatiello, S.; Viggiani, M.T.; Amoruso, A.; et al. A comparison of the nutritional status between adult celiac patients on a long-term, strictly gluten-free diet and healthy subjects. *Eur. J. Clin. Nutr.* **2016**, *70*, 23–27. [CrossRef] [PubMed]
31. Field, R.; Field, T.; Pourkazemi, F.; Rooney, K. Low-carbohydrate and ketogenic diets: A scoping review of neurological and inflammatory outcomes in human studies and their relevance to chronic pain. *Nutr. Res. Rev.* **2022**. [CrossRef]

nutrients

MDPI

Review

Food, Dietary Patterns, or Is Eating Behavior to Blame? Analyzing the Nutritional Aspects of Functional Dyspepsia

Charalampia Amerikanou [1], Stamatia-Angeliki Kleftaki [1], Evdokia Valsamidou [1], Eirini Chroni [1], Theodora Biagki [1], Demetra Sigala [2], Konstantinos Koutoulogenis [1], Panagiotis Anapliotis [1], Aristea Gioxari [2] and Andriana C. Kaliora [1,*]

[1] Department of Nutrition and Dietetics, School of Health Science and Education, Harokopio University, 70 El. Venizelou Ave, 17676 Athens, Greece; amerikanou@windowslive.com (C.A.); matina.kleftaki@gmail.com (S.-A.K.)

[2] Department of Nutritional Science and Dietetics, School of Health Science, University of the Peloponnese, Antikalamos, 24100 Kalamata-Messinia, Greece

* Correspondence: akaliora@hua.gr; Tel.: +30-210-954-9226

Abstract: Functional dyspepsia is a gastrointestinal disorder characterized by postprandial fullness, early satiation, epigastric pain, and epigastric burning. The pathophysiology of the disease is not fully elucidated and there is no permanent cure, although some therapies (drugs or herbal remedies) try to reduce the symptoms. Diet plays a critical role in either the reduction or the exacerbation of functional dyspepsia symptoms; therefore dietary management is considered to be of high importance. Several foods have been suggested to be associated with worsening functional dyspepsia, such as fatty and spicy foods, soft drinks, and others, and other foods are thought to alleviate symptoms, such as apples, rice, bread, olive oil, yogurt, and others. Although an association between functional dyspepsia and irregular eating habits (abnormal meal frequency, skipping meals, late-night snacking, dining out, etc.) has been established, not many dietary patterns have been reported as potential factors that influence the severity of functional dyspepsia. A higher adherence to Western diets and a lower adherence to FODMAPs diets and healthy patterns, such as the Mediterranean diet, can contribute to the worsening of symptoms. More research is needed on the role of specific foods, dietary patterns, or specific eating habits in the management of functional dyspepsia.

Keywords: functional dyspepsia; epigastric pain syndrome; postprandial distress syndrome; foods; dietary patterns; eating behavior; nutrition; diet

Citation: Amerikanou, C.; Kleftaki, S.-A.; Valsamidou, E.; Chroni, E.; Biagki, T.; Sigala, D.; Koutoulogenis, K.; Anapliotis, P.; Gioxari, A.; Kaliora, A.C. Food, Dietary Patterns, or Is Eating Behavior to Blame? Analyzing the Nutritional Aspects of Functional Dyspepsia. *Nutrients* **2023**, *15*, 1544. https://doi.org/10.3390/nu15061544

Academic Editor: Ruggiero Francavilla

Received: 9 March 2023
Revised: 20 March 2023
Accepted: 21 March 2023
Published: 22 March 2023

1. Introduction

Functional dyspepsia (FD) is considered to be one of the most common disorders in clinical practice [1]. It has a high prevalence that affects 10–30% of adults and 3.5–27% of children worldwide [2]. Despite its high prevalence, there are major uncertainties regarding its definition, pathophysiology, diagnosis, treatment, and prognosis [1].

1.1. Pathophysiology

Although the etiology of the disorder has not been fully elucidated, the main pathophysiological mechanisms that have been proposed throughout the years include motility alterations and psychosocial factors. The disruption of the microbiota–gut–brain axis, with abnormal central modulation, visceral hypersensitivity, and increased mucosal permeability contribute to the pathophysiology of FD. Increased intestinal permeability, immune activation, and gut dysbiosis caused by stress, which in turn affect the nervous system, suggests the concept of an impaired bidirectional communication of the "brain–gut axis" in FD [3]. In addition, acute enteric infections lead to colonic inflammation, recruitment of eosinophils and mast cells, lymphoid follicles, and duodenal mucosal bacterial loads, which affect the symptomatology of FD patients. Increased levels of inflammatory cytokines in the colonic mucosa are associated with anxiety and depression, which are related to the gut–brain axis [4]. Finally, genetic factors (such as polymorphisms in the genes related to gastrointestinal mobility and immune function), Helicobacter pylori infection, and impaired duodenal mucosal barrier function have been linked with worse FD symptoms [5–7] (Figure 1).

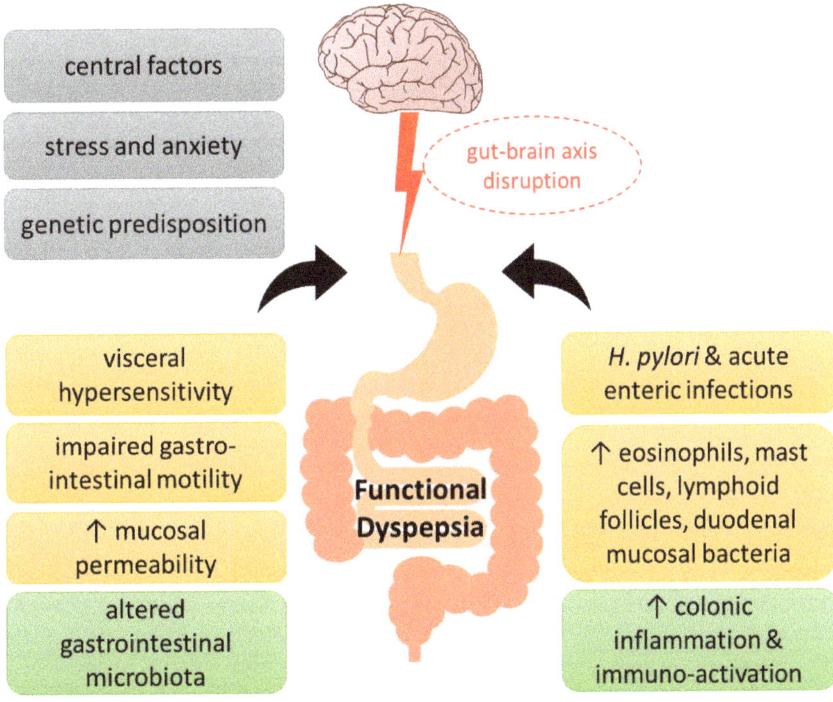

Figure 1. Pathogenesis of functional dyspepsia (FD). A series of pathogenic factors have been proposed for FD, including central nervous system abnormalities and genetic predisposition, as well as psychological factors, which have been suggested to interfere with the gut–brain axis function. Visceral hypersensitivity, impaired gastrointestinal motility, increased epithelial barrier permeability of the duodenal mucosa and infections, such as *Helicobacter pylori*, have been associated with altered intestinal flora in FD towards immune activation, immune cell infiltration, and low-grade inflammation.

1.2. Clinical Manifestations

According to the Rome IV criteria for the diagnosis of functional dyspepsia, the main clinical symptoms include bothersome postprandial fullness, early satiation, epigastric pain, and/or epigastric burning along with the absence of any structural disease that may explain the symptoms. Furthermore, the above symptomatology impairs the patient's quality of life and emotional health, and creates significant financial burden due to increased medical expenses and reduced work productivity. [8]. FD symptoms must be present for a minimum of 3 days a week during the last 3 months, they must be chronic, and start at least 6 months before diagnosis [9]. FD diagnosis includes an evaluation of the clinical history, physical examination, minimal laboratory tests, and a normal upper endoscopy. It is further categorized into epigastric pain syndrome (EPS) and eating-related postprandial distress syndrome (PDS). PDS is defined by bothersome postprandial fullness, that can affect typical activities, and/or bothersome early satiation, that can prevent the completion of a regular-sized meal. EPS is defined by bothersome epigastric pain and/or epigastric burning, both severe enough to disturb usual activities. The Rome IV classification involves not only PDS and EPS, but also their overlap (PDS-EPS overlapped syndrome), which is observed more frequently in hospital than in the general population [8].

1.3. Medicines

As there is no standard treatment for FD, research on effective therapies is ongoing, but still needs further confirmation. The acid-suppressive therapy with proton pump inhibitors (PPIs) is the most common treatment method [10]. Treatment with the tetracyclic antidepressant mirtazapine improves the quality of life, and buspirone, a serotonin-1A receptor agonist, can alleviate FD symptoms [11]. Prokinetics facilitate the gastric emptying rate [12], while amitriptyline, a neuromodulator, seems to be less effective for the treatment of FD [13]. Finally, the antibiotic rifaximin can change the duodenal microenvironment and reduce FD symptoms [14].

1.4. Herbal Remedies

Several herbal remedies have been proven effective and safe in FD with comparable outcomes with conventional treatments, and can serve as complementary and alternative medicine, especially when first line therapeutic approaches fail or are inaccessible to patients [9]. Some herbal oils improve PDS and EPS, and improve gastrointestinal symptom rating scale (GSRS) numbers and quality of life scores [15]. Herbal treatments show anti-inflammatory effects and contribute to an improvement in the function of gut microbiota, immune system, central stimuli, and intestinal motility in FD [16]. A systematic review and meta-analysis of 23 randomized controlled trials (RCTs) comparing the effectiveness of herbal treatments versus a placebo or other standard treatments for FD found that the majority of participants (>60%) in the herbal treatment group experienced an improvement in symptomatology and quality of life, compared to participants in the placebo group [17]. Chinese herbal medicines have been considered an effective alternative to prokinetics, according to a meta-analysis of 28 RCTs showing that Chinese herbal remedies were more effective than prokinetics at reducing the overall symptoms [18]. A combination of three herbs (*Trachyspermum ammi* L., *Anethum graveolens* L., and Zataria multiflora Boiss) may be important in the treatment of FD, as the essential oils were proven more effective than omeprazole [15]. Similarly, the Japanese Yukgunja-tang, also known as Rikkunshito, is a mixture of eight herbs that is frequently prescribed in FD [19], and it was proven more effective in the total clinical efficacy rate in a meta-analysis of 10 studies with 1246 patients, when combined with Western medicine over the use of Western medicine alone [20]. Additionally, perilla/ginger nutraceuticals have been shown to ameliorate some FD symptoms, such as epigastric pain, heartburn, and gastric reflux, with minor adverse events [21]. Artichoke leaf extract supplementation resulted in a greater amelioration of the multiple correspondence analysis scale compared to a placebo [22], while ginger accelerated gastric emptying [23]. The use of peppermint and caraway oil, a combination with unique properties, showed a statistically significant effect in the global improvement of FD symptoms in a meta-analysis of five RCTs [24]. A unique Greek herbal remedy known as Chios mastic gum has been shown to alleviate the symptoms of FD when taken daily for three weeks over a placebo [25]. The Hong Kong index of dyspepsia was used to assess the efficacy of the mastic treatment.

The current literature review was conducted with the aim of mapping the development of research on the nutritional aspects of FD during the period beginning on 1 January 2010 and ending on 31 December 2022. The research studies used fulfilled the following criteria: (1) they assessed the impact of foods, dietary patterns, eating behaviors, and botanicals on FD, including the relief of FD symptoms as the main outcome, (2) the literature was published in English, and (3) published from 2010 onwards. Duplicate/in vitro/animal studies, book chapters, study protocols, case reports, comments, and letters were excluded, as well as studies that involved multifactorial lifestyle interventions, in which the effects of dietary factors could not be distinguished from either genetic or lifestyle factors. To be accurate and trustworthy, the literature search was carried out using the PubMed-MEDLINE and Scopus databases. The search strategy included the following keywords: "Functional dyspepsia", "Epigastric pain syndrome", "Postprandial distress syndrome" "Nutrition", "Diet", "Foods", "Dietary patterns", "Dietary intervention", "Nutrients", "Macronutrients", and "Micronutrients". References from the extracted articles and reviews were also used to complete the data bank. The relevance of the studies was based on the title, abstract, and the full manuscript, which was reviewed. The search for duplicate publications was performed using an electronic database, and then the full text of each potentially relevant study was reviewed to ensure that it was consistent with the search criteria. A specific table was constructed by the same independent reviewers to facilitate the data extraction and selection, including the following information: article title, authors' names, year of publication, participants' characteristics, study design, duration of intervention, type of intervention, study outcomes, major findings, and limitations. Additionally, the studies were grouped according to their design (i.e., food-based intervention and dietary counseling interventions), and secondly according to the outcome.

2. Certain Foods as Inducers or Suppressants of Functional Dyspepsia Symptoms

Food consumption is the main triggering factor for dyspepsia in most FD patients. Several studies have reported that the consumption of specific foods may trigger or suppress FD symptoms. Most commonly reported triggering foods include fatty and spicy foods, soft drinks, wheat products, products containing caffeine, and alcohol [9]. Studies investigating foods that may trigger or alleviate symptoms of FD are presented in Table 1.

In the study of Akhondi-Meybodi and colleagues, three hundred and eighty four patients with FD, who had previously undergone endoscopy, were examined for their response to 114 foods in terms of relief or aggravation of their symptoms. Symptoms were most aggravated by sausage and bologna, pickles, vinegar, soft drinks, grains, tea, salt, pizza, watermelon, red pepper, and macaroni, as well as soft drinks, and acidic fruits [26]. Moreover, the foods that most frequently led to an alleviation of symptoms were apples, rice, rock candy, bread, caraway seeds, dates, honey, yogurt, quince, and walnuts. The evidence that particularly spicy foods stimulate FD symptoms has been confirmed by other studies. A spicy food's burning sensation is caused by capsaicin found in high concentrations in chili peppers. Patients with FD who consume capsaicin-containing foods experience more symptoms than healthy controls or those who consume placebos [27,28]. It has been reported that transient receptor potential vanilloid-1 receptors (TRPV1) interact with capsaicin to cause the burning sensations, however the G315 polymorphism of the TRPV1 gene is inversely correlated with FD [29]. The prevalence of FD subtypes did not vary with the consumption of spicy foods or TRPV1 genotypes in a comparison of symptom generation according to these two factors, and TRPV1 polymorphisms were not associated with scores on symptom severity questionnaires, but eating spicy food was associated with higher scores for retching and stomach fullness [30]. When comparing the nutritional habits of 168 adults with FD to 135 healthy control subjects, spicy, but also fatty foods, and carbonated drinks were the most frequently reported food items to cause symptoms. In postprandial distress syndrome FD, symptoms were more likely to be brought on by carbonated beverages and legumes [31]. Similarly, the consumption of spicy, hot, raw, or cold foods, and dairy foods or products has been associated with FD symptoms, while tea was associated with FD prevention [32]. Canned foods, fast foods, and alcoholic beverages have also been linked with FD symptoms [33].

Patients with FD commonly report food hypersensitivities to wheat and gluten, a protein found in wheat. According to an Australian population-based study [34], 29% of people with FD avoided gluten, and self-reported gluten sensitivity had a significant association with FD diagnosis. In some cases, gluten-free diets have helped FD patients to improve their symptoms. Furthermore, in a randomized double-blind placebo-controlled trial in FD patients, the application of a gluten-free diet resulted in an improvement of symptoms in 35% of patients, yet only 18% of those patients were confirmed to have this sensitivity to gluten, with symptoms reoccurring following a blind gluten challenge [35]. This suggests that other components in wheat-based foods, such as fructans, may induce FD symptoms.

In sixty patients with FD, a 1-month retrospective food consumption frequency questionnaire was used to examine fifty-one foods that may stimulate FD symptoms [36]. The consumption of broccoli, radish, celery, green olives, and olive oil was lower in patients with increased postprandial fullness. Participants who had more pain in the stomach reported a lower consumption of dried fruits, green olives, butter, fast food, and alcohol, but the consumption of sunflower oil was higher. Overall, foods high in fat are often blamed by patients with FD for making their symptoms worse. Unknown pathophysiological mechanisms are responsible for high-fat foods' ability to cause FD symptoms. However, cholecystokinin seems to be an important mediator [37]. Consuming fatty foods can slow down gastric emptying, disrupt gastric motility, and make dyspeptic patients feel fuller after meals [38]. Despite the known evidence regarding fatty foods, almonds do not appear to worsen FD symptoms as expected [38]. This effect may be related to the high content of tryptophan, a precursor of serotonin. Serotonin (5-hydroxytryptamine) is a key neurotransmitter involved in the regulation of gastrointestinal motility and sensory function. Indeed, stimulation of 5-HT1 serotonergic receptors induces smooth gastric muscle contractions that enhances gastric emptying, and appears to improve abdominal symptoms in patients [39]. However, when examining the brain activity using functional magnetic resonance imaging (fMRI) following the consumption of high- and low-fat foods with accurate or inaccurate fat information, researchers discovered that FD patients displayed more pronounced FD symptoms than healthy controls. These symptoms were less relieved following the consumption of high-fat yogurt than low-fat yogurt, regardless of the actual fat content. This suggests that low-fat foods may have a placebo effect or that high-fat foods may have a nocebo effect on symptom expression [40]. Interestingly, extra-virgin olive oil enriched with probiotics or antioxidants incorporated blindly into the regular diet of subjects with FD for seven days, induced a significant improvement in dyspeptic symptoms in those receiving the probiotic- or antioxidant-enriched oil diet, with the probiotic-enriched oil having a greater impact [41].

The relationship between coffee consumption and gastroesophageal reflux has been widely investigated [42–44]. Coffee has been considered a beverage that should be avoided, as it has been shown to worsen the symptoms of FD [8,26]. Chinese people with FD have been reported to be more likely to have a preference for coffee [45]. In the study by Correia et al., the effects of removing and substituting caffeinated or decaffeinated coffee with a non-caffeinated coffee alternative in the diet of fifty-one patients with FD were investigated [46]. Using a self-reported questionnaire, this descriptive, quasi-experimental pre/post intervention study looked at the relationship between functional dyspepsia and non-caffeinated coffee consumption. A statistically significant reduction in FD symptoms was reported following a month of consuming the coffee substitute.

Table 1. Studies investigating foods that may trigger or alleviate symptoms of FD.

a/a	Ref. No	Foods/Nutrients Investigated	Design	Results	FD Symptoms
1	[26]	Commonly consumed foods	• Observational study; • Three-hundred and eighty-four patients with an FD diagnosis (Rome III diagnostic criteria), aged 39.16 ± 14 years, 39.6% men.	• Sausage and bologna, pickled foods and fruits, vinegar, soft drinks, grains, tea, salty foods, pizza, watermelon, red pepper, macaroni (pasta), and fatty oils; • Apples, rice, rock candy, bread, caraway seeds, dates, honey, yogurt, quinces, almonds, and walnuts.	↑ ↓
2	[27]	Capsaicin containing capsule	• Randomized, double-blind, placebo-controlled trial; • Seventy-three patients with FD (Rome II diagnostic criteria), aged 39.3 ± 11.0 years, 26% men; • Intervention group ($n = 42$) received a capsaicin containing capsule (0.75 mg), and placebo group ($n = 31$).	• Capsaicin vs. placebo group: ✓ Positive test in upper gastrointestinal symptoms; ✓ Median symptom score.	↑
3	[28]	Capsaicin containing capsule	• Case-control study: 61 healthy controls and 54 FD patients (Rome II diagnostic criteria) received a capsaicin containing capsule (0.75 mg); • Placebo case-control study: 19 healthy controls and 13 FD patients (Rome II diagnostic criteria) received a placebo capsule.	• FD patients vs. controls and vs. placebo: ✓ Positive test in upper gastrointestinal symptoms; ✓ Median symptom score.	↑
4	[29]	Capsaicin	• Case-control study • 98 subjects with no upper abdominal symptoms and 109 patients with FD (Rome III diagnostic criteria)	• Transient receptor potential vanilloid-1 receptors (TRPV1) in FD; • The G315 polymorphism of the TRPV1 gene is inversely correlated with FD.	↑
5	[30]	Spicy foods	• Observational study; • One-hundred ant twenty-one FD patients (Rome III diagnostic criteria) of whom 35 carried TRPV1 CC and 28 carried GG genotypes.	• Stomach fullness and retching regardless of genotype.	↑
6	[31]	Commonly consumed foods	• Case-control study; • A total of 168 adults with FD (Rome III diagnostic criteria) and 135 healthy controls; • FD patients were categorized into epigastric pain syndrome (EP-FD), postprandial distress syndrome (PS-FD), and mixed (MX-FD) subgroups.	• FD patients vs. controls: ✓ fried and fatty foods, hot spices, and carbonated drinks. • Carbonated drinks in PS-FD group vs. other subgroups.	↑ ↑

Table 1. *Cont.*

a/a	Ref. No	Foods/Nutrients Investigated	Design	Results	FD Symptoms
7	[32]	Commonly consumed foods	• Case-control study; • Seven-hundred and fifty-nine university students categorized into the FD (Rome III diagnostic criteria) group (n = 128) and healthy group.	• FD patients vs. controls: ✓ Spicy, hot, raw, or cold foods and dairy. • Tea was associated with FD prevention.	↑
8	[33]	Commonly consumed foods	• Observational study; • One-hundred and eighty-four subjects participated in a 4-month study; • FD (Rome III diagnostic criteria) was present in 7.6%, and gastroesophageal reflux disease was present in 31.0%.	• Canned foods, fast foods, and alcoholic beverages in FD.	↑
9	[34]	Wheat	• Observational study; • A total of 3542 people were randomly selected from the Australian population.	• Self-reported wheat sensitivity in FD (Rome III diagnostic criteria).	↑
10	[35]	Gluten	• Observational study: 77 patients with refractory FD followed a gluten-free diet for 6 weeks; • Patients with ≥30% improvement (n = 27) participated in a randomized double-blind placebo-controlled crossover trial: the intervention group (n = 14) received gluten-free muffins for one week and the control group received gluten muffins for one week.	• Gluten free diet (observational study): ✓ Gastrointestinal symptoms; - Sixty-five percent did not respond, while 35% cases showed gastrointestinal symptoms improvement. • Gluten challenge (randomized trial): - Symptoms recurred in five cases suggesting the presence of non-celiac gluten sensitivity.	↔ ↔
11	[36]	List of foods that may stimulate FD	• Observational study; • Sixty patients with FD followed a gluten-free diet for 6 weeks (Rome IV diagnostic criteria).	• Consumption of broccoli, radish, celery, green olives, and olive oil in subjects with postprandial fullness; • Consumption of alcohol, dried fruits, green olives, butter, and fast food in subjects with stomach pain; • Consumption of sunflower oil in subjects with stomach pain.	↓ ↓ ↓

Table 1. *Cont.*

a/a	Ref. No	Foods/Nutrients Investigated	Design	Results	FD Symptoms
12	[38]	Commonly consumed foods	• Case-control study; • Forty-one patients with FD (30 women, 11 men; mean age: 46 ± 12 years) and 30 healthy volunteers (25 women, five men; mean age: 35 ± 12 years).	• FD patients vs. controls: ✓ Carbonated drinks, fried foods, red meat, sausage, coffee, pasta (macaroni, lasagna), milk, cheese, sweets, pepper, bananas, pineapple, cucumber, orange, beans, bread, and spicy foods; ✓ Heartburn: pepper and coffee; ✓ Bloating: carbonated drinks, onions, beans, and bananas; ✓ Epigastric burning: coffee, cheese, onions, pepper, milk, chocolate, and pineapple; ✓ Fullness: red meat, bananas, bread, cakes, pasta, sausage, fried foods, beans, mayonnaise, milk, chocolate, eggs, sweets, and oranges.	↑ ↑ ↑ ↑ ↑
13	[40]	Four different yogurts differentiated in fat composition label	• Case-control cross-over study; • Twelve FD (Rome III diagnostic criteria) patients (five men, aged 46.46 ± 5.64 years) and 14 age- and body mass index-controlled healthy subjects (five men, aged 45.79 ± 4.71 years); • Subjects consumed four different yogurts during four separate visits: high-fat yogurt with "high fat" label (HH), high-fat yogurt with "low fat" label (HL), low-fat yogurt with "high fat" label (LH), and low-fat yogurt with "low fat" label (LL).	• FD patients vs. controls: ✓ Burning, discomfort, pain, bloating, nausea, and fullness at baseline; ✓ Satiation, discomfort, burning, and abdominal pain for high fat label vs. low fat label; ✓ Amplitude of low-frequency fluctuations (ALFFs) regardless of the type of yogurt consumed; ✓ Functional connectivity from the insula to the occipital cortex (I-O) after high fat ingestion; ✓ Functional connectivity from the insula to the occipital cortex (I-O) after low fat ingestion; ✓ Functional connectivity from the insula to the precuneus (I-P) after ingestion of low fat–labeled yogurt. • In FD patients: ✓ I-O functional connectivity was negatively correlated with nausea; ✓ I-P functional connectivity was negatively correlated with FD symptom intensity, food craving, and depression.	↑ ↑ ↑ ↑ ↓ ↑

Table 1. *Cont.*

a/a	Ref. No	Foods/Nutrients Investigated	Design	Results	FD Symptoms
14	[41]	Extra virgin olive oil enriched with antioxidants or probiotics	• Randomized controlled trial; • Eight FD patients (Rome III diagnostic criteria); • Each subject received two vials of 9 mL (equal to the daily food requirement) per day either of: (a) extra virgin olive oil, (b) extra virgin olive oil enriched with antioxidants, or (c) oil enriched with probiotics, to be added to the meals for 7 days.	• Probiotic olive oil vs. plain olive oil: ✓ Nausea; ✓ Pain/discomfort. • Probiotic vs. antioxidant rich olive oil: ✓ Belching; ✓ Postprandial gastric distension and fullness. • Probiotic olive oil was more effective than antioxidant olive oil.	 ↓ ↓ ↓ ↓
15	[45]	Commonly consumed foods	• Observational study; • A total of 1304 adults residents were recruited: 165 had existing organic dyspepsia, 203 were diagnosed with FD (Rome III diagnostic criteria), and the other 936 were healthy controls.	• FD patients vs. controls: ✓ Consumption of fatty foods, sweets, and coffee.	 ↑
16	[46]	Non-caffeinated coffee substitute	• A quantitative study; • Fifty-one patients, aged 29–83 years, and diagnosed with FD; • Each participant received a commercially available 7-ounce bottle of a non-caffeinated coffee substitute, consisting of roasted barley as the main ingredient with roasted malt barley, roasted chicory, and roasted rye; • Participants were instructed to substitute their usual daily coffee consumption with the non-caffeinated coffee substitute for 1 month.	• Post vs. pre-intervention: ✓ Dyspepsia symptoms; ✓ Reflux, indigestion, diarrhea, and constipation.	 ↓ ↓

3. Dietary Patterns and Eating Behaviors; Proportions, Variety, Frequency, and Cooking Habits

The dietary patterns and behavioral parameters examined constitute a multidimensional construct, including diet composition spanning from alimentary selection to the subsequent development of healthy or disordered eating habits, and ranging from meal frequency and regularity to cooking techniques. To the best of our knowledge, only a small percentage of studies provide a solution for to what extent do specific dietary behaviors influence and ultimately cause susceptibility to dyspeptic symptoms (Table 2).

Despite the difficulty of adhering to a FODMAP diet, Staudacher et al. noticed that individuals with underlying symptoms of FD and coexisting irritable bowel syndrome, and who followed a diet limited to FODMAPs, showed a higher reduction in the epigastric and total symptom score in comparison with those who received personalized dietary counseling at the discretion of the dietitians who participated in the research [47]. Similarly to Staudacher et al., Goyal and his team studied the effects of a low FODMAP diet in a panel of people with persistent FD, and subsequently they compared them to a slightly restricted standard diet that included an avoidance to spices, soft drinks, tea, coffee, and alcohol. Both interventions showed a significant remission of symptom severity and quality of life improvement, although one sub-group analysis of volunteers with postprandial distress syndrome or bloating were better responders to a low FODMAP regime [48].

Schnabel et al., within the NutriNet-Santé prospective observational cohort study, examined the interaction of consuming ultra-processed foods in people with various functional gastrointestinal disorders. Results showed that subjects suffering from FD and concurrent irritable bowel syndrome have a higher risk of being affected because of the increased daily amount of ultra-processed foods consumed [49]. In 2034 children and adolescents with functional gastrointestinal disorders, including FD, approximately 90% of the subjects reported fast food consumption, and there was a significantly higher prevalence of a history of functional disorder in fast food consumers, compared to non-consumers [50]. The increased risk of functional disorders was associated with the regular fast food intake. Ultra-processed and fast foods consist of the core of the Westernized dietary pattern. The popularity of ultra-processed and fast foods is partly due to their inexpensive and easy to find nature. However, the high fat content, particularly the trans fatty acid content and the presence of additives or reaction products owed to processing [51–54] may be linked with the augmentation of FD symptoms. Similarly, the prevalence of FD in Asia has increased during the last decades due to the change towards a more Westernized diet, higher in fat- and lower in carbohydrate-rich foods, whereas the spicy foods in Asian diets are associated with a higher risk of developing FD [55].

Table 2. Studies investigating dietary patterns and eating behaviors associated with FD.

a/a	Ref. No	Dietary Pattern/Eating Behavior	Design	Results	FD Symptoms
1	[47]	Low FODMAP	• Randomized-controlled study; • Fifty-nine patients with FD (Rome IV diagnostic criteria); • Individuals received either low FODMAP (n = 40) or standard dietary advice as per clinical judgment from the consulting dietitian (n = 19).	• Low FODMAP vs. control: ✓ Epigastric score; ✓ Total symptom score.	 ↓ ↓
2	[49]	Ultra-processed food rich diet	• Observational study; • A total of 44,551 adults (> 45 years) from the French NutriNet-Santé Study.	• Dietary factors associated with FD: ✓ Ultra-processed food rich diet.	 ↑
3	[50]	Fast food diets	• Observational study; • A total of 2034 adolescents (aged 12–19 years) from the Nutrition and Health Survey in Taiwan (NAHSIT) with functional gastrointestinal disorders (Rome III diagnostic criteria).	• Dietary factors associated with FD development: ✓ Fast foods; ✓ Low fiber intake and frozen desserts in the diet.	 ↑ ↑
4	[56]	Fruit and vegetable intake	• Observational study; • A total of 3362 middle-age participants of whom 14.5% was diagnosed with FD (Rome III diagnostic criteria).	• Dietary factors associated with FD development: ✓ Fruit consumption; ✓ Vegetable consumption. • Fruit consumption was associated with a lower risk of early satiation.	 ↓ ↔
5	[57]	Mediterranean diet	• Observational study; • A toatl of 1134 subjects (age 17–83 years); • Seven-hundred and nineteen (63.4%) were healthy controls, 172 (13.3%) patients had irritable bowel syndrome (IBS), and 243 (23.3%) had FD (Rome III diagnostic criteria).	• Dietary factors associated with FD development: ✓ Low adherence to a Mediterranean diet.	 ↑
6	[31]	Eating behavior	• Case-control study; • A total of 168 adults with FD (Rome III diagnostic criteria) and 135 healthy controls; • FD patients were categorized into epigastric pain syndrome (EP-FD), postprandial distress syndrome (PS-FD), and mixed (MX-FD) subgroups.	• Symptomatology in FD patients vs. controls: ✓ Meal frequency; ✓ Snacking; ✓ Intra-meal fluid consumption.	 ↔ ↔ ↑
7	[36]	Eating behavior	• Observational study; • Sixty patients with FD followed a gluten-free diet for 6 weeks (Rome IV diagnostic criteria).	• Symptomatology in FD patients vs. controls: ✓ Meal frequency; ✓ Roasting.	 ↔ ↑

Table 2. *Cont.*

a/a	Ref. No	Dietary Pattern/Eating Behavior	Design	Results	FD Symptoms
8	[45]	Eating behavior	• Observational study: • A total of 1304 adults residents were recruited: 165 had existing organic dyspepsia, 203 were diagnosed with FD (Rome III diagnostic criteria), and the other 936 were healthy controls.	• Dietary habits in FD vs. controls: ✓ Irregular mealtime, meal frequency, night snacking, skipping breakfast, and dining out.	↑
9	[58]	Meal frequency	• Observational study; • A total of 4763 individuals from the general adult population.	• Dietary factors associated with FD development: ✓ >3 meals/day; ✓ Three to five snacks/day.	↓ ↓
10	[59]	Eating behavior	• Observational study: • A total of 8923 Japanese university students: 168 subjects had FD (Rome III diagnostic criteria) and 8745 were healthy controls.	• Dietary factors associated with FD development: ✓ Skipping breakfast/lunch; ✓ Skipping dinner, extra meals (snacks), or midnight snacks; ✓ Frequency of meals.	↑ ↔ ↓
11	[38]	Eating behavior	• Case-control study; • Forty-one patients with FD (30 women, 11 men; mean age: 46 ± 12 years) and 30 healthy volunteers (25 women, 5 men; mean age: 35 ± 12 years).	• Dietary factors associated with FD development: ✓ Overnight fast; ✓ Frequency of meals; ✓ Daytime fast, fast eating, or going to bed soon after the meal.	↑ ↓ ↔
12	[60]	Eating behavior	• Observational study; • Eighty-nine adult women: 11 subjects had FD (Rome III diagnostic criteria), six subjects with gastroesophageal reflux disease, and 72 healthy controls.	• Dietary factors associated with FD development: ✓ Fast eating.	↑

Quite the opposite, a dietary pattern that appears to be closely related to FD symptom amelioration is one that contains fruits and vegetables. According to the cross-sectional study by Tabibian et al., people on a diet rich in fruit have a 32% lower risk of FD, as well as a lower risk of early satiation and postprandial fullness in comparison with those on a low fruit diet. Moreover, a high vegetable intake seems to have a beneficial effect on FD, however, only in male subjects [56]. The same pattern was observed by Zito and his associates, who also emphasized through their results the importance of adopting a balanced diet, such as the Mediterranean diet due to its preventive function against FD. Specifically, a low adherence to the Mediterranean diet is associated with FD mainly in younger people [57].

Regarding the eating habits of FD subjects, a confusion about the frequency of main meals and snacks per day prevails in the bibliography. Göktaş et al. did not manage to demonstrate a significant difference in the frequency of main meals among FD and control subjects, as both groups had three main meals per day (68.5% and 70.4%, respectively). Likewise for snack consumption, researchers found no difference in their frequency between the two groups [31]. These findings corroborate those of Çolak et al., who recently extrapolated the conclusion that meal frequency had no impact on the triggering of FD symptoms among FD subjects [36]. Moreover, in a study involving 1139 volunteers, of whom 936 were healthy subjects and 203 had been diagnosed with FD, Xu et al. [45] demonstrated that the vast majority of the latter had by far more unhealthy nutritional habits than the control group (75.86% versus 37.50%, respectively; $p < 0.001$). It appears that the number of meals during the day shows an inversely independent relationship with the occurrence of FD. Hassanzadeh et al. [58] concluded that regularly consuming at least three main meals per day was not only indissolubly related with a 52% lower chance of developing FD compared to eating one meal, but also was inversely associated with the prevalence of early satiety. Similar findings were observed in Yamamoto et al.'s [59] cross-sectional study, where the risk of FD in subjects who ate one, two, and three meals daily was 4.8%, 2.2%, and 1.7%, respectively.

Correspondingly, the literature review shows that, three to five snacks per day are found to be a determinant factor for low FD incidence data, inhibiting alongside both the occurrence of postprandial discomfort (42%) and epigastric pain symptoms (43%) [58].

Concerning late-night snacking, the findings are controversial, with Xu et al. arguing that it is an independent triggering factor of postprandial fullness, while Yamamoto et al. state that there is no correlation with the development of FD. Carvalho et al. claim that the duration of the overnight fast was significantly longer in subjects with FD than in the controls, as the former tended to eat dinner earlier. On the contrary, this difference is equated to daytime fasting [38].

Based on the current literature, another causative agent for the generation of dyspeptic symptoms is the skipping of one or even more main meals during the day, and depending on which meal (breakfast, lunch, dinner), there are different impacts on FD and its subtypes. In particular, Yamamoto et al. claim that FD subjects were more prone to skip lunch and dinner than the controls, and the habit of omitting breakfast and/or lunch was independently inversely related to a high incidence of FD symptoms (breakfast: adjusted OR, 1.60 [95% CI, 1.10–2.32] and lunch: adjusted OR, 2.52 [95% CI, 1.04–5.18]). Contrarily, researchers did not prove any relation between FD prevalence and skipping dinner [59]. A more comprehensive description can be found in Xu et al.'s study, that not only confirms the analogous findings, but also reveals the close association between skipping breakfast and PDS (18/203) and EPS (13/203) subtypes, but not with their overlapping type (9/203) [45].

Apropos irregular meals and the tendency of dining out, which are positively connected to the generation of FD symptoms, investigators claim that both correlated with PDS, but only an erratic dinner time was related to EPS and the overlapping subtype [45]. Furthermore, in respect to the possible correlation between sleep and the development of FD symptoms, researchers did not prove any statistically significant difference between FD and control subjects who go to bed following a meal [38].

In regard to meal time or chewing efficiency, and their role in triggering FD symptoms, Çolak et al. [36] claim that this assertion has no merit, although some clinical studies demonstrate that subjects with FD are more likely to self-report themselves as fast eaters than the controls [38,60]. In the same cross-sectional study, the findings indicated that roasting was the most usual cooking method in subjects experiencing the postprandial fullness symptoms [36]. Furthermore, compared to both PDS and EPS subtypes, as well as the control group, the overlapping group appeared to have a significantly higher intra-meal fluid consumption [31] (Figure 2).

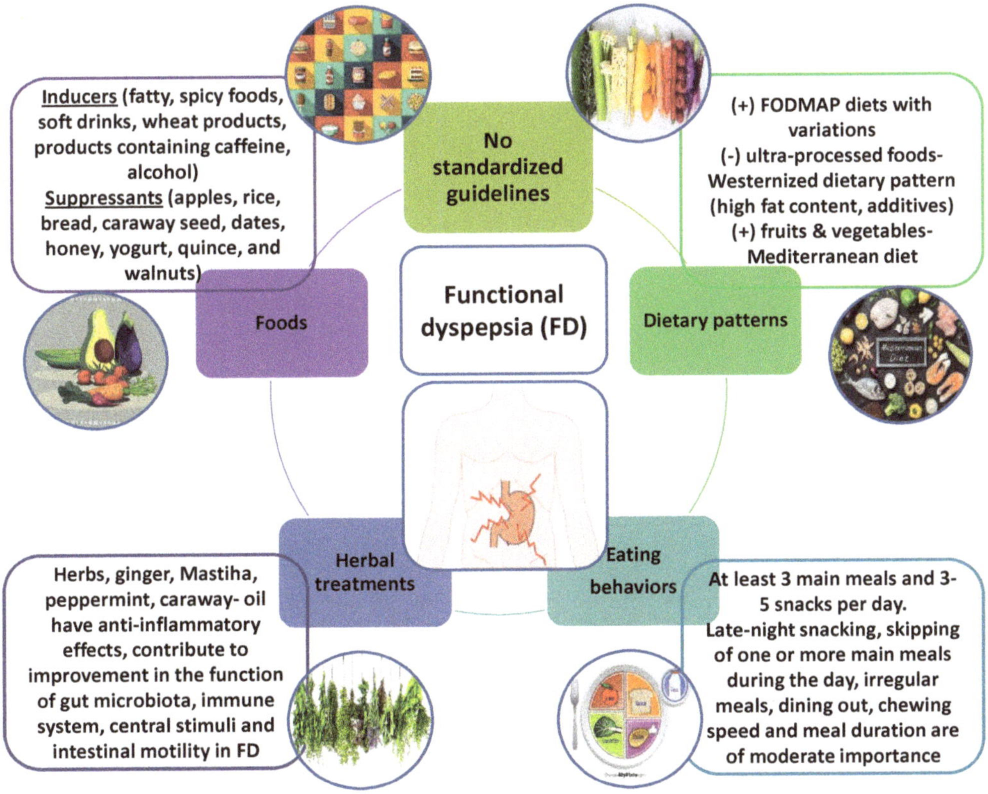

Figure 2. Functional dyspepsia (FD) is considered one of the most common disorders in clinical practice. Standardized guidelines for the nutritional approach and eating habits of patients with FD do not exist. However, assessing the impact of foods, dietary patterns, eating behaviors and botanicals on FD can help, on the one hand, to elucidate the pathophysiology of the disease, and on the other hand, to improve palliative treatments.

4. Conclusions

Well-structured and standardized guidelines for the nutritional approach and eating habits of patients with FD do not exist, and randomized controlled trials are few, while most available evidence comes from observational studies. Hence, although the contribution of specific foods identified as triggers by FD patients varies, some causal relationships between specific foods and symptoms have been demonstrated. The retrospective nature of most studies and the lack of a standardized method for verifying the food-symptom association accounts for the difficulty in accumulating a definitive list of foods to avoid or foods to favor. Some of the most frequently reported foods to avoid are fatty foods, processed foods, and wheat products. Alike, processed foods and overall the Western dietary pattern, which includes fatty foods, as these are considered inducers of FD symptoms. Eating small and frequent meals may be a reasonable suggestion to reduce symptoms, however the evidence is inadequate and inconsistent. In the future, the improvement in our understanding of the pathophysiological mechanisms underlying FD might lead to well-designed and target-driven clinical trials on the effects of foods, dietary patterns, or specific eating habits in the management of FD.

Author Contributions: All authors contributed equally to writing this review article. All authors have read and agreed to the published version of the manuscript.

Funding: This work received funding by the Chios Mastiha Research and Development Center.

Institutional Review Board Statement: Not applicable.

Informed Consent Statement: Not applicable.

Data Availability Statement: No new data were created.

Conflicts of Interest: The authors declare no conflict of interest.

References

1. Wauters, L.; Dickman, R.; Drug, V.; Mulak, A.; Serra, J.; Enck, P.; Tack, J.; Accarino, A.; Barbara, G.; Bor, S.; et al. United European Gastroenterology (UEG) and European Society for Neurogastroenterology and Motility (ESNM) consensus on functional dyspepsia. *United Eur. Gastroenterol. J.* **2021**, *9*, 307–331. [CrossRef] [PubMed]
2. Drago, L.; Meroni, G.; Pistone, D.; Pasquale, L.; Milazzo, G.; Monica, F.; Aragona, S.; Ficano, L.; Vassallo, R. Evaluation of main functional dyspepsia symptoms after probiotic administration in patients receiving conventional pharmacological therapies. *J. Int. Med. Res.* **2021**, *49*, 0300060520982657. [CrossRef]
3. Wauters, L.; Talley, N.J.; Walker, M.M.; Tack, J.; Vanuytsel, T. Novel concepts in the pathophysiology and treatment of functional dyspepsia. *Gut* **2019**, *69*, 591–600. [CrossRef] [PubMed]
4. Holtmann, G.; Shah, A.; Morrison, M. Pathophysiology of Functional Gastrointestinal Disorders: A Holistic Overview. *Dig. Dis.* **2017**, *35* (Suppl. S1), 5–13. [CrossRef]
5. Wauters, L.; Li, H.; Talley, N.J. Editorial: Disruption of the Microbiota-Gut-Brain Axis in Functional Dyspepsia and Gastroparesis: Mechanisms and Clinical Implications. *Front. Neurosci.* **2022**, *16*, 941810. [CrossRef] [PubMed]
6. Komori, K.; Ihara, E.; Minoda, Y.; Ogino, H.; Sasaki, T.; Fujiwara, M.; Oda, Y.; Ogawa, Y. The Altered Mucosal Barrier Function in the Duodenum Plays a Role in the Pathogenesis of Functional Dyspepsia. *Dig. Dis. Sci.* **2019**, *64*, 3228–3239. [CrossRef]
7. Wauters, L.; Burns, G.; Ceulemans, M.; Walker, M.M.; Vanuytsel, T.; Keely, S.; Talley, N.J. Duodenal inflammation: An emerging target for functional dyspepsia? *Expert Opin. Ther. Targets* **2020**, *24*, 511–523. [CrossRef]
8. Stanghellini, V.; Chan, F.K.L.; Hasler, W.L.; Malagelada, J.R.; Suzuki, H.; Tack, J.; Talley, N.J. Gastroduodenal Disorders. *Gastroenterology* **2016**, *150*, 1380–1392. [CrossRef]
9. Duboc, H.; Latrache, S.; Nebunu, N.; Coffin, B. The Role of Diet in Functional Dyspepsia Management. *Front. Psychiatry* **2020**, *11*, 23. [CrossRef]
10. Moayyedi, P.M.; Lacy, B.E.; Andrews, C.N.; Enns, R.A.; Howden, C.W.; Vakil, N. ACG and CAG Clinical Guideline: Management of Dyspepsia. *Am. J. Gastroenterol.* **2017**, *112*, 988–1013. [CrossRef]
11. Tack, J.; Janssen, P.; Masaoka, T.; Farré, R.; Van Oudenhove, L. Efficacy of Buspirone, a Fundus-Relaxing Drug, in Patients With Functional Dyspepsia. *Clin. Gastroenterol. Hepatol.* **2012**, *10*, 1239–1245. [CrossRef]
12. Pittayanon, R.; Yuan, Y.; Bollegala, N.P.; Khanna, R.; Lacy, B.E.; Andrews, C.N.; Leontiadis, G.I.; Moayyedi, P. Prokinetics for Functional Dyspepsia. *Am. J. Gastroenterol.* **2019**, *114*, 233–243. [CrossRef]
13. Talley, N.J.; Locke, G.R.; Saito, Y.A.; Almazar, A.E.; Bouras, E.P.; Howden, C.W.; Lacy, B.E.; DiBaise, J.K.; Prather, C.M.; Abraham, B.P.; et al. Effect of Amitriptyline and Escitalopram on Functional Dyspepsia: A Multicenter, Randomized Controlled Study. *Gastroenterology* **2015**, *149*, 340–349.e2. [CrossRef]
14. Meyrat, P.; Safroneeva, E.; Schoepfer, A.M. Rifaximin treatment for the irritable bowel syndrome with a positive lactulose hydrogen breath test improves symptoms for at least 3 months. *Aliment. Pharmacol.Ther.* **2012**, *36*, 1084–1093. [CrossRef]
15. Bordbar, G.; Miri, M.B.; Omidi, M.; Shoja, S.; Akhavan, M. Comparison of a Novel Herbal Medicine and Omeprazole in the Treatment of Functional Dyspepsia: A Randomized Double-Blinded Clinical Trial. *Gastroenterol. Res. Pract.* **2020**, *2020*, 5152736. [CrossRef] [PubMed]
16. Kim, Y.S.; Kim, J.-W.; Ha, N.-Y.; Kim, J.; Ryu, H.S. Herbal Therapies in Functional Gastrointestinal Disorders: A Narrative Review and Clinical Implication. *Front. Psychiatry* **2020**, *11*, 601. [CrossRef] [PubMed]
17. Heiran, A.; Bagheri Lankarani, K.; Bradley, R.; Simab, A.; Pasalar, M. Efficacy of herbal treatments for functional dyspepsia: A systematic review and meta-analysis of randomized clinical trials. *Phytother. Res.* **2021**, *36*, 686–704. [CrossRef]

18. Ho, L.; Zhong, C.C.W.; Wong, C.H.L.; Wu, J.C.Y.; Chan, K.K.H.; Wu, I.X.Y.; Leung, T.H.; Chung, V.C.H. Chinese herbal medicine for functional dyspepsia: A network meta-analysis of prokinetic-controlled randomised trials. *Chin. Med.* **2021**, *16*, 140. [CrossRef]
19. Yoon, J.Y.; Ko, S.-J.; Park, J.-W.; Cha, J.M. Complementary and alternative medicine for functional dyspepsia: An Asian perspective. *Medicine* **2022**, *101*, e30077. [CrossRef]
20. Ko, S.; Park, J.; Kim, M.; Kim, J.; Park, J. Effects of the herbal medicine Rikkunshito, for functional dyspepsia: A systematic review and meta-analysis. *J. Gastroenterol. Hepatol.* **2021**, *36*, 64–74. [CrossRef] [PubMed]
21. Di Pierro, F.; Giovannone, M.; Saponara, M.; Ivaldi, L. Effectiveness of a nutraceutical supplement containing highly standardized perilla and ginger extracts in patients with functional dyspepsia. *Minerva Gastroenterol. Dietol.* **2020**, *66*, 35–40. [CrossRef] [PubMed]
22. Giacosa, A.; Guido, D.; Grassi, M.; Riva, A.; Morazzoni, P.; Bombardelli, E.; Perna, S.; Faliva, M.A.; Rondanelli, M. The Effect of Ginger (Zingiber officinalis) and Artichoke (Cynara cardunculus) Extract Supplementation on Functional Dyspepsia: A Randomised, Double-Blind, and Placebo-Controlled Clinical Trial. *Evid.-Based Complement. Altern. Med.* **2015**, *2015*, 915087. [CrossRef] [PubMed]
23. Hu, M.-L. Effect of ginger on gastric motility and symptoms of functional dyspepsia. *World J. Gastroenterol.* **2011**, *17*, 105. [CrossRef]
24. Li, J.; Lv, L.; Zhang, J.; Xu, L.; Zeng, E.; Zhang, Z.; Wang, F.; Tang, X. A Combination of Peppermint Oil and Caraway Oil for the Treatment of Functional Dyspepsia: A Systematic Review and Meta-Analysis. *Evid. -Based Complement. Altern. Med.* **2019**, *2019*, 7654947. [CrossRef]
25. Dabos, K.J.; Sfika, E.; Vlatta, L.J.; Frantzi, D.; Amygdalos, G.I.; Giannikopoulos, G. Is Chios mastic gum effective in the treatment of functional dyspepsia? A prospective randomised double-blind placebo controlled trial. *J. Ethnopharmacol.* **2010**, *127*, 205–209. [CrossRef]
26. Akhondi-Meybodi, M.; Aghaei, M.A.; Hashemian, Z. The role of diet in the management of non-ulcer dyspepsia. *Middle East J. Dig. Dis.* **2015**, *7*, 19–24.
27. Führer, M.; Vogelsang, H.; Hammer, J. A Placebo Controlled Trial of an Oral Capsaicin Load in Patients With Functional Dyspepsia. *Gastroenterology* **2011**, *23*, 918-e397. [CrossRef]
28. Hammer, J.; Führer, M.; Pipal, L.; Matiasek, J. Hypersensitivity for capsaicin in patients with functional dyspepsia. *Neurogastroenterol. Motil.* **2008**, *20*, 125–133. [CrossRef] [PubMed]
29. Tahara, T.; Shibata, T.; Nakamura, M.; Yamashita, H.; Yoshioka, D.; Hirata, I.; Arisawa, T. Homozygous TRPV1 315C Influences the Susceptibility to Functional Dyspepsia. *J. Clin. Gastroenterol.* **2010**, *44*, e1–e7. [CrossRef]
30. Lee, S.Y.; Masaoka, T.; Han, H.S.; Matsuzaki, J.; Hong, M.J.; Fukuhara, S.; Choi, H.S.; Suzuki, H. A prospective study on symptom generation according to spicy food intake and TRPV1 genotypes in functional dyspepsia patients. *Neurogastroenterol. Motil.* **2016**, *28*, 1401–1408. [CrossRef]
31. Göktaş, Z.; Köklü, S.; Dikmen, D.; Öztürk, Ö.; Yılmaz, B.; Asıl, M.; Korkmaz, H.; Tuna, Y.; Kekilli, M.; Karamanoğlu Aksoy, E.; et al. Nutritional habits in functional dyspepsia and its subgroups: A comparative study. *Scand. J. Gastroenterol.* **2016**, *51*, 903–907. [CrossRef] [PubMed]
32. Huang, Z.-p.; Wang, K.; Duan, Y.-h.; Yang, G. Correlation between lifestyle and social factors in functional dyspepsia among college freshmen. *J. Int. Med. Res.* **2020**, *48*, 1–8. [CrossRef]
33. Chirila, I.; Morariu, I.D.; Barboi, O.B.; Drug, V.L. The role of diet in the overlap between gastroesophageal reflux disease and functional dyspepsia. *Turk. J. Gastroenterol.* **2016**, *27*, 73–80. [CrossRef] [PubMed]
34. Potter, M.D.E.; Walker, M.M.; Jones, M.P.; Koloski, N.A.; Keely, S.; Talley, N.J. Wheat Intolerance and Chronic Gastrointestinal Symptoms in an Australian Population-based Study: Association Between Wheat Sensitivity, Celiac Disease and Functional Gastrointestinal Disorders. *Am. J. Gastroenterol.* **2018**, *113*, 1036–1044. [CrossRef]
35. Shahbazkhani, B.; Fanaeian, M.M.; Farahvash, M.J.; Aletaha, N.; Alborzi, F.; Elli, L.; Shahbazkhani, A.; Zebardast, J.; Rostami-Nejad, M. Prevalence of Non-Celiac Gluten Sensitivity in Patients with Refractory Functional Dyspepsia: A Randomized Double-blind Placebo Controlled Trial. *Sci. Rep.* **2020**, *10*, 2401. [CrossRef] [PubMed]
36. Colak, H.; Gunes, F.E.; Ozen Alahdab, Y.; Karakoyun, B. Investigation of Eating Habits in Patients with Functional Dyspepsia. *Turk. J. Gastroenterol.* **2022**, *5*, 11. [CrossRef]
37. Azadbakht, L.; Khodarahm, M. Dietary fat intake and functional dyspepsia. *Adv. Biomed. Res.* **2016**, *5*, 76. [CrossRef]
38. Carvalho, R.V.B.; Lorena, S.L.S.; de Souza Almeida, J.R.; Mesquita, M.A. Food Intolerance, Diet Composition, and Eating Patterns in Functional Dyspepsia Patients. *Dig. Dis. Sci.* **2009**, *55*, 60–65. [CrossRef]
39. Mawe, G.M.; Hoffman, J.M. Serotonin signalling in the gut—Functions, dysfunctions and therapeutic targets. *Nat. Rev. Gastroenterol. Hepatol.* **2013**, *10*, 473–486. [CrossRef]
40. Lee, I.-S.; Kullmann, S.; Scheffler, K.; Preissl, H.; Enck, P. Fat label compared with fat content: Gastrointestinal symptoms and brain activity in functional dyspepsia patients and healthy controls. *Am. J. Clin. Nutr.* **2018**, *108*, 127–135. [CrossRef]
41. Ianiro, G.; Pizzoferrato, M.; Franceschi, F.; Tarullo, A.; Luisi, T.; Gasbarrini, G. Effect of an extra-virgin olive oil enriched with probiotics or antioxidants on functional dyspepsia: A pilot study. *Eur. Rev. Med. Pharmacol. Sci.* **2013**, *17*, 2085–2090. [PubMed]
42. Mehta, R.S.; Song, M.; Staller, K.; Chan, A.T. Association Between Beverage Intake and Incidence of Gastroesophageal Reflux Symptoms. *Clin. Gastroenterol. Hepatol.* **2020**, *18*, 2226–2233.e4. [CrossRef] [PubMed]

43. Yuan, L.-Z.; Yi, P.; Wang, G.-S.; Tan, S.-Y.; Huang, G.-M.; Qi, L.-Z.; Jia, Y.; Wang, F. Lifestyle intervention for gastroesophageal reflux disease: A national multicenter survey of lifestyle factor effects on gastroesophageal reflux disease in China. *Ther. Adv. Gastroenterol.* **2019**, *12*, 1–12. [CrossRef]
44. Wang, C.-C.; Wei, T.-Y.; Hsueh, P.-H.; Wen, S.-H.; Chen, C.-L. The role of tea and coffee in the development of gastroesophageal reflux disease. *Tzu Chi Med. J.* **2019**, *31*, 169. [CrossRef] [PubMed]
45. Xu, J.-H.; Lai, Y.; Zhuang, L.-P.; Huang, C.-Z.; Li, C.-Q.; Chen, Q.-K.; Yu, T. Certain Dietary Habits Contribute to the Functional Dyspepsia in South China Rural Area. *Med. Sci. Monit.* **2017**, *23*, 3942–3951. [CrossRef]
46. Correia, H.; Peneiras, S.; Levchook, N.; Peneiras, E.; Levchook, T.; Nayyar, J. Effects of a non-caffeinated coffee substitute on functional dyspepsia. *Clin. Nutr. ESPEN* **2020**, *41*, 412–416. [CrossRef]
47. Staudacher, H.M.; Nevin, A.N.; Duff, C.; Kendall, B.J.; Holtmann, G.J. Epigastric symptom response to low FODMAP dietary advice compared with standard dietetic advice in individuals with functional dyspepsia. *Neurogastroenterol. Motil.* **2021**, *33*, e14148. [CrossRef]
48. Goyal, O.; Nohria, S.; Batta, S.; Dhaliwal, A.; Goyal, P.; Sood, A. Low FODMAP diet versus traditional dietary advice for functional dyspepsia: A randomized controlled trial. *J. Gastroenterol. Hepatol.* **2021**, *37*, 301–309. [CrossRef]
49. Schnabel, L.; Kesse-Guyot, E.; Allès, B.; Touvier, M.; Srour, B.; Hercberg, S.; Buscail, C.; Julia, C. Association Between Ultra-processed Food Consumption and Risk of Mortality Among Middle-aged Adults in France. *JAMA Intern. Med.* **2019**, *179*, 490. [CrossRef]
50. Shau, J.P.; Chen, P.H.; Chan, C.F.; Hsu, Y.C.; Wu, T.C.; James, F.E.; Pan, W.H. Fast foods–are they a risk factor for functional gastrointestinal disorders? *Asia Pac J Clin Nutr.* **2016**, *25*, 393–401. [CrossRef]
51. Feinle-Bisset, C.; Azpiroz, F. Dietary Lipids and Functional Gastrointestinal Disorders. *Am. J. Gastroenterol.* **2013**, *108*, 737–747. [CrossRef]
52. Poti, J.M.; Mendez, M.A.; Ng, S.W.; Popkin, B.M. Is the degree of food processing and convenience linked with the nutritional quality of foods purchased by US households? *Am. J. Clin. Nutr.* **2015**, *101*, 1251–1262. [CrossRef]
53. Csáki, K.F. Synthetic surfactant food additives can cause intestinal barrier dysfunction. *Med. Hypotheses* **2011**, *76*, 676–681. [CrossRef] [PubMed]
54. Tessier, F.J.; Birlouez-Aragon, I. Health effects of dietary Maillard reaction products: The results of ICARE and other studies. *Amino Acids* **2010**, *42*, 1119–1131. [CrossRef] [PubMed]
55. Yap, P.; Mahadeva, S.; Goh, K.-L. The Influence of Cultural Habits on the Changing Pattern of Functional Dyspepsia. *Dig. Dis.* **2014**, *32*, 217–221. [CrossRef] [PubMed]
56. Tabibian, S.; Hajhashemy, Z.; Shaabani, P.; Saneei, P.; Keshteli, A.H.; Esmaillzadeh, A.; Adibi, P. The relationship between fruit and vegetable intake with functional dyspepsia in adults. *Neurogastroenterol. Motil.* **2021**, *33*, e14129. [CrossRef]
57. Zito, F.P.; Polese, B.; Vozzella, L.; Gala, A.; Genovese, D.; Verlezza, V.; Medugno, F.; Santini, A.; Barrea, L.; Cargiolli, M.; et al. Good adherence to mediterranean diet can prevent gastrointestinal symptoms: A survey from Southern Italy. *World J. Gastrointest. Pharmacol. Ther.* **2016**, *7*, 564. [CrossRef]
58. Hassanzadeh, S.; Saneei, P.; Keshteli, A.H.; Daghaghzadeh, H.; Esmaillzadeh, A.; Adibi, P. Meal frequency in relation to prevalence of functional dyspepsia among Iranian adults. *Nutrition* **2016**, *32*, 242–248. [CrossRef]
59. Yamamoto, Y.; Furukawa, S.; Watanabe, J.; Kato, A.; Kusumoto, K.; Miyake, T.; Takeshita, E.; Ikeda, Y.; Yamamoto, N.; Kohara, K.; et al. Association Between Eating Behavior, Frequency of Meals, and Functional Dyspepsia in Young Japanese Population. *J. Neurogastroenterol. Motil.* **2022**, *28*, 418–423. [CrossRef]
60. Sinn, D.H.; Shin, D.H.; Lim, S.W.; Kim, K.-M.; Son, H.J.; Kim, J.J.; Rhee, J.C.; Rhee, P.-L. The Speed of Eating and Functional Dyspepsia in Young Women. *Gut Liver* **2010**, *4*, 173–178. [CrossRef]

MDPI
St. Alban-Anlage 66
4052 Basel
Switzerland
www.mdpi.com

Nutrients Editorial Office
E-mail: nutrients@mdpi.com
www.mdpi.com/journal/nutrients

www.ingramcontent.com/pod-product-compliance
Lightning Source LLC
Chambersburg PA
CBHW040752250926
51850CB00007BB/1473
* 9 7 8 3 7 2 5 8 0 0 5 7 5 *